数据生态：
MySQL
复制技术与生产实践

罗小波 沈刚 / 著

电子工业出版社
Publishing House of Electronics Industry
北京·BEIJING

内 容 简 介

全书共分为 3 篇：基础篇、方案篇和参考篇。按照"基本原理"→"生产实践"→"更多参考"的逻辑顺序讲述，书中配了大量的原理与方案示意图，力求用通俗易懂的语言、直观明了的示意图、完整的知识涵盖面将 MySQL 复制技术讲透。

其中，基础篇侧重介绍主从复制的原理和复制技术的演进，方案篇侧重介绍主从复制技术在生产环境中的应用方案，参考篇侧重介绍二进制日志的基本组成及主从复制中常见对象复制的安全性等。

本书适用初、中、高级 MySQL DBA、数据库架构师及相关开发人员阅读。

未经许可，不得以任何方式复制或抄袭本书之部分或全部内容。
版权所有，侵权必究。

图书在版编目（CIP）数据

数据生态: MySQL 复制技术与生产实践/罗小波，沈刚著. —北京：电子工业出版社，2020.10
ISBN 978-7-121-39714-1

Ⅰ.①数… Ⅱ.①罗… ②沈… Ⅲ.①SQL 语言－程序设计 Ⅳ.①TP311.132.3

中国版本图书馆 CIP 数据核字（2020）第 189281 号

责任编辑：张春雨
印　　刷：三河市双峰印刷装订有限公司
装　　订：三河市双峰印刷装订有限公司
出版发行：电子工业出版社
　　　　　北京市海淀区万寿路 173 信箱　　　邮编：100036
开　　本：787×1092　1/16　　印张：28.5　　字数：658 千字
版　　次：2020 年 10 月第 1 版
印　　次：2020 年 10 月第 1 次印刷
定　　价：119.00 元

凡所购买电子工业出版社图书有缺损问题，请向购买书店调换。若书店售缺，请与本社发行部联系，联系及邮购电话：（010）88254888，88258888。
质量投诉请发邮件至 zlts@phei.com.cn，盗版侵权举报请发邮件至 dbqq@phei.com.cn。
本书咨询联系方式：010-51260888-819，faq@phei.com.cn。

推荐序一

MySQL 在其 3.23 版本时就颇具前瞻性地推出主从复制（replication）特性，距今约有 20 年了。正是这一举措，使得 MySQL 赶上了互联网 1.0 时代的发展大势。利用复制特性可以很方便地实现数据库架构的扩展及读写分离功能，以提升架构承载的容量。

我想，从这个角度来说，复制是 MySQL 最主要的特性一点都不为过。

本书专注于 MySQL 复制特性，从基本概念和原理着手，再到实践操作、复制架构的方案、常见故障案例等方面，也涵盖了传统复制架构下最新的半同步复制、多源复制、多线程复制等内容，可谓 MySQL 复制特性的全面指南。

相信我，本书对 MySQL 复制特性的阐述既有广度也不失深度，无论是对于在校生还是技术专家，本书都值得细读。

叶金荣

Oracle ACE Director(MySQL)

推荐序二

在开源国产数据库崛起的今天，罗小波和沈刚所著的这本佳作《数据生态：MySQL 复制技术与生产实践》，无疑将为 MySQL 在各行业的推广和使用做出贡献，这也是像我这样的从商业数据库转到开源数据库的从业者的福音。

MySQL 能够成为"最流行的开源数据库"，其复制技术起了巨大的作用。Sharing Nothing 架构、横向扩展、高可用、容灾、数据集成与聚合，这一系列名词所代表的架构和应用场景都与复制技术有关。多应用场景意味着与复制相关的技术灵活而复杂：异步复制、增强半同步复制、语句级复制、行级复制、按位点复制、GTID 复制、多源复制、级联复制、多线程复制、双主架构、一主多从架构、延迟复制、读写分离等。不同的业务场景，使用不同的复制架构，该如何正确地构建 MySQL 复制拓扑、如何有效地监控和正确地维护，这些都是架构师和 DBA 要解决的问题。

小波他们这本书系统讲解了 MySQL 复制技术的知识点，涵盖上面所提到的各种场景和架构相关的问题。我特别佩服小波，佩服他在技术研究和写作上的系统性，在书中对高层架构和实践细节的知识讲述得细致而全面。毫无疑问，这些知识将有助于 DBA 构建企业级 MySQL 运行环境，在数据库可扩展性和高可用性等方面，做到正确地实施和监控，以及管理和维护。

本书以专题的形式切入，力求将 MySQL 主从复制的原理和实践讲透。掌握 MySQL 复制技术后，再了解和掌握生态圈中的第三方产品，包括 MHA、Orchestrator、Replication Manager、Otter、Canal、ProxySQL 等，来补充 MySQL 架构，就不再是难事。

由于篇幅原因，这本书没有涉及 MySQL Group Replication（简称 MGR，即组复制），所以很期待小波的下一本书能够对 MySQL 复制技术的另一个分支"组复制"进行全面、细致、深入的讲解。

再次感谢小波给社区和行业带来的佳作。

熊军

云和恩墨产品研发副总裁

推荐序三

很高兴能够为大家推荐罗小波老师与他同事沈刚的新作《数据生态：MySQL 复制技术与生产实践》。这本书汇集了罗老师他们近年来的学习成果和工作经验，是关于 MySQL 生态的又一力作。

与概要性地介绍 MySQL 的图书不同，这本书详细介绍了 MySQL 复制技术及其在生产实践中的具体应用和操作，是一本专题类的图书。

熟悉 MySQL 的人都知道，MySQL 的复制技术是其核心技术之一，是灵活运用 MySQL 的基础。从 3.23 版本开始，MySQL 推出了异步复制功能，之后不断进化，推出半同步复制、无损半同步复制，以及目前最新的组复制功能。其基本原理几乎是一致的——利用二进制日志文件在数据库服务器之间的传播和数据回放，实现多台数据库服务器之间的数据同步。一个合格的 MySQL 从业人员，必须掌握复制技术的基础知识，熟悉复制技术的各种解决方案，并灵活利用它们去满足生产系统中的各种需求。

这本书的"基础篇"介绍了复制技术的原理，对相关概念进行了详尽的解读；"方案篇"则介绍了不同方案的应用与实际操作；最后的"参考篇"对二进制日志等内容进行了详细的说明。这本书详尽阐述了 MySQL 复制技术。

随着国家加强对数据库技术安全可控的要求，使用开源的 MySQL 数据库已经成为一种趋势，开源代表着代码安全、数据库可控。各行各业不断地推进 MySQL 的部署，MySQL 从业人员也由互联网行业扩散到传统行业。利用复制技术可以实现 MySQL 的高可用性及高扩展性，是保障 MySQL 数据安全和进行数据库扩展的必要技术。安全可控对于传统行业至关重要，因此极力推荐 MySQL 从业者，特别是传统行业的 DBA 阅读此书。

徐铁韬

Oracle 公司 MySQL 全球事业部亚太区解决方案高级工程师

微信公众号"MySQL 解决方案工程师"运营者和内容作者

前　言

写书的出发点

2019 年 11 月，我们撰写了《千金良方：MySQL 性能优化金字塔法则》一书。从那以后，身边不断有人问我一个问题："写技术类的书不怎么赚钱，为什么还要写？"刚开始，我还认真回答，但提问者听到回答之后大多仍然表示不解，后来问的人多了，我索性回答："为了赚名气！"这个答案简单、粗暴、有效！的确，通过《千金良方：MySQL 性能优化金字塔法则》一书，我们小"赚"了些名气，不过对于为什么写书这个问题，这不是我的全部答案。现在，借本书的前言，我将自己全部的想法写出来，希望能完整、全面地回答这个问题。

- 督促自己有计划地学习

2018 年 12 月，我们有幸参加了电子工业出版社在北京举办的作译者聚会。聚会上有两位嘉宾的话令我印象深刻。一位来自阿里巴巴的安全专家说他已经撰写十余本书了，另一位来自美团的某团队负责人说他规划要写十余本书。"听君一席话，胜读十年书"，他们完全颠覆了我之前对于写书这件事的看法。在此之前，我一直以为，技术图书是写作者不断沉淀工作中所学的知识，由量变到质变的产物。能写一本书，说明某个方面知识的积累已经达到一定厚度，非常不易。在那一刻，我才知道，原来写书这件事，还可以刻意规划，也难怪大拿们能够做到连续数十载一年写一本书！于是，从那时开始，我们也开始尝试有计划地学习、有计划地写书。

- 化解繁重的工作和技能精进之间的矛盾

在 2018 年 12 月的作译者聚会之后不久，我们再次受邀参加电子工业出版社在杭州举办的作译者聚会。这次聚会上，一位嘉宾的话又让我获益匪浅，他说"视野的高度决定了做事的高度"。在他的影响下，我阅读了吴军博士的《见识》一书，其中"西瓜与芝麻"和"不做伪工作者"的故事令人印象深刻。这让我联想到，自己及身边的同事、朋友们，也未尝不是经常做着捡芝麻的事情，沉迷于低效率的勤奋而不自知，由于工作时长问题，相当多的人也缺少精进的时间和精力。

不少朋友曾经问我如何从零开始学习 MySQL，这些人中有刚走出校园的实习生、程序员，也有想从其他数据库开发转向 MySQL 开发的开发者。对于这一问题，我一时间不知如何作答，但问的人多了，一来我不好意思总说自己也不知道如何学习 MySQL，二来突然发觉这或许是一个需要有人站出来解决的问题。于是我开始思考，想象作为一个 MySQL "小白"应该如何入门。

经过不断的思考，我慢慢有了一些答案。例如，可以尝试系统地研究某个知识点，将其研究透彻，然后分享出来供大家一起学习，以便帮助那些没有足够精力和时间精进的朋友们快速进步，让那些可能走弯路的朋友们少走弯路。或许这就像吴军博士在《文明之光》中所描述的那样，当农耕文明发展到一定阶段，赖以生存的食物不再短缺的时候，就能够腾出一部分人力不再从事农耕劳作，转而专门从事满足新需求的工作，如此这般，人类文明便可以不断向前进步。

- 系统地学习和研究某个课题

借用吴军博士的话说，一本书可以看作系统学习和研究某个课题后的答卷，就好比在大学里写论文一样，它能够验证你对课题的学习和研究是否有成效，让大家都可以看见、分享你的学习和研究成果。

- 锻炼写作能力

写作是一项技能，需要反复地刻意练习，而写书的过程通常要求高、周期长，这是一个非常好的练习机会。

掌握正确的学习和研究方法

要系统地学习和研究一个课题，按照由浅入深的顺序，我们可以将其过程大致分为如下三个阶段：

- 第一阶段：认识整体

对于一个复杂课题，可以先从整体上搞清楚它的知识体系组成框架，搞清楚其中各个组成部分（知识模块）的大致脉络，从而从全局上建立起初步认知，以便为下一阶段选择知识模块进行深入研究做好铺垫。

- 第二阶段：逐一深入

基于第一阶段的整体认识，可以优先选择一些工作中需要用到的知识模块，或者感兴趣的知识模块，作为子课题逐一进行深入的系统研究。

- 第三阶段：回归整体

由于人的精力有限，在第二阶段，深入研究各知识模板期间，一些知识模块可能会被遗忘，因此需要回归整体，结合自己的验证与理解，建立牢固的知识体系。

在学习和研究的过程中要勤动手记录，可将零散的知识整理为博客文章，博客文章积累到一定数量之后亦可整理为图书，持续地做这样的知识积累工作，直到形成完善的知识体系，日后可供自己与他人使用。

2019 年 11 月出版的《千金良方：MySQL 性能优化金字塔法则》，可以说是我们在第二阶段中对一个子课题（MySQL 性能优化）的答卷，而本书则可以说是我们对另一个子课题（MySQL 主从复制）的答卷。对于 MySQL 来说，主从复制是一个非常重要的知识模块，而且据我所知，还有非常多从事 MySQL 相关工作的同行们，对 MySQL 主从复制的原理、应用场景等知识掌握得并不全面，甚至对其一知半解的人也不在少数。因此，在本书中，我们将其作为重点展开介绍。不过，遗憾的是，在学习和研究的第一阶段，我们还没有撰写相关图书，但我们正在积极筹备中，在不久的将来会为广大读者朋友们交上一份满意的"答卷"！

MySQL 的数据生态

MySQL 的二进制日志记录了一个数据库实例内数据的变更，这些内容是按照时间的先后顺序记录的。根据具体的二进制日志格式选项设置，可以记录数据库实例内执行的原始 SQL 语句文本，也可以记录数据库实例内执行 SQL 语句时产生的数据变更的行记录值。二进制日志可以满足类似如下一些应用场景：

- MySQL Server 崩溃之后对其进行恢复时，二进制日志作为事务的协调者。

对于一个未完成提交的事务，MySQL Server 崩溃之后进行恢复时，会在二进制日志中检查是否存在对应的内容，如果存在，则事务可以重新提交，如果不存在（或没有启用二进制日志记录功能），则必须回滚事务，将该事务修改的记录进行回滚。

- 在不同实例间进行数据同步（主从复制、组复制）。

主库中的数据变更被记录到二进制日志，然后主库将二进制日志发送给从库，从库使用这些二进制日志进行回放，模拟主库中执行的操作，从而实现主从库之间的数据同步。

- 数据异地容灾。

可以在异地数据中心增加一台服务器，将其配置为新的从库，从主库中获取二进制日志进行回放，达到数据异地容灾的目的；也可以由应用程序模拟一个从库，从主库中获取二进制日志，进行解析、处理，然后将数据存放在容灾专用的系统中。

- 恢复被误删除的数据（通常称为"数据闪回"）。

通常，对于未提交的事务，事务的 ACID 特性能够保证数据的一致性，直接使用事务提供的回滚功能即可实现对误删除数据的恢复，不受事务控制的修改语句除外，因为不受事务控制的语句修改的数据无法执行回滚。但事务一旦完成提交，就无法再对数据进行回滚，这时可以对二进制日志中记录的值与条件进行反转，生成新的 SQL 语句（要求二进制日志以 row 格式记录，必须记录全镜像，且只支持增/删/改语句的反向操作），然后重新在数据库实例中执行新的 SQL 语句，从而恢复被误删除的数据。

- 数据在异构数据库之间流转。

二进制日志中记录的是数据的逻辑变更，可以从中提取出数据的纯文本和字段的顺序，然后通过应用程序做一些处理，数据就可以在异构数据库之间流转了。这是其他大多数数据库软件不具备的特性。

二进制日志独有的特性是 MySQL 数据流动与循环的基石，在不同应用场景下形成了独特的"数据生态"，这也是本书名字的由来。

读者对象

无论你是 MySQL 的初学者、数据库架构师及相关开发人员、非 MySQL 数据库 DBA，还是中高级的 MySQL DBA，认真阅读此书，相信或多或少都能有所收获。

- MySQL 的初学者、MySQL 相关开发人员、非 MySQL 数据库的 DBA，可以从头开始完整学习 MySQL 的复制技术。
- 数据库架构师、中高级 MySQL DBA，可以借助本书对 MySQL 复制技术的相关知识进行查漏补缺，扫除盲点。

如何阅读本书

全书分为"基础篇""方案篇""参考篇"，其中：

- "基础篇"对 MySQL 主从复制技术的用途、概念、基本原理及演进等进行全方位的介绍。
- "方案篇"对 MySQL 主从复制技术在生产中的应用场景、复制拓扑的生命管理周期、高可用切换与主库故障转移等进行全方位的介绍。
- "参考篇"对 MySQL 二进制日志的基本组成结构、常见的复制对象在主从复制拓扑中的流转过程等进行全方位的介绍。

对于初学者、MySQL 相关开发人员、非 MySQL 数据库的 DBA 而言，如果时间充足，建议从头到尾依次学习本书内容。

数据库架构师、中高级 MySQL DBA 则可以通过目录快速查找所需内容。

如果你具备一定的 MySQL 源码阅读能力，或者想要挑战更高难度，可以结合简书平台高鹏的专栏"深入理解主从原理 32 讲"中的系列文章进行学习。对该专栏中的内容，高鹏也进行了整理，并整理成了《深入理解 MySQL 主从原理》一书，有需要的读者朋友可自行购买。

作者分工

本书作者各自负责的章节：

- 罗小波：负责撰写"基础篇""方案篇"，以及"参考篇"中的第 28 章。
- 沈刚：负责撰写"参考篇"中除第 28 章以外的其他章节，以及博文视点官网的附录资源。

致谢

首先，非常感谢给本书作序的大拿：叶金荣老师、熊军、徐轶韬。感谢撰写封底推荐语的大拿：林晓斌（丁奇）、高鹏（八怪）、温正湖、杨奇龙、熊中哲、李春。感谢他们的认可与支持！

其次，非常感谢为本书校稿的朋友们：刘云、董红禹、高鹏。感谢他们不辞辛苦，反复咀嚼文字，努力寻找书稿中的纰漏，帮助提升阅读体验！

再次，非常感谢参与本书命名讨论的朋友们：李春、董红禹、徐婷、杜蓉、符隆美、孙黎！

最后，非常感谢帮忙为本书进行宣传的叶金荣老师、大力配合我们推动图书出版事宜的电子工业出版社编辑符隆美，以及其他负责内容审核、校对、排版的编辑们！

目录

基 础 篇

第 1 章 复制的概述 2
    1.1 适用场景 2
    1.2 数据同步方法 3
    1.3 数据同步类型 4
    1.4 复制格式 4

第 2 章 复制的基本原理 6
    2.1 概述 6
    2.2 细节 7

第 3 章 复制格式详解 10
    3.1 复制格式概述 10
    3.2 复制格式明细 11
        3.2.1 基于 statement 和基于 row 的复制的优缺点 11
        3.2.2 使用 row 格式的二进制日志进行复制 14
    3.3 如何确定与记录复制中的安全和不安全语句 15

第 4 章 传统复制与 GTID 复制 18
    4.1 传统复制 18
    4.2 GTID 复制 19
        4.2.1 GTID 的格式和存储 19
        4.2.2 GTID 的生命周期 23
        4.2.3 GTID 自动定位 26
        4.2.4 GTID 复制模式的限制 27

第 5 章 半同步复制 29
    5.1 半同步复制的原理 29

5.2 半同步复制的管理接口 33
5.3 半同步复制的监控 33
5.4 半同步复制的注意要点 34

第 6 章 多线程复制 36

6.1 单线程复制原理 36
6.2 DATABASE 多线程复制 38
  6.2.1 原理 38
  6.2.2 系统变量的配置 40
6.3 LOGICAL_CLOCK 多线程复制 40
  6.3.1 原理 40
  6.3.2 系统变量的配置 42
6.4 WRITESET 多线程复制 42
  6.4.1 原理 42
  6.4.2 系统变量的配置 50

第 7 章 多源复制 51

7.1 复制通道 51
7.2 单通道操作命令 52
7.3 复制语句的向前兼容性 53
7.4 启动选项和复制通道选项 53
7.5 复制通道的命名约定 55

第 8 章 从库中继日志和状态日志 56

8.1 中继日志和状态日志概述 56
8.2 从库中继日志 57
8.3 从库状态日志 58

第 9 章 通过 PERFORMANCE_SCHEMA 库检查复制信息 66

9.1 PERFORMANCE_SCHEMA 库中的复制信息记录表概述 66
9.2 PERFORMANCE_SCHEMA 库中的复制信息记录表详解 70
  9.2.1 replication_applier_configuration 表 70
  9.2.2 replication_applier_status 表 71
  9.2.3 replication_applier_status_by_coordinator 表 72
  9.2.4 replication_applier_status_by_worker 表 74
  9.2.5 replication_connection_configuration 表 76
  9.2.6 replication_connection_status 表 78

9.2.7　replication_group_member_stats 表 80
　　9.2.8　replication_group_members 表 81

第 10 章　通过其他方式检查复制信息 83

　10.1　复制状态变量 83
　10.2　复制心跳信息 83
　10.3　SHOW SLAVE STATUS 语句输出信息详解 85
　10.4　通过 SHOW PROCESSLIST 语句查看复制线程状态 92
　10.5　SHOW MASTER STATUS 语句输出详解 93
　10.6　SHOW SLAVE HOSTS 语句 93

第 11 章　MySQL 复制延迟 Seconds_Behind_Master 究竟是如何计算的 96

　11.1　"口口相传"的计算方法 96
　11.2　探寻"正确"的计算方法 97
　11.3　验证 100
　　11.3.1　我们想确认什么 100
　　11.3.2　提前确认一些信息 100
　　11.3.3　执行验证 102
　11.4　小结 103

第 12 章　如何保证从库在意外中止后安全恢复 104

　12.1　从库的崩溃与恢复概述 104
　12.2　从库的崩溃与恢复详解 104
　　12.2.1　单线程复制的安全恢复 104
　　12.2.2　多线程复制的安全恢复 105

第 13 章　MySQL Server 复制过滤 108

　13.1　MySQL Server 复制过滤规则概述 108
　13.2　库级别复制过滤选项的评估 109
　13.3　表级别复制过滤选项的评估 113
　13.4　复制过滤规则的应用 115

方　案　篇

第 14 章　搭建异步复制 120

　14.1　操作环境信息 120
　14.2　全新初始化场景 121
　　14.2.1　传统复制 121

    14.2.2 GTID 复制 124
  14.3 已有数据场景 126
    14.3.1 传统复制 127
    14.3.2 GTID 复制 134
  14.4 变量模板 140
    14.4.1 传统复制模式的变量模板 140
    14.4.2 GTID 复制模式的变量模板 144

第 15 章 搭建半同步复制 146
  15.1 半同步复制插件的安装和配置环境要求 146
  15.2 半同步复制插件的安装和配置 147
    15.2.1 关键步骤 147
    15.2.2 详细过程 148
  15.3 半同步复制工作状态的验证 152

第 16 章 通过扩展从库以提高复制性能 155
  16.1 操作环境 155
  16.2 横向扩展 156
    16.2.1 扩展从库的简要步骤 157
    16.2.2 扩展从库的详细过程 158
    16.2.3 配置从库的读负载均衡 164
  16.3 提高复制性能 165

第 17 章 复制模式的切换 167
  17.1 操作环境信息 167
  17.2 复制模式的相关概念 167
  17.3 传统复制在线变更为 GTID 复制 173
    17.3.1 简要步骤 173
    17.3.2 详细过程 173
  17.4 GTID 复制在线变更为传统复制 179
    17.4.1 简要步骤 179
    17.4.2 详细过程 180
  17.5 GTID 复制离线变更为传统复制 186
    17.5.1 简要步骤 186
    17.5.2 详细过程 187
  17.6 传统复制离线变更为 GTID 复制 191
    17.6.1 简要步骤 191

| | | 17.6.2 详细过程 | 191 |

第 18 章 复制拓扑的在线调整 ... 196
18.1 操作环境信息 ... 196
18.2 传统复制模式下的复制拓扑在线调整 ... 197
    18.2.1 并行复制变更为串行复制 ... 198
    18.2.2 串行复制变更为并行复制 ... 206
18.3 GTID 复制模式下的复制拓扑在线调整 ... 211
    18.3.1 并行复制变更为串行复制 ... 212
    18.3.2 串行复制变更为并行复制 ... 214

第 19 章 主从实例的例行切换 ... 218
19.1 操作环境信息 ... 218
19.2 在线切换 ... 219
    19.2.1 基于账号删除的在线切换 ... 220
    19.2.2 基于修改连接数的在线切换 ... 232

第 20 章 数据库故障转移 ... 239
20.1 操作环境信息 ... 240
20.2 主库故障转移的关键步骤 ... 241
20.3 主库故障转移的详细过程 ... 243
    20.3.1 环境的准备 ... 243
    20.3.2 执行步骤 ... 245

第 21 章 搭建多源复制 ... 255
21.1 操作环境信息 ... 255
21.2 基于传统复制的多源复制 ... 256
    21.2.1 传统复制模式下的单线程多源复制 ... 256
    21.2.2 传统复制模式下的多线程多源复制 ... 261
21.3 基于 GTID 复制的多源复制 ... 263
    21.3.1 GTID 复制模式下的单线程多源复制 ... 264
    21.3.2 GTID 复制模式下的多线程多源复制 ... 267
21.4 多源复制拓扑中复制相关的操作语句变化 ... 268

第 22 章 MySQL 版本升级 ... 273
22.1 MySQL 版本之间的复制兼容性 ... 273
22.2 升级复制的设置 ... 274

第 23 章　将不同数据库的数据复制到不同实例......276
23.1　操作环境信息......276
23.2　通过设置复制过滤规则将不同数据库的数据复制到不同实例......277
23.2.1　通过只读选项配置复制过滤规则......278
23.2.2　通过动态语句配置复制过滤规则......286

第 24 章　发生数据误操作之后的处理方案......293
24.1　操作环境信息......294
24.2　主库发生误操作后的数据恢复......295
24.2.1　通过延迟复制恢复数据......296
24.2.2　通过闪回工具恢复数据......302
24.3　从库发生误操作后的数据恢复......309
24.3.1　通过修改系统变量 slave_exec_mode 恢复数据......309
24.3.2　通过 GTID 特性注入空事务恢复数据......316

第 25 章　常用复制故障排除方案......323
25.1　确认故障现象......323
25.2　信息收集与故障排查......323
25.3　复制故障的修复......325
25.4　无法解决的问题......326

参　考　篇

第 26 章　二进制日志文件的基本组成......328
26.1　什么是二进制日志......328
26.2　二进制日志的组成......328
26.3　二进制日志内容解析......329
26.3.1　基于 row 的复制的二进制日志内容解析......330
26.3.2　基于 statement 的复制的二进制日志内容解析......344
26.4　小结......351

第 27 章　常规 DDL 操作解析......352
27.1　操作环境信息......352
27.2　常规 DDL 操作示例......353
27.3　二进制日志内容解析......356
27.4　小结......359

第 28 章　为何二进制日志中同一个事务的事件时间点会乱序 360
28.1　操作环境信息 360
28.2　验证前的准备 360
28.3　验证过程 361

第 29 章　复制 AUTO_INCREMENT 字段 367
29.1　操作环境信息 367
29.2　复制 AUTO_INCREMENT 字段的操作示例 367
29.3　对二进制日志的解析及解释 368
29.3.1　基于 row 的复制中 AUTO_INCREMENT 字段的复制 368
29.3.2　基于 statement 的复制中 AUTO_INCREMENT 字段的复制 369
29.3.3　混合复制中 AUTO_INCREMENT 字段的复制 371
29.4　使用 AUTO_INCREMENT 字段时的注意事项 371
29.5　小结 372

第 30 章　复制 CREATE ... IF NOT EXISTS 语句 373
30.1　操作环境信息 373
30.2　复制 CREATE ... IF NOT EXISTS 语句的操作演示 374
30.3　二进制日志解析结果的解释 375
30.4　小结 376

第 31 章　复制 CREATE TABLE ... SELECT 语句 377
31.1　操作环境信息 377
31.2　复制 CREATE TABLE ... SELECT 语句的操作示例 377
31.3　二进制日志的解析及解释 380
31.3.1　statement 格式二进制日志的解析及解释 380
31.3.2　row 格式和 mixed 格式二进制日志的解析及解释 381
31.4　使用 CREATE TABLE ... SELECT 语句时的注意事项 384
31.5　小结 384

第 32 章　在主从复制中使用不同的表定义 385
32.1　操作环境信息 385
32.2　主从库的表字段数不同时如何复制 386
32.2.1　源表字段数多于目标表字段数 386
32.2.2　目标表字段数多于源表字段数 388
32.3　不同类型字段的复制 389
32.3.1　属性提升 390

         32.3.2 有损转换与无损转换 ...390
    32.4 小结 ...391

第 33 章 复制中的调用功能 ...392
    33.1 操作环境信息 ...392
    33.2 复制中的调用功能操作示例 ...392
         33.2.1 在 READ-COMMITTED 隔离级别、基于 row 的复制场景下数据库的操作记录 ...393
         33.2.2 在 READ-COMMITTED 隔离级别、基于 statement 的复制场景下数据库的操作记录 ...396
    33.3 二进制日志的解析及解释 ...396
         33.3.1 row 和 mixed 格式二进制日志的解析及解释 ...396
         33.3.2 statement 格式二进制日志的解析及解释 ...402
    33.4 小结 ...407

第 34 章 复制 LIMIT 子句 ...408
    34.1 操作环境信息 ...408
    34.2 复制 LIMIT 子句的操作示例 ...408
    34.3 二进制日志的解析及解释 ...409
         34.3.1 statement 格式二进制日志的解析及解释 ...410
         34.3.2 row 格式和 mixed 格式二进制日志的解析及解释 ...411
    34.4 小结 ...414

第 35 章 复制 LOAD DATA 语句 ...415
    35.1 操作环境信息 ...415
    35.2 复制 LOAD DATA 语句的操作示例 ...416
         35.2.1 准备演示数据 ...416
         35.2.2 LOAD DATA 语句的操作 ...417
    35.3 二进制日志的解析及解释 ...417
         35.3.1 statement 格式二进制日志的解析及解释 ...417
         35.3.2 row 格式和 mixed 格式二进制日志的解析及解释 ...419
    35.4 小结 ...420

第 36 章 系统变量 max_allowed_packet 对复制的影响 ...421
    36.1 系统变量简介 ...421
    36.2 操作环境信息 ...422
    36.3 max_allowed_packet 对复制的影响操作示例 ...423

　　　　36.3.1　max_allowed_packet 对主库的影响 423
　　　　36.3.2　max_allowed_packet 对从库的影响 424
　　36.4　小结 427

第 37 章　复制临时表 429
　　37.1　操作环境信息 429
　　37.2　复制临时表的操作示例 429
　　　　37.2.1　基于 statement 的复制且隔离级别为 REPEATABLE-READ 430
　　　　37.2.2　基于 row 的复制且隔离级别为 REPEATABLE-READ 432
　　　　37.2.3　混合复制且隔离级别为 REPEATABLE-READ 433
　　　　37.2.4　使用临时表时如何安全关闭从库 433
　　37.3　与临时表相关的其他注意事项 435
　　37.4　小结 435

第 38 章　复制中的事务不一致问题 436
　　38.1　事务不一致的场景类型 436
　　38.2　事务不一致的原因 437
　　38.3　事务不一致的后果 437
　　38.4　小结 438

基 础 篇

第 1 章 复制的概述

简单来说,"复制"就是将来自一个 MySQL Server(这里指 master 角色,即主库)的数据变更,通过其逻辑的二进制日志(binlog)传输到其他的一个或多个 MySQL Server(这里指 slave 角色,即从库)中,其他 MySQL Server 通过应用(回放)这些逻辑的二进制日志来完成数据的同步。这些 MySQL Server 之间的逻辑关系,我们称为"复制拓扑"(也可以称为"复制架构")。

默认情况下,复制是异步的,即主库将二进制日志传输到从库之后,并不关心从库是否成功收到。从库是否收到这些二进制日志,不影响主库的任何读/写访问;而从库的复制线程也可以随意暂停或停止,并不影响主库的读/写访问。通常,异步模式能够发挥数据库的最高性能,但数据安全性却得不到很好的保证,如果对数据安全性的要求较高,可以考虑使用半同步复制。

另外,默认情况下,复制的数据是针对整个实例的(排除部分系统表),你可以根据自身需求选择是否需要复制整个实例的数据,是只复制某些库,还是只复制某些表的数据等。接下来将简要介绍 MySQL 中复制拓扑的一些适用场景,以及与复制相关的概念。

1.1 适用场景

MySQL 主从复制拓扑可用于如下场景:

1. 横向扩展(Scale-Out)

横向扩展是指在多个从库之间进行读负载均衡,以提高读性能。在此扩展方案中,所有数据变更在主库上执行,读负载可以分摊到一个或者多个从库上。可以把之前在主库上的读负载剥离出来,以承载更多的写请求,另外,如果读负载越来越大,可以通过扩展从库来提高读性能。

2. 数据安全性(Data Security)

数据安全性在很大程度上需要靠数据副本来保证。在这里,副本可以理解为我们通常所说的备份。在 MySQL 中,为了保证二进制日志的位置和数据的一致性,通常在执行备份时需要阻塞写操作,防止因数据变更而造成备份过程中所获取的二进制日志位置与数据不一致。

大多数时候，我们使用的 MySQL 主从复制拓扑是这样的：一主 N 从、两主 N 从（N 大于或等于 1）。其中，主库提供读与写或者只写服务，从库提供只读服务（多个从库通常会提供读负载均衡以分摊单个从库的读压力）。而主库通常只有一个（在有两个主库的复制拓扑中，另一个主库通常用于提供快速的高可用切换），提供只读服务的从库通常有多个。因此，虽然在主库上执行备份能够尽可能保证备份数据的实时性（即拥有最新的数据），但是由于在备份过程中需要通过阻塞写操作来保证数据的一致性，因此会严重影响应用的写可用性。对于主库来说，从库有多个，所以如果在从库上执行备份，对只读应用的可用性影响就要小很多（从库的复制机制本身也支持断点续传）。也就是说，在执行备份操作时，选择使用从库而不使用主库是一个更好的替代方案，这样可以尽量减少对主库性能以及数据安全性的影响。

3．分析（Analytics）

在主库上运行 OLTP（联机事务处理）应用，而 OLAP（联机分析处理）应用可以在从库上运行，避免在主库上运行 OLAP 应用对主库性能造成影响。

4．远程数据分发（Long-Distance Data Distribution）

可以使用复制特性为远程站点（例如分公司或者分地域的应用中心）创建数据的本地副本，那些对数据实时性没有要求的应用可以访问本地副本，剩下的一小部分对实时性有要求的应用访问主库。远程站点既可以作为灾备中心，也可以用于实现跨地域访问以分摊负载，以及实现就近访问，加快访问速度。

5．在线滚动升级（Online Rolling Upgrade）

先升级从库，然后执行主从切换，再升级主库，从而实现数据库应用不停机的在线升级（这里说的"不停机"并不是指应用完全无感知，在执行主从切换时，应用需要重新连接数据库）。

6．高可用切换（Failover）

当主库发生故障时，可以把从库提升为主库，从而快速恢复服务。

1.2 数据同步方法

MySQL 5.7 支持以下两种数据同步方法（也可以称为复制模式或复制方法）：

- 传统复制：也可以称为基于二进制日志文件和位置的复制，在从库中配置复制时，要求指定从主库中获取的二进制日志文件（binlog file）和位置（binlog position），以便从库中的复制线程启动时，能够以指定的二进制日志文件和位置为起点，持续读取主库中的二进制日志，并在从库中应用，从而达到数据同步的目的。
- 基于 GTID 的复制（本书中为了表述上的方便，简称为 GTID 复制）：GTID（全局事务标识符）是新的事务性复制方法，利用 GTID 可以自动在主库中寻找需要复制的二进制日志记录，因此不需要关心日志文件或位置，极大地简化了许多常见的复制任务。使用 GTID 复制可确保主库和从库之间的一致性。有关 GTID 的更多细节，详见 4.2 节 "GTID 复制"。

1.3 数据同步类型

MySQL 支持如下 4 种不同类型的数据同步：

- 异步复制：是最早出现的复制技术，MySQL 内置支持，不需要额外安装插件。其中，一个实例充当主库，一个或多个其他实例充当从库，与同步复制形成对比。
- 半同步复制：MySQL 5.7（从 MySQL 5.5 开始支持半同步复制）除内置了异步复制之外，还支持半同步复制。使用半同步复制时，主库的会话在提交事务之前，会等待至少一个从库返回收到二进制日志的 ACK 消息（确认接收，并将事务的事件记录到从库的中继日志中）。
- 延迟复制：MySQL 5.7 还支持延迟复制（从 MySQL 5.6 开始支持），使得从库可以故意滞后于主库至少一段指定的时间，以便在出现误操作时，有时间对误操作的数据进行补救。
- 同步复制：指的是需要保证写操作完全同步到其他数据节点，而不仅仅是二进制日志被其他节点接收。例如 NDB Cluster，它是一种集群架构，所有节点都可以进行读/写，而且每个节点上发生的写操作都会实时同步到其他节点。对于需要同步复制的场景，可以使用 NDB Cluster，或者其他类似的开源解决方案（虚拟同步，非完全同步），例如 Percona XtraDB Cluster（PXC）、MariaDB Galera Cluster（MGC）、MySQL Group Replication（MGR）。

1.4 复制格式

MySQL 支持两种二进制日志格式（二进制日志记录的是数据变更的逻辑）：一种是记录语句原始文本（即 statement 格式，语句文本按原样记录到二进制日志），另一种是记录数据的逐行变更（即 row 格式，将一个语句记录到二进制日志中，可能产生多行数据变更记录）。就具体的复制格式而言，可以分为三种，由系统变量 binlog_format 进行设置，有效值为：statement（对应下文中的 SBR）、row（对应下文中的 RBR）、mixed（mixed 格式，混合使用 statement 和 row 格式，对应下文中的 MBR）。这三种格式的详情如下：

- 基于 statement 的复制（Statement Based Replication，SBR）：当系统变量 binlog_format 设置为 statement 时表示使用 SBR。SBR 复制的是整个 SQL 语句的原始文本，日志量较小，但容易出现主从库数据不一致。对于 SBR，当执行的某个语句被判定为不安全时，是否允许其执行还取决于事务的隔离级别。
- 基于 row 的复制（Row Based Replication，RBR）：当系统变量 binlog_format 设置为 row 时表示使用 RBR。RBR 复制的是发生更改的数据行的实际记录（原始语句会被转换为发生变更的行数据记录），日志量较大，但可以保证主从库数据的一致性。
- 混合复制（Mixed Based Replication，MBR）：系统变量 binlog_format 设置为 mixed 时表示使用 MBR。MBR 实际上是由 MySQL 自行判断的，即在不影响数据一致性的情况下，使用 SBR；如果可能影响数据一致性，则自动转换为 RBR。

提示：本章中复制相关的概念较多，容易混淆，这里做一个总结。MySQL 中的复制，按照数据同步方法划分，可分为传统复制和 GTID 复制；按照数据同步类型划分，可分为异步复制、半同步复制、延迟复制和同步复制；按照复制格式划分，可分为基于 statement 的复制（SBR）、基于 row 的复制（RBR）和混合复制（MBR）。

有关复制的格式的更多信息，详见第 3 章"复制格式详解"。

第 2 章 复制的基本原理

第 1 章简单介绍了 MySQL 复制(技术)相关的基本概念,本章将详细介绍其基本原理。MySQL 的复制技术自诞生以来,随着各种各样的应用场景对数据安全性及复制性能的要求不断提高,也在不断迭代与优化。要深刻理解 MySQL 的复制技术,就要从它的基本原理说起[1],下面将对此展开介绍。

2.1 概述

复制是基于主库(master)二进制日志中写入的关于该数据库的所有更改(更新、删除等)的日志记录来实现的。从库(slave)利用这些二进制日志中的事件记录进行回放来同步数据。但对于不修改任何数据的查询语句,主库不会记录二进制日志(例如 SELECT 语句),所以,对于查询语句,从库自然也没有需要回放的内容。

连接到主库的每个从库都会复制一份主库二进制日志的副本。从库利用来自主库的二进制日志进行回放,就可以模拟主库中的数据变更操作,例如,建表、修改表结构等 DDL(Data Definition Language,数据定义语言)操作以及 DML(Data Manipulation Language,数据操纵语言)操作。

提示:对于主库中的二进制日志同步到从库时发起请求的方向,在此有必要做一些说明。二进制日志的同步,既有从库主动请求的情况,也有主库主动推送的情况。

- 对于复制线程在主从之间新建立连接或重新建立连接的情况,不是主库主动推送二进制日志(因为这时主库并不知道需要发送哪些二进制日志给新建立连接的从库),而是从库主动向主库请求所需的二进制日志(从库向主库注册连接时,携带了从库自身所需二进制日志的位置信息)。主从之间的复制连接始终是从库先发起请求的,就算主库主动断开了从库的连接,重新建立连接时也是从库重试的,而不是主库。
- 如果复制线程已经在主从之间建立连接,而且从库已经完全接收建立连接时请求的二进制日志内容,后续的增量二进制日志是由主库主动推送给从库的,而不是从库

---

[1] 注:从 MySQL 5.7 开始,支持能够兼顾数据一致性和数据库高可用性的组复制拓扑,由于本书只围绕主从复制拓扑展开,所以后续提及的"复制技术"仅针对"主从复制拓扑"。

主动向主库请求的（因为这时主库随时都可能写入新的内容，从库难以即时感知，也没必要即时感知）。例如，主库在某段时间内没有写入任何数据，从库已经完全接收了最新的二进制日志内容，当主库再次写入新的数据时产生的新二进制日志，主库会直接发送给从库，而不需要发送一个通知给从库 I/O 线程，然后让从库来拉取（主库在产生新的二进制日志时，只会通知主库自身的 Binlog Dump 线程，而不会通知从库 I/O 线程，当主库的 Binlog Dump 线程收到通知之后，会直接把新产生的二进制日志发送给从库 I/O 线程）。

每个从库的复制操作都是独立运行的，即每个从库对从主库拉取的二进制日志的回放都是独立发生的，互不影响。每个从库都按照自己的进度读取主库二进制日志（从主库中同步的二进制日志保存在每个从库自身的中继日志中）并进行回放来更新自身的数据，而且每个从库都可以独立地随意启动/停止自己的 I/O 线程和 SQL 线程。

主库和从库都会定期输出在复制过程中的状态，可以通过相关的复制状态变量来监控这些状态的变化。从库在进行回放之前，会先将从主库接收的二进制日志写入本地的中继日志，中继日志中还记录了主库二进制日志文件位置的有关信息。

2.2 细节

MySQL 的复制功能用三个线程来实现：一个线程在主库上（Binlog Dump 线程），两个线程在从库上（I/O 线程和 SQL 线程），如图 2-1 所示。

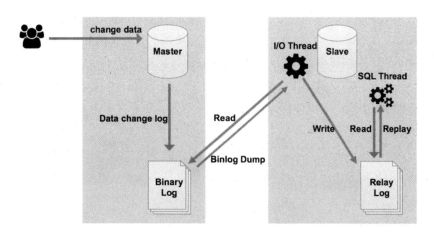

图 2-1

结合图 2-1 从左往右看，复制的原理大致如下：

（1）用户提交对数据的修改，然后 Master（主库）把所有数据库变更写进 Binary Log（二进制日志），主库通过 Binlog Dump 线程把二进制日志内容推送给 Slave（从库），从库被动接收数据，不是主动去获取，除非是新建连接。

- 当从库正常连接到主库时，在主库中使用 SHOW PROCESSLIST 语句可以查看到标识为"Binlog Dump"的线程。

- Binlog Dump 线程在读取二进制日志中要发送到从库的每个事件时，会获取二进制日志上的锁。一旦读取完事件，即使事件还未发送到从库，二进制日志上的锁也会被释放。

（2）在从库上执行 START SLAVE 语句时，已经使用 CHANGE MASTER TO 语句配置好复制信息，从库会创建一个 I/O 线程，该线程连接到主库并请求主库为其发送所需的二进制日志（从库向主库注册连接时，里面携带了请求的二进制日志的位置，该位置表示从库所请求的二进制日志的起点）。

- 从库 I/O 线程与主库的 Binlog Dump 线程成功建立连接之后，从库 I/O 线程接收主库 Binlog Dump 线程发送的二进制日志，并将它们写入从库本地的 Relay Log（中继日志文件）。

- I/O 线程的状态可以通过 SHOW SLAVE STATUS 语句输出的 Slave_IO_running 字段值来查看，或者通过 SHOW STATUS 语句输出的 Slave_running 状态变量查看（该状态变量在 MySQL 5.7 中已弃用，MySQL 8.0 已将其移除）。

（3）从库 SQL 线程读取并解析中继日志中的内容，按照读取的顺序进行回放（二进制日志中存放的事务顺序就是主库中事务的提交顺序），并将数据变更写入本地数据库文件中，这样就实现了数据在主从数据库（实例）之间的同步。

每一对主从关系中，都有三个线程。主库可以连接多个从库，而且会为每一个连接成功的从库创建一个 Binlog Dump 线程；从库也可以连接多个主库（多源复制），而且会为每一个连接成功的主库创建自己的 I/O 线程和 SQL 线程。

从库使用两个线程分别将主库的二进制日志读取到本地，并应用这些日志以实现主从之间的数据同步。因此，即便 SQL 线程执行语句缓慢，也不会影响 I/O 线程读取日志的速度（正常情况下，I/O 线程不会成为性能瓶颈，除非网络出现问题）。如果 SQL 线程已经应用完所有的中继日志，就表明 I/O 线程已经获取了主库中所有的二进制日志。

SHOW PROCESSLIST 语句提供的信息可以告诉你主库和从库上有关复制的一些状态，如下所示：

```
# 在主库上执行 SHOW PROCESSLIST 语句可以看到类似如下的信息：线程 Id 为 2 的线程就是 Binlog Dump 线程，其中 State 字段值为 "Has sent all binlog to slave; waiting for binlog to be updated"，表示主库已经发送所有的数据更新日志，正在等待新的数据更新写入二进制日志；如果在主库中看不到 Binlog Dump 线程，则说明复制未进行，即当前没有任何从库与主库建立了连接
mysql> show processlist\G
*************************** 1. row ***************************
   Id: 2
 User: root
 Host: localhost:32931
   db: NULL
```

```
    Command: Binlog Dump
       Time: 94
      State: Has sent all binlog to slave; waiting for binlog to be updated
       Info: NULL
```

在从库上执行 SHOW PROCESSLIST 语句可以看到类似如下的信息。线程 Id 为 10 的就是 I/O 线程，State 字段值为 "Waiting for master to send event"，表示从库已经读取了主库 Binlog Dump 线程发送的所有二进制日志，正在等待主库发送新的二进制日志。线程 Id 为 11 的就是 SQL 线程，State 字段值为 "Has read all relay log; waiting for the slave I/O thread to update it"，表示从库已经应用所有中继日志中的更新，正在等待 I/O 线程写入新的中继日志

```
mysql> show processlist\G
*************************** 1. row ***************************
     Id: 10
   User: system user
   Host:
     db: NULL
Command: Connect
   Time: 11
  State: Waiting for master to send event
   Info: NULL
*************************** 2. row ***************************
     Id: 11
   User: system user
   Host:
     db: NULL
Command: Connect
   Time: 11
  State: Has read all relay log; waiting for the slave I/O thread to update it
   Info: NULL
```

关于查看复制状态的更完整的介绍，见第 9 章 "通过 PERFORMANCE_SCHEMA 库检查复制信息" 和第 10 章 "通过其他方式检查复制信息"。

第 3 章 复制格式详解

通过前面的两章,我们了解了复制技术的使用场景及其基本原理与实现,知道复制是通过二进制日志记录在主从库之间的流转来实现的。第 1 章对二进制日志的记录格式做了简要的介绍,本章将详细阐述复制格式。

3.1 复制格式概述

复制功能之所以能够正常工作,是因为写入二进制日志的事件是从主库读取,然后在从库上回放的。根据事件的类型,事件以不同的格式被记录在二进制日志中。复制格式由系统变量 binlog_format 控制(主要针对 DML 语句生效)。根据主库中记录的二进制日志格式以及系统变量 binlog_format 的不同值,在配置使用时可以将复制划分为如下几种格式(在第 1 章中有所提及,详见 1.4 节"复制格式"):

- 使用 statement 格式的二进制日志时,主库会将 SQL 语句文本写入二进制日志。在从库上执行 SQL 语句,然后将主库的数据变更应用到从库中,这称为基于 statement(语句)的复制,简称为 SBR。
- 使用 row 格式的二进制日志时,主库会将产生的事件(一组事件)写入二进制日志,以事件来表示数据的变更。将这些表示数据变更的事件复制到从库,然后在从库中应用这些事件,把主库数据同步到从库,这称为基于 row(行)的复制,简称为 RBR。
- 还可以使用 statement 和 row 的混合(mixed)格式的二进制日志,具体为 statement 还是 row 格式,由记录的内容决定。默认使用的是 statement 格式,根据语句以及使用的存储引擎,在特殊情况下会自动切换到 row 格式。这种使用 mixed 格式二进制日志的复制,简称为 MBR。

在 MySQL 5.7.7 之前,默认的二进制日志采用 statement 格式。在 MySQL 5.7.7 及更高的版本中,默认的二进制日志变更为 row 格式。MySQL NDB Cluster 7.5 中默认的二进制日志为 mixed 格式。要注意,MySQL NDB Cluster 的复制始终使用基于 row 的格式,NDB 存储引擎与基于 statement 的复制不兼容。

通过系统变量 binlog_format 来控制二进制日志的格式时,可以在会话(session)或全局(global)级别动态修改其值。在会话级别修改时,修改的值只对当前会话生效,会话断开即失效，而且修改的值对其他会话不可见；在全局级别修改时，修改的值对修改之后新建立的所有客户端连接生效，对之前已建立的客户端连接不生效（包括执行全局级别修改操作的连接本身）。动态修改的值在数据库进程重启后会丢失，如果要对这个值进行持久化，就需要在配置文件中进行设置。

注意：在某些情况下不能动态修改二进制日志格式，否则容易导致复制失败。例如，事务内不允许修改会话级别的二进制日志格式。

要修改系统变量 binlog_format 在全局级别和会话级别的值，用户必须拥有 SUPER 权限。通常，对于大部分会话而言，修改系统变量的值不需要用户具有 SUPER 权限，但在某些会话中修改它们可能会在会话之外产生影响（例如，系统变量 binlog_format、sql_log_bin 和 sql_log_off），因此用户需要拥有 SUPER 权限。

基于 statement 和基于 row 的复制各自有不同的问题和限制。有关它们的优缺点对比详见 3.2 节。

使用基于 statement 的复制，可能会遇到复制存储过程或触发器的问题（在主从数据库上各自执行这些语句会导致主从库的数据不一致），可以通过使用基于 row 的复制来避免这些问题。更多信息详见 3.2 节。

3.2 复制格式明细

3.2.1 基于 statement 和基于 row 的复制的优缺点

每一种二进制日志格式都有优点和缺点。对于大多数用户而言，混合复制是兼具数据完整性和较高性能的最佳选择。但是，执行某些任务时，需要根据实际情况来选择使用 statement 或者 row 二进制日志格式，本节对不同复制格式的优缺点进行对比，供读者在决策时参考。

1. 基于 statement 的复制的优点
- 技术成熟。
- 写入日志文件的数据较少。当更新或删除操作涉及多行时，可以大大减少存储空间，在利用二进制日志备份与恢复数据时也可以快速完成。
- 日志文件中包含所有的数据变更的原始语句，可用于数据库审计。

2. 基于 statement 的复制的缺点
- 一些执行结果不确定的 DML 语句，不能使用基于 statement 的复制，否则可能会造成主从库的数据不一致。
- UDF（用户自定义函数，即用户创建的函数）和存储过程执行的结果也不确定，因为具体的返回值受传入的参数值的影响。

- 在 DML 语句中，使用不带 ORDER BY 的 LIMIT 子句时，由于在主从库之间执行的排序结果可能不同，所以执行结果是不确定的（如果使用混合复制，会自动使用 row 格式记录执行 DML 语句后对数据所做的变更，而不是记录 DML 语句本身）。
- 使用 statement 格式的日志时，一些内置的函数无法正确复制，如下：
  - LOAD_FILE()
  - UUID()、UUID_SHORT()
  - USER()
  - FOUND_ROWS()
  - SYSDATE()（主库和从库都使用 --sysdate-is-now 选项启动时适用）
  - GET_LOCK()
  - IS_FREE_LOCK()
  - IS_USED_LOCK()
  - MASTER_POS_WAIT()
  - RAND()
  - RELEASE_LOCK()
  - SLEEP()
  - VERSION()

注意：无法正确复制不一定就不允许执行，在 READ-COMMITTED（读提交）隔离级别下，某些函数不允许执行，而在 REPEATABLE-READ（可重复读）隔离级别下却允许执行，但不一定保证能够正确复制。凡是在执行语句时产生了警告信息的，都需要留意。

- INSERT INTO ... SELECT 语句在基于 statement 的复制中需要的行级锁比基于 row 的复制多。
- 未使用索引的 UPDATE 语句需要进行表扫描，基于 statement 的复制可能比基于 row 的复制锁定的行数更多。
- 对于复杂语句，必须在主从库之间先评估数据的一致性（DML 语句），基于 row 的复制则不存在这个风险，因为主库的二进制日志只记录发生数据变更的行，而从库执行这些二进制日志时也只会执行发生数据变更的行，而不是执行实际的复杂语句本身。

提示：

使用基于 statement 的复制时：

- 从 MySQL 5.7 开始，类似 NOW() 的函数（这里指的是一些从系统变量 timestamp 获取时间值的函数）可以正确地在主从库之间进行复制（对于这些时间函数，在 MySQL 5.6 及其之前的版本在 REPEATABLE-READ 隔离级别下也是允许执行的，但在 READ-COMMITTED 隔离级别下基本不允许执行）。因为二进制日志中每个 Query_log_event（一个事件类型）都会记录时间戳（例如，SET TIMESTAMP=1555828207//;），所以对于使用时间戳的一些函数，可以在二进制日志中直接记录 SQL 语句文本，而且可以确保主从库的一致性。

- 如果碰到无法正确复制的语句，在 REPEATABLE-READ 隔离级别下将发出警告信息，并正常执行语句，但在 READ-COMMITTED 和 READ-UNCOMMITTED（读未提交）隔离级别下不允许执行。例如警告信息："Note (Code 1592): Unsafe statement written to the binary log using statement format since BINLOG_FORMAT = STATEMENT. Statement is unsafe because it uses a system function that may return a different value on the slave."，可以使用 SHOW WARNINGS 语句查看。

3. 基于 row 的复制的优点

- 可以正确复制所有数据的变更，这是最安全的复制格式。

注意：会更新 MySQL 系统库数据的 GRANT、REVOKE、TRIGGER、PROCEDURE、VIEW 等操作，都是使用 statement 格式复制到从库的。而 CREATE TABLE ... SELECT 之类的语句的复制，会被拆分为两个步骤：建表操作使用 statement 格式的日志记录；涉及数据插入操作时，会使用 row 格式的日志记录。但 GTID 复制模式不允许执行 CREATE TABLE ... SELECT 语句，因为两步操作会导致产生两个不同的 GTID（在 GTID 机制下，二进制日志中的每一个操作都会生成一个单独的 GTID）。从逻辑上来说，这显然是不合理的，所以启用 GTID 之后，GTID 的使用限制中有不允许执行该语句这一条。

- 对于以下类型的语句，从库需要的行锁更少，实现了更高的并发性：

    - INSERT INTO ... SELECT

    - 使用了 AUTO_INCREMENT（自增字段）的 INSERT 语句（这里指的是 INSERT 语句在对定义了自增字段的表执行插入数据时，不指定自增字段名和自增字段值，让其自动分配）。

    - 在 UPDATE 或 DELETE 语句中，WHERE 条件字段未使用索引时，可能导致全表扫描，但大多数被扫描的行实际上都不会被修改，只有满足 WHERE 条件值的行才会真正被修改。采用 statement 格式的二进制日志中记录的是原始 SQL 语句，这时如果该语句无法使用索引，则会扫描并锁定全表的所有数据；而如果采用 row 格式，则二进制日志中记录的是逐行数据变更，从库在回放这些二进制日志时也逐行回放，不会锁住所有行。这得益于从 MySQL 5.6 开始引入的一个新特性：在 row 格式下，如果表存在主键或唯一索引，那么可以通过特殊的优化算法找到能够唯一标志行的主键值或唯一索引值，从而避免对不需要修改的行加锁。查找算法由系统变量 slave_rows_search_algorithms 进行设置。关于该特性，可参考高鹏的"复制"专栏，登录简书网站搜索"第 24 节：从库数据查找和参数 slave_rows_search_algorithms"。

提示：综上所述，对于任何 INSERT、UPDATE 或 DELETE 语句，从库需要的行锁可能都会更少。

4. 基于 row 的复制的缺点

- 生成更多的二进制日志数据，因为基于 row 的复制会将每行数据的变更都写入二进制日志。利用二进制日志进行备份和恢复的时间也会更长。此外，二进制日志的文件锁也会因为需要更长的时间来写入数据而被持有更久的时间，这可能会影响数据库的并发能力。可以使用系统变量 binlog_row_image = minimal 来减少二进制日志的写入量。

- 如果要生成大字段的 BLOB 值，使用基于 row 的复制比使用基于 statement 的复制耗费的时间更长，因为前者记录了 BLOB 字段的具体值，而不是生成数据的语句。

- 无法直接看到从库中执行的语句，但是可以使用 mysqlbinlog 工具的--base64-output=decode-rows 和--verbose 选项进行查看，或者在主库中启用系统变量 binlog_rows_query_log_events，它会在二进制日志中写入一个 Rows_query_log_event 类型的事件来记录原始的语句文本，可以使用 mysqlbinlog 工具的-vv 选项来查看。

- 对于使用 MyISAM 存储引擎的表，当 INSERT 语句操作多行数据，在从库中重放该 INSERT 语句时，可能需要更多的表级锁，即在基于 row 的复制中，MyISAM 引擎的并发性能会受到很大影响。

3.2.2 使用 row 格式的二进制日志进行复制

所使用的二进制日志格式不同，在二进制日志文件中记录的日志量及其写入时间也各不相同。在实际使用场景中，需要根据应用程序和环境来选择。

在基于 row 的复制中，不会复制临时表，临时表只能被创建临时表的线程访问，因此没有必要记录到二进制日志中。但如果使用基于 statement 的复制，则二进制日志中会记录对临时表的操作语句，而实际上把它们记录到二进制日志中也没有什么用处。

注意：从 MySQL 5.7.25 开始，MySQL 会跟踪创建每个临时表时生效的二进制日志格式。如果是 statement 格式，则当客户端会话连接断开时，会记录 DROP TEMPORARY TABLE IF EXISTS 语句；如果是 row 格式，则不会记录该语句。在之前的版本中，无论二进制日志被设置为何种格式，当客户端连接断开时都会在二进制日志中记录 DROP TEMPORARY TABLE IF EXISTS 语句，以确保主从库都会删除该临时表。

在基于 row 的复制中，当修改的行数较多时，可能会将行数据的变更拆分到多个事件中，因此在主库中的非事务表修改多行，在从库中进行重放时会更频繁地持有表级锁。如果是不同的表，可能一定程度上能增加并发性；如果是相同的表，则不能增加并发性。

由于基于 row 的复制是将行数据的变更都记录到二进制日志中，因此日志量可能会迅速增加，而且在从库中进行重放时可能需要更长的时间，所以当应用程序访问从库时，应用开发人员需要清楚访问从库可能出现数据延迟的情况。

通过 mysqlbinlog 工具解析二进制日志，可以看到其中使用了 BINLOG 语句（这里指的是二进制日志文件中用于记录 Base64 编码的 BINLOG 语句）来显示 row 格式的行数据变更信息。此语句将事件内容显示为不容易读懂的 Base64 编码字符串，可以结合使用 mysqlbinlog 工具的--base64-output=decode-rows 和--verbose 选项，将这个字符串解析为更适合人阅读的格式，以便更容易使用二进制日志来恢复误删除的数据或从故障中恢复。

当系统变量 slave_exec_mode 设置为 IDEMPOTENT 时，通常仅对 MySQL NDB Cluster 复制有用（在 NDB 存储引擎中，该系统变量的默认值为 IDEMPOTENT，使用其他引擎时，其默认值为 STRICT）。如果 slave_exec_mode 设置为 IDEMPOTENT，当找不到行记录或者主键冲突时，会自动跳过发生错误的事件，这就意味着从库上最终并没有应用这些发生错误的事件数据，主从库之间的数据一致性会被破坏。

不能在查询语句中使用 Server ID 来过滤复制内容（这里指的是在 DML 语句的 WHERE 条件值中使用@@server_id 系统变量），因为在 row 格式的二进制日志中会将其转换为具体的值记录下来，而不是记录@@server_id 变量字符串。

如果需要使用 Server ID 来过滤复制内容，可以在从库中配置复制时，使用 CHANGE MASTER TO 语句的 IGNORE_SERVER_IDS 选项指定需要过滤的 Server ID（对于 row 格式和 statement 格式的二进制日志的复制过滤都支持此方法）。不建议在 DML 语句中使用包含 WHERE @@server_id <> id_value 的子句来过滤复制内容，例如，WHERE @@server_id <> 1，因为对于 row 格式的二进制日志，这样的语句不能正常进行复制过滤。如果确实需要在语句中使用系统变量 server_id 过滤语句，请使用 statement 格式的二进制日志。

关于数据库级别的复制选项，在 row 格式和 statement 格式的二进制日志中，--replicate-do-db、--replicate-ignore-db 和--replicate-rewrite-db 选项的效果差别很大。通常情况下，不建议使用数据库级的复制选项，而应该用表级复制选项，例如，--replicate-do-table 和--replicate-ignore-table。

使用 row 格式的二进制日志时，如果从库在更新非事务性表时停止了复制线程，则从库中可能发生数据不一致（因为非事务表数据无法回滚）。因此，建议在使用基于 row 的复制时，所有表都使用事务存储引擎（例如 InnoDB）。另外，无论使用何种格式的复制和存储引擎，建议在关闭从库 MySQL 进程之前，使用 STOP SLAVE 或 STOP SLAVE SQL_THREAD 先停止复制线程，因为这样能避免复制时出现的一些问题。

3.3 如何确定与记录复制中的安全和不安全语句

MySQL 复制中语句是否"安全"是指是否可以使用基于 statement 的格式（这里指的是在二进制日志文件中实际记录的内容为 statement 格式，不是指设置系统变量 binlog_format = statement）正确复制语句，如果能正确复制，则认为语句是安全的，否则就认为是不安全的。

- 某些执行结果不确定的函数被视为不安全（详见下文）。
- 使用浮点数的函数（执行结果与硬件相关）的语句被认为不安全。

根据语句是否被认为是安全的，以及二进制日志的格式（即系统变量 binlog_format 的当前值），对语句有不同的处理方式。

- 使用 row 格式的日志时，对安全和不安全语句的处理没有区别。
- 使用 mixed 格式的日志时，被视为不安全的语句在记录到二进制日志时会自动转换为 row 格式，被视为安全的语句在记录到二进制日志时会使用 statement 格式。
- 使用 statement 格式的日志时，对标记为不安全的语句会生成警告，甚至拒绝执行，被标记安全的语句则被正常记录。

每个被标记为不安全的语句，MySQL 都会生成一个警告。在早期版本中，如果在主库上执行大量不安全的语句（这里指的是会触发警告的语句），可能会导致错误日志文件过大。为了防止这种情况的发生，MySQL 5.5.27 及其之后的 5.5 发行版、MySQL 5.6.7 及其之后的 5.6 发行版、MySQL 5.7 及其以上发行版提供了一种警告抑制机制：在任意 50 秒的时间段内，当 ER_BINLOG_UNSAFE_STATEMENT 发出超过 50 次警告时，就会启用警告抑制。

当某种警告达到 50 次时，在最后 S 秒内重复 N 次的最后一个警告将被写入错误日志中，即一旦触发警告抑制机制，则对于 50 秒内超过 50 次的警告，只将最后一次记录到错误日志中。如果警告持续保持该频率，则警告抑制持续有效，一旦警告频率低于此阈值，则所有的警告信息将都正常记录到错误日志中。警告抑制不会影响确定语句的安全性，也不会影响向客户端发送警告信息，MySQL 客户端仍然会收到所有的警告信息。

在 statement 格式的日志中，包含某些函数的语句被认为是不安全的，因为在主从数据库中执行的结果可能不相同，在 READ-UNCOMMITTED 和 READ-COMMITTED 隔离级别下不允许执行，REPEATABLE-READ 或 SERIALIZABLE（串行）隔离级别允许执行，但是会收到警告信息。这些函数包括：FOUND_ROWS()、GET_LOCK()、IS_FREE_LOCK()、IS_USED_LOCK()、LOAD_FILE()、MASTER_POS_WAIT()、PASSWORD()、RAND()、RELEASE_LOCK()、ROW_COUNT()、SESSION_USER()、SLEEP()、SYSDATE()、SYSTEM_USER()、USER()、UUID()、UUID_SHORT()。

以下一些函数虽然执行结果也不确定，但是它们被视为安全的（实际上这些函数中除了使用系统变量 timestamp 获取时间戳的时间函数之外，其他大多数函数在主从数据库中的执行结果并不一致，在 READ-UNCOMMITTED 和 READ-COMMITTED 隔离级别下这些函数都不允许执行，在 REPEATABLE-READ 或 SERIALIZABLE 隔离级别下允许执行，但不会收到警告信息，使用这些函数时需要留意主从数据的一致性）。这些函数包括：CONNECTION_ID()、CURDATE()、CURRENT_DATE()、CURRENT_TIME()、CURRENT_TIMESTAMP()、CURTIME()、LAST_INSERT_ID()、LOCALTIME()、LOCALTIMESTAMP()、NOW()、UNIX_TIMESTAMP()、UTC_DATE()、UTC_TIME()、UTC_TIMESTAMP()。

执行语句中如果有对系统变量的引用，使用 statement 格式的日志时，将无法正确复制大多数的系统变量。

对于 UDF，由于无法控制 UDF 的作用，因此必须假设在 UDF 中执行的语句是不安全的。

Fulltext plugin（全文索引插件）在不同的 MySQL Server 上的行为可能不同，在不同的语句中执行结果也可能不相同，因此，跟 Fulltext plugin 相关的所有语句都被视为不安全。

使用触发器或存储程序 UPDATE 一个具有 AUTO_INCREMENT 字段的表时，被认为不安全，因为更新行的顺序在主从数据库中可能不相同。另外，如果一个表具有复合主键，且复合主键包含一个 AUTO_INCREMENT 字段，而该字段又不是这个复合主键的第一字段时，那么对该表的 INSERT 操作也被认为不安全。

INSERT INTO ... ON DUPLICATE KEY UPDATE 语句在具有多个唯一约束（主键 + 唯一索引 = 多个唯一约束）的表中执行时，被认为不安全。因为它对存储引擎检查索引键的顺序很敏感，MySQL Server 更新行的选择依赖于唯一索引的检查顺序，而该顺序是不确定的，所以在基于 statement 的复制中，该语句也会被标记为不安全。

使用 LIMIT 子句执行 UPDATE 操作时，未使用 ORDER BY 来指定检索行的顺序的语句，被视为不安全。

主从库之间的系统日志表的内容可能不相同，当访问或引用日志表时，可能返回不同的结果。

在同一个事务中，混合事务引擎与非事务引擎的读/写操作被认为不安全。

在事务中，对内部使用的一些记录表的所有读/写操作都被认为是不安全的。

LOAD DATA 被视为不安全，当 binlog_format = mixed 时，语句将以 row 格式记录。要注意：当 binlog_format = statement 时（而且是 REPEATABLE-READ 或者 SERIALIZABLE 隔离级别），LOAD DATA 不会生成警告，这一点与其他不安全的语句不同。

如果在主库上并行提交的两个 XA（一种分布式事务的协议名称）事务在从库上按相反的顺序执行，那么基于 statement 的复制可能会发生不安全的锁依赖关系，这可能导致从库发生死锁，进而导致复制失败。

- 当设置 binlog_format = statement 时，XA 事务中的 DML 语句被标记为不安全，并生成警告。
- 当设置 binlog_format = mixed 或 binlog_format = row 时，XA 事务中的 DML 语句将使用 row 格式的日志记录，不存在该问题。

提示：
- 更多关于二进制日志的内容，可参考第 26 章 "二进制日志文件的基本组成"。
- 更多关于隔离级别的内容，可参考《千金良方：MySQL 性能优化金字塔法则》的第 19 章 "事务概念基础"。

第 4 章　传统复制与 GTID 复制

MySQL 5.6 之前的版本只支持传统复制，即"基于二进制日志文件（binlog file）和位置（binlog pos）的复制"。在该复制模式中，复制拓扑的初始化配置和变更、复制的高可用切换等操作都需要找到正确的二进制日志文件和位置，否则就无法正确复制。然而，寻找该位置信息的过程所涉及的操作步骤较为烦琐，于是在 MySQL 5.6 及其之后的版本中，出现了基于 GTID 的复制（为了表述上的方便，书中也简称为 GTID 复制）模式。它利用 GTID 自动定位的特性，不再需要二进制日志的位置信息，也就省去了寻找这些信息所需的烦琐步骤，极大地简化了复制拓扑的初始化配置和变更，以及复制的高可用切换等操作。本章将对这两种复制模式做基本的介绍。

4.1　传统复制

传统复制是指基于二进制日志文件和位置的复制，实际上它更像是在一些特定操作下（例如，新搭建复制拓扑、调整复制拓扑、主从切换等），告诉从库如何找到正确的复制位置的一种解决方案（GTID 复制模式与此类似，但它把手动定位变更为自动定位）。每个从库使用基于二进制日志文件和位置的复制配置时，需要在使用 CHANGE MASTER TO 语句时，指定需要连接的主库 IP 地址、账号、密码、端口、二进制日志文件名和二进制日志的位置。这些信息可以保存在从库的磁盘文件中，也可以保存在从库的配置信息表中。有关细节详见第 8 章"从库中继日志和状态日志"。

第 2 章对复制的基本原理做了详细介绍，这里我们根据本章的需要，再简要阐述一遍复制的原理。

- 作为主库的 MySQL 实例（数据源）将数据变更作为"事件"写入二进制日志文件。二进制日志中记录的数据变更信息可能会以不同的日志记录格式存储（具体看系统变量 binlog_format 如何设置）。作为从库的 MySQL 实例读取主库的二进制日志，并在从库的本地数据库上应用主库二进制日志中的事件（日志重放）。

- 每个从库都会收到主库二进制日志的全部副本内容。每个从库自行决定如何执行其中的内容。除非另行指定一些过滤规则，否则在从库上会应用主库二进制日志中的所有事件。

- 每个从库都会记录当前对应主库的二进制日志文件位置、二进制日志文件名，以及在此文件中它已从主库读取和处理的位置（即 SQL 线程和 I/O 线程的位置）。每个从库独立地应用主库的二进制日志，相互之间不产生影响并各自记录自身应用到的位置，而且就算有从库与主库的连接发生断开或重连，也不会影响主库的操作（这里主要指主库的可用性）。

在同一个复制架构组内，所有实例的 Server ID 须唯一，以便唯一标识一个实例机器的二进制日志来源。这样做可以避免在较为复杂的复制架构中，重复传输与应用二进制日志（例如，环形复制和双主复制）。GTID 复制模式也是如此，下文将不再赘述。

关于基于二进制日志文件和位置的复制拓扑的搭建步骤，详见第 14 章"搭建异步复制"。

4.2　GTID 复制

GTID（Global Transaction Identifier，全局事务标识符）复制，即基于 GTID 实现的复制，指的是基于事务的复制。使用 GTID 时，在某个 MySQL Server（后文为了表述上的方便，也简称为 Server）上提交的事务可以被任意从库应用识别与跟踪，使用 GTID 复制在搭建新从库或者因故障转移到新主库时，会自动根据 GTID 来定位对应的二进制日志文件和位置，更准确地说，是自动寻找从库缺失的 GTID SET 对应的二进制日志记录，极大地降低了这些任务的复杂度。

由于 GTID 复制是完全基于事务的，因此也更容易确定主库和从库的数据是否一致，只要在主库上提交的所有事务在从库上也成功提交，就能保证两者之间的数据一致性。在一个持续正常运行的主从复制拓扑中，主从库之间的 GTID SET 一致就可以粗略地认为主从库的数据是一致的，人为的误操作除外。另外，为了获得最佳效果，建议使用基于 row 格式的复制。

GTID SET 信息在主库与从库中都会保存。这意味着可以通过 GTID SET 信息来追踪二进制日志的来源。此外，一旦在给定 Server 中提交过某个 GTID 的事务，则该 Server 将忽略后续提交的相同 GTID 的事务。因此，主库上提交的事务在从库上只能应用一次，之后碰到重复的 GTID 时会自动跳过整个事务，这有助于保证主从库数据一致。

提示：在 MySQL 5.6.x 中，如果使用 GTID 复制，则从库必须使用系统变量 log_bin 来启用二进制日志记录功能，系统变量 log_slave_updates 将主库的二进制日志记录到从库自身的二进制日志中。因为 GTID 需要被持久化，MySQL 5.7.x 及其之后的版本新增了一张 InnoDB 存储引擎的 mysql.gtid_executed 表来持久化 GTID 信息，所以可以不启用系统变量 log_bin 和 log_slave_updates。

4.2.1　GTID 的格式和存储

4.2.1.1　GTID 的格式

在主从复制拓扑中，按照使用规范，主库提供读/写操作，从库提供只读操作，所以源 Server 通常是主库的角色。GTID 是在源 Server 上创建的具有唯一性的标识符，并与源

Server 上提交的每个事务相关联。此标识符在给定的复制拓扑中的所有 Server 上都是唯一的。

通过 GTID 可以区分事务的来源（通过 GTID 组成中的 UUID 可以区分事务是由哪个 Server 提交的，关于 GTID 组成的介绍详见下文）。当主库事务被提交并将二进制日志写入二进制日志文件中时，会为其分配新的 GTID，保证事务的 GTID 单调递增且生成的数字之间没有间隔。如果事务未写入二进制日志文件（例如，事务被过滤，或者事务是只读的），则在源 Server 上不会为其分配 GTID。

在从库上应用主库的二进制日志时会保留主库事务的 GTID，即使从库在复制事务时进行了过滤，主库事务的 GTID 也会在从库中持久化，即数据可能已被过滤，但是 GTID 仍然会被记录下来。mysql 系统库下的系统表 gtid_executed 用于保存从库中应用的所有事务已分配的 GTID。在 MySQL 5.7 中，当启用系统变量 log_bin 和 log_slave_updates 时，表 mysql.gtid_executed 中的 GTID SET 不包括最后一个正在使用的二进制日志文件中的 GTID。

从库在应用主库的二进制日志时，碰到具有相同 GTID 的事务时会跳过，也就是说，主库的事务在从库上的应用不会超过 1 次（即不会重复执行），这可以保证主从库数据的一致性。一旦在给定 Server 上提交了给定 GTID 的事务，则该 Server 将不会尝试执行具有相同 GTID 的任何事务，也不会引发错误信息且不会执行事务中的语句 。

- 如果某个给定 GTID 的事务正在某个 Server 上执行，且该事务处于活跃状态（未提交也未回滚），则在该 Server 上尝试执行具有相同 GTID 的并行事务时都将被阻塞。如果这个活跃的事务被回滚，则第一个被阻塞的并行事务会继续尝试使用该 GTID 来提交，在此期间，如果其他并行会话尝试使用相同的 GTID 来提交事务，则会被阻塞。如果给定 GTID 的事务被提交，则使用了相同 GTID 的其他并行事务的所有语句都会被跳过。

GTID 由用冒号"："分隔的 UUID 和 TID 构成，即 GTID = source_id:transaction_id。

- source_id：标识事务的源 Server，通常为主库的 server_uuid。
- transaction_id：是由在主库上提交事务的顺序确定的序列号。例如，第一个事务的 transaction_id 为 1（事务的 GTID 不能使用 0 作为序列号），假如在该 Server 上提交的事务已经到了第 23 个，则第 23 的事务的 GTID 为 3E11FA47-71CA-11E1-9E33-C80AA9429562:23。

二进制日志中事务的 GTID 可以在 mysqlbinlog 工具的输出中查看，带 GTID 的复制状态可以通过 PERFORMANCE_SCHEMA 系统库下的复制状态表查看，例如，replication_applier_status_by_worker 表。更多信息详见第 9 章"通过 PERFORMANCE_SCHEMA 库检查复制状态"。

GTID SET 指的是由一个或多个 GTID 列表，或一个 GTID 范围组成的集合。GTID SET 在 MySQL Server 中有几种使用方式。例如：

- 系统变量 gtid_executed 和 gtid_purged 存储的值需要使用 GTID SET。
- START SLAVE 子句 UNTIL SQL_BEFORE_GTIDS 和 UNTIL SQL_AFTER_GTIDS 需要使用 GTID SET。其中，前者用于指定从库重放事务到 GTID SET 中的第一个 GTID 即停止，后者用于指定重放到 GTID SET 的最后一个 GTID 之后停止。
- 内置函数 GTID_SUBSET()和 GTID_SUBTRACT()的输入参数需要使用 GTID SET。

来自同一个实例的一系列 GTID 可以合并成单个表达式。例如：

```
# 来自同一个MySQL实例的第1~5个事务,其中server_uuid是3E11FA47-71CA-11E1-9E33-C80AA9429562,
表达式如下：
3E11FA47-71CA-11E1-9E33-C80AA9429562:1-5

# 来自同一个MySQL实例的第1~3个事务、第11个事务、第47~49个事务，表达式如下：
3E11FA47-71CA-11E1-9E33-C80AA9429562:1-3:11:47-49

# 来自server_uuid为2174B383-5441-11E8-B90A-C80AA9429562实例的第1~3个事务,server_uuid
为24DA167-0C0C-11E8-8442-00059A3C7B00实例的第1~19个事务，表达式如下：
2174B383-5441-11E8-B90A-C80AA9429562:1-3,24DA167-0C0C-11E8-8442-00059A3C7B00:1-19

# 提示：有多个GTID SET时，按server_uuid值中的字母顺序排列，使用逗号分隔。GTID中用于表示事务序
号的数字也会合并，合并之后按升序排序，不连续的数字之间，使用冒号分隔
```

4.2.1.2 GTID 的存储

上文提到，从 MySQL 5.7 开始，在 mysql 系统库下提供一张 InnoDB 引擎的 gtid_executed 表来持久化存储 GTID。该表中的每一行数据都包含事务的始发实例 UUID，以及对应的 GTID SET 的起始和结束事务 ID；对于仅引用单个 GTID 的行，起始和结束事务 ID 相同。

安装或升级 MySQL Server 时，如果该表不存在，则 MySQL Server 会自动使用类似如下的 CREATE TABLE 语句创建 mysql.gtid_executed 表（注意，请勿自行尝试创建与修改此表，这里列出建表 DDL 语句仅仅是为了便于说明）：

```
CREATE TABLE gtid_executed(
    source_uuid CHAR(36) NOT NULL,
    interval_start BIGINT(20) NOT NULL,
    interval_end BIGINT(20) NOT NULL,
    PRIMARY KEY(source_uuid, interval_start)
)
```

mysql.gtid_executed 表仅供 MySQL Server 内部使用。当从库禁用二进制日志功能或关闭系统变量 log_slave_updates 时，每个事务的 GTID 都会被实时记录到该表。另外，在 MySQL Server 中执行 FLUSH LOG 语句时，二进制日志会执行切换，上一个二进制日志文件中的 GTID 将会被记录到该表中。当二进制日志被丢失时，该表中的 GTID 记录不会丢失，例如，在 MySQL Server 中对二进制日志执行清理时，被清理的二进制日志包含的 GTID 在该表中仍然会保留，但当执行 RESET MASTER 语句时，该表中的内容会被重置（清空）。

仅当系统变量 gtid_mode 被设置为 ON 或 ON_PERMISSIVE 时，GTID 才存储在 mysql.gtid_executed 表中。是否在该表中实时存储 GTID，取决于系统变量 log_bin 和 log_slave_updates 的值，详情如下：

- 如果禁用二进制日志记录（注意，禁用二进制日志需要注释掉系统变量 log_bin，而不是将其设置为 OFF），或者禁用 log_slave_updates（设置为 OFF 或 0），则 MySQL Server 在每个事务提交时将事务的 GTID 一并记录到该表。

- 如果启用了二进制日志记录（启用了系统变量 log_bin 和 log_slave_updates。注意，在启动 MySQL Server 时，为系统变量 log_bin 指定了具体值，就表示启用二进制日志记录功能，而不是通过将其设置为"ON"字符串来启用二进制日志记录功能，如果这样做，最终会让所有的二进制日志文件的前缀都变成"ON"字符串），则只在切换二进制日志或关闭 MySQL Server 时，将先前二进制日志的所有事务的 GTID 写入 mysql.gtid_executed 表。这种情况适用于主库或启用了二进制日志记录的从库。如果 MySQL Server 意外停止，则当前二进制日志文件中的 GTID SET 不会保存在 mysql.gtid_executed 表中。在 MySQL Server 重新启动期间，将扫描二进制文件并将这些 GTID 记录到表中（注意，如果 MySQL Server 重新启动时禁用了二进制日志记录功能，则在重新启动期间不会扫描二进制日志中的 GTID，也就是说二进制日志中的 GTID 无法被记录下来，也就无法启动复制），但不会记录所有已执行事务的 GTID，最后一个二进制日志文件中的 GTID 不会被记录。GTID 的完整记录由系统变量 gtid_executed 的全局值提供，始终使用全局变量@@global.gtid_executed（该变量在事务每次提交后更新）来表示 MySQL Server 的最新 GTID 状态，并且在查询该全局变量的值时不会查询 mysql.gtid_executed 表中的数据。

以下是一个 mysql.gtid_executed 表压缩示例。

```
# 随着时间的推移，mysql.gtid_executed 表记录了大量数据，而且很多行都源自同一个 MySQL Server 的 GTID
记录，类似如下：
+--------------------------------+----------------+--------------+
| source_uuid                    | interval_start | interval_end |
|--------------------------------+----------------+--------------|
| 3E11FA47-71CA-11E1-9E33-C80AA9429562 | 37 | 37 |
| 3E11FA47-71CA-11E1-9E33-C80AA9429562 | 38 | 38 |
| 3E11FA47-71CA-11E1-9E33-C80AA9429562 | 39 | 39 |
| 3E11FA47-71CA-11E1-9E33-C80AA9429562 | 40 | 40 |
| 3E11FA47-71CA-11E1-9E33-C80AA9429562 | 41 | 41 |
| 3E11FA47-71CA-11E1-9E33-C80AA9429562 | 42 | 42 |
| 3E11FA47-71CA-11E1-9E33-C80AA9429562 | 43 | 43 |
...

# 为了节省空间，MySQL Server 定期压缩 mysql.gtid_executed 表，将每个这样的行合并成一个 GTID 范围
的单行，类似如下：
+--------------------------------+----------------+--------------+
| source_uuid                    | interval_start | interval_end |
```

```
|----------------------------------+-------------+-----------|
| 3E11FA47-71CA-11E1-9E33-C80AA9429562 | 37 | 43 |
...
```

通过设置系统变量 gtid_executed_compression_period，可以在每次执行压缩表之前控制表中允许记录的事务数，即每隔多少个事务压缩一次，从而控制压缩率。此变量的默认值为 1000，意味着在默认情况下每 1000 个事务之后执行表的压缩。将 gtid_executed_compression_period 设置为 0 可以防止执行压缩，如果要关闭压缩功能，请留意磁盘空间是否充足。

启用二进制日志记录时，不会使用 gtid_executed_compression_period 的值，即不根据该变量定义的事务数量来执行压缩，而是在每次执行二进制日志切换时压缩 mysql.gtid_executed 表。

mysql.gtid_executed 表的压缩由名为 thread/sql/compress_gtid_table 的专用前台线程执行。此线程未在 SHOW PROCESSLIST 的输出中列出，但可以通过查询 performance_schema.threads 表来查看，类似如下内容：

```
mysql> select * from performance_schema.threads where name like '%gtid%'\G
*************************** 1. row ***************************
         THREAD_ID: 26
              NAME: thread/sql/compress_gtid_table
              TYPE: FOREGROUND
     PROCESSLIST_ID: 1
......
```

线程 thread/sql/compress_gtid_table 通常会休眠，直到执行了系统变量 gtid_executed_compression_period 定义的数量的事务时被唤醒，执行对 mysql.gtid_executed 表的压缩，然后继续休眠，直到下一次执行的事务达到 gtid_executed_compression_period 定义的数量时再被唤醒执行压缩，无限重复此循环。如果禁用二进制日志记录的同时将此值设置为 0，则意味着线程始终处于休眠状态且永不会被唤醒

4.2.2 GTID 的生命周期

GTID 的生命周期包括以下几个阶段：

（1）在主库上执行事务并提交。事务在提交时会被分配一个 GTID，该 GTID 由主库的 UUID 和主库上尚未使用的最小非零事务序列号组成。随后，此 GTID 被写入主库的二进制日志（在二进制日志的 GTID 事件中记录，该事件排在事务的 binlog event group 的最前面）。如果未将事务写入二进制日志（例如，因为主库中过滤了对某个数据库操作的事务，即不写入二进制日志；或者在主库中某个事务是只读的），则不会为其分配 GTID。

（2）如果为事务分配了 GTID，则在提交事务时，通过二进制日志中的 Gtid_log_event 将 GTID 写入二进制日志做原子保留。如果二进制日志执行了切换或 Server 关闭，则会将二进制日志文件中涉及的所有事务的 GTID 写入 mysql.gtid_executed 表中保存。

（3）如果为事务分配了 GTID，则在提交事务之后不久通过非原子方式将 GTID 添加到

系统变量 gtid_executed（@@global.gtid_executed）的 GTID SET 中，系统变量 gtid_executed 的 GTID SET 值在复制中用来标记已提交数据的状态，即表示当前数据库中的数据对应着哪些事务。当启用二进制日志记录（主库必须启用）后，系统变量 gtid_executed 中的 GTID SET 是已提交事务的完整记录，但 mysql.gtid_executed 表中记录的 GTID SET 不完整。在启用系统变量 log_bin 和 log_slave_updates 时，该表中只记录除最后一个正在使用的二进制日志之外（最新的 GTID 保存在当前活跃的二进制日志文件中，也即这里所说的最后一个二进制日志文件），其他所有更早的二进制日志中所涉及的 GTID SET。

（4）将二进制日志数据传输到从库并存储到中继日志之后，从库读取 GTID 并将读取到的 GTID 设置为其系统变量 gtid_next 的值，以告诉从库必须使用此 GTID 记录下一个事务。注意，从库是在会话上下文中设置系统变量 gtid_next 的值的，而不是在全局级别设置的。

（5）从库验证在 gtid_next 中获得的 GTID，是否已有线程获得该 GTID 的所有权，以便处理事务。首先，读取和检查复制事务的 GTID，在处理被读取 GTID 的事务之前，不仅需要保证在从库上此前该 GTID 没有被其他事务应用过，而且还需要保证当前没有其他会话已经读取此 GTID 但尚未提交事务，即持有该 GTID 且处于活跃状态的事务。因此，如果多个客户端尝试同时应用同一事务，即拥有同一个 GTID 的事务，则 Server 只允许其中一个执行，从而解决此问题。从库的系统变量 gtid_owned（@@global.gtid_owned）显示当前正在使用的每个 GTID 以及拥有它的线程的 ID。因此，如果某个 GTID 已经被使用，当某个线程再次处理相同 GTID 的事务时不会引发错误，而是会使用自动跳过功能来忽略该事务。

（6）如果 GTID 尚未被使用，则从库会应用该事务。由于在重放主库二进制日志时，二进制日志的 GTID EVENT 中的语句 "SET @@SESSION.GTID_NEXT='...';" 会将从库的系统变量 gtid_next 设置为该语句指定的 GTID，因此从库不会尝试为此事务生成新的 GTID。

（7）如果在从库上启用了二进制日志记录且启用了系统变量 log_slave_updates，则从库在将二进制日志回放完之后，也会将这些日志中涉及的 GTID 写入自身的二进制日志文件做原子保留。在二进制日志执行切换或关闭 Server 时，Server 会将二进制日志文件中所有事务的 GTID 写入 mysql.gtid_executed 表中保存。

（8）如果在从库上禁用二进制日志记录或者禁用了系统变量 log_slave_updates，则通过实时将 GTID 直接写入 mysql.gtid_executed 表来原子性地保留 GTID。MySQL 在每个事务中附加一条语句，将 GTID 插入 mysql.gtid_executed 表中。在这种情况下，mysql.gtid_executed 表是从库上应用的事务的完整记录。请注意，在 MySQL 5.7 中，将 GTID 插入 mysql.gtid_executed 表中的操作，对于 DML 语句是原子操作，但对于 DDL 语句则不是原子操作（因为 DDL 语句不受事务控制）。因此，如果 Server 在执行 DDL 语句时意外退出，则 GTID 的状态可能会变得不一致。但从 MySQL 8.0 开始，DDL 语句和 DML 语句都支持原子操作。

（9）在从库上提交复制事务后不久，GTID 通过非原子化的方式将其添加到从库的系统变量 gtid_executed（@@global.gtid_executed）中。对于主库，此 GTID SET 包含所有已提交的事务集。对于从库，如果禁用二进制日志记录，就无法使用二进制日志文件来持久化

GTID，但会将其实时记录到 mysql.gtid_executed 表中。如果启用了二进制日志记录，则 mysql.gtid_executed 表并不会实时记录事务的 GTID，也就是说，这时该表中记录的 GTID 并不完整，最新的 GTID 不在该表中，而是在最后一个二进制日志文件中。但是要注意，系统变量 gtid_executed 中的 GTID SET 是完整的，它也是在这种情况下唯一完整记录 GTID 的地方。

在主库上完全过滤掉的客户端事务（例如，主库使用复制过滤参数指定过滤某个数据库）由于未分配 GTID，因此不会被添加到系统变量 gtid_executed 中，也不会被添加到 mysql.gtid_executed 表中。但是，在从库上为了保证主从库数据的一致性，即使使用复制过滤参数过滤了某些事务，这些被过滤的事务对应的 GTID 也必须被持久化（需要确保在从库上启用了二进制日志记录且启用了系统变量 log_slave_updates）。从库在将这些被过滤的事务写入自身的二进制日志时，通过 Gtid_log_event 事件写入仅包含 BEGIN 和 COMMIT 语句的空事务，而不包含被过滤的事务对应的数据变更日志。如果禁用二进制日志记录或禁用系统变量 log_slave_updates，则被从库过滤的事务的 GTID 将被实时写入 mysql.gtid_executed 表。在从库上为被过滤的事务保留 GTID，可以确保与主库拥有一致的 GTID SET。另外，当从库重新连接到主库时，还可以确保不会重复拉取被过滤的事务。

在多线程复制的从库（slave_parallel_workers > 0）上，可以并行应用事务，因此复制的事务可以无序提交，除非设置 slave_preserve_commit_order = 1。当发生这种情况时，系统变量 gtid_executed 中的 GTID SET 将包含多个 GTID 范围，它们之间存在间隙。通常，在主库或单线程复制的从库上，GTID 的数值是单调递增的，数值之间没有间隙。对于多线程复制，从库上的 GTID 间隙仅发生在最近应用的事务中，并在复制过程中会填充。当使用 STOP SLAVE 语句正常停止复制线程时，会先等待应用完成正在执行的事务，然后才会真正停止复制线程，这样就能够填补 GTID 间隙。如果发生异常关闭，例如 Server 故障或使用 KILL 语句停止复制线程，则 GTID 可能会出现间隙。

客户端可以在执行事务之前将变量@@session.gtid_next 设置为有效的 GTID 值来模拟复制事务。mysqlbinlog 解析二进制日志文件时生成的 GTID 事件中也是使用此技术生成 GTID 的，这样客户端就可以重放该解析内容以保留 GTID。通过客户端提交的模拟复制事务，或者将解析的二进制日志文本内容直接导入某个数据库实例，完全等同于通过复制线程提交的复制事务，并且事后也无法区分它们。

系统变量 gtid_purged（@@global.gtid_purged）中的 GTID SET 包含已在 Server 上提交，但在 Server 上的任何二进制日志文件中都不存在的所有事务的 GTID。

- 在从库上禁用二进制日志记录时，该系统变量的值表示从库上所提交的所有复制事务的 GTID。
- 启用二进制日志记录时，该系统变量的值表示已写入二进制日志且已被清除的二进制日志文件中包含的事务的 GTID。
- 通过语句 SET @@global.gtid_purged 可以为系统变量@@global.gtid_purged 明确设

置一个 GTID SET。

Server 在启动时，将对系统变量 gtid_purged 中的 GTID SET 进行初始化。每个二进制日志文件中的内容都以事件 Previous_gtids_log_event 开头，该事件包含先前二进制日志文件中的所有 GTID SET，由前一个文件的 Previous_gtids_log_event 事件中的 GTID 与前一个文件中每个 Gtid_log_event 事件的 GTID 组成。最旧和最新的二进制日志文件的 Previous_gtids_log_event 事件的内容用于计算 Server 启动时系统变量 gtid_purged 对应的 GTID SET。

4.2.3　GTID 自动定位

GTID 自动定位功能替代了在传统复制中手工指定用于确定主从库之间数据流的起始点、停止点和恢复点的文件偏移位置的工作。使用 GTID 时，从库需要与主库同步的所有信息（这里主要指的是主库中数据变更产生的二进制日志记录）都直接从复制数据流中自动获取。

要使用 GTID 复制启动从库，请不要在 CHANGE MASTER TO 语句中包含 MASTER_LOG_FILE 或 MASTER_LOG_POS 选项，这些选项指定二进制日志文件的名称和文件中的起始位置。对于 GTID，你需要启用 MASTER_AUTO_POSITION 选项。有关 GTID 复制的配置，以及启动主库和从库的完整说明，详见第 14 章 "搭建异步复制"。

默认情况下，MASTER_AUTO_POSITION 选项被禁用。如果在从库上启用了多源复制，则需要为每个复制通道设置该选项。如果将该选项启用之后再禁用，会使从库恢复为基于文件的复制。在这种情况下，你必须指定 MASTER_LOG_FILE 或 MASTER_LOG_POS 选项中的一个或两个（从 GTID 复制模式切换到传统复制模式时，建议同时指定 RELAY LOG 相关的选项，否则还未来得及重放的中继日志会被清理，容易出现复制故障）。

当复制从库启用 GTID（gtid_mode=on|on_permissive|off_permissive）并启用 MASTER_AUTO_POSITION 选项时，将激活自动定位以连接主库。主库必也须设置 gtid_mode = on 才能使从库连接成功。在初始握手通信中，从库发送一个 GTID SET，其中包含从库已经收到、已提交或两者都已完成的事务。此 GTID SET 等于系统变量 gtid_executed（@@global.gtid_executed）中的 GTID SET 与接收事务时记录在 performance_schema.replication_connection_status 表中的 received_transaction_set 字段表示的 GTID 的并集（received_transaction_set 字段值可以使用 select received_transaction_set from performance_schema.replication_connection_status 语句来查询）。

主库通过发送其二进制日志中记录的事务来响应从库，但这些事务的 GTID 未包含在从库发送的 GTID SET 中，即这些 GTID 未包含在从库请求连接主库时注册的 GTID SET 中，因为向主库注册的这些 GTID 代表在从库中已经被应用，或者存在于从库的中继日志中，不需要重复发送，以确保主库只发送从库中未接收或未提交的 GTID 对应的事务。如果从库从多个主库接收事务，则此时自动跳过事务的功能可确保事务不会被重复应用。

如果主库应发送给从库的任何事务已从主库的二进制日志中清除，则主库将错误信息 ER_MASTER_HAS_PURGED_REQUIRED_GTIDS 发送给从库，而且复制线程启动失败或报错中止。此时，从库无法自动从此错误中恢复，需要从另一个数据源恢复丢失的事务或者使用最新的备份（可在主库上实时备份）来恢复一个新的从库，替换旧的从库。然后，可以考虑是否修改主库上二进制日志文件的保留时长，以避免二进制日志被过早清理。

如果在事务传输期间，主库发现从库接收或提交了与主库 UUID 相同的事务，但主库本身没有它们的记录，则主库将错误信息 ER_SLAVE_HAS_MORE_GTIDS_THAN_MASTER 发送给从库，并且复制线程启动失败或报错终止。如果主库在没有设置系统变量 sync_binlog = 1 的情况下碰到电源故障或操作系统崩溃，而且丢失了尚未在主库中落盘但已被从库接收的已提交事务的二进制日志时，就会发生这种情况。这可能导致主从 Server 具有相同 GTID 但是记录的数据却完全不同（主库中丢失的 GTID 被重新分配给客户端提交新的数据），即产生了主从库数据不一致。此时，正确的恢复方法是手动检查主库和从库的数据是否出现了不一致。如果确实出现不一致，而且主库上缺少事务，则可以将主库设置为从库，从其他拥有该缺失事务的 Server 中同步所缺少的事务；或者从复制拓扑中删除主库或从库，重建搭建复制拓扑。当成功恢复主库之后，根据需要选择是否将主库角色切换回去。

使用 GTID 复制时，关于跳过复制错误的步骤详见第 24 章"发生数据误操作之后的处理方案"，关于复制模式的变更步骤详见第 17 章"复制模式的切换"。

4.2.4　GTID 复制模式的限制

由于 GTID 复制依赖于事务，因此在使用时不支持 MySQL 中的某些功能。下面是使用 GTID 复制时的一些限制。

1. 更新操作涉及非事务引擎

- 在同一个事务中，不能同时操作支持事务（InnoDB）和不支持事务（MyISAM）的引擎。这是由于同时操作这两类引擎时可能导致将多个 GTID 被分配给同一个事务。
- 主从数据库 Server 中相同的表使用不同的存储引擎时（其中，一个 Server 使用事务表，另一个 Server 使用非事务表），如果在非事务表上定义了触发器，可能导致事务与 GTID 之间的一一对应关系被破坏。

2. CREATE TABLE ... SELECT 语句

使用 GTID 复制时，CREATE TABLE ...语句不允许同时使用 SELECT 语句。当 binlog_format 设置为 statement 时，CREATE TABLE ... SELECT 语句是作为一个整体且只分配一个 GTID 的事务形式被记录在二进制日志中的，但如果使用 row 格式的二进制日志，则该语句将被记录为具有两个 GTID 的两个事务。如果主库使用 statement 格式的二进制日志，而从库使用 row 格式的二进制日志，则从库将无法正确处理事务。

3. 临时表

使用 GTID 时（这里指的是系统变量 enforce_gtid_consistency 设置为 ON 时），事务、存储过程、存储函数和触发器内不支持 CREATE TEMPORARY TABLE 和 DROP TEMPORARY TABLE 语句，不过可以在这些对象之外执行 CREATE TEMPORARY TABLE 和 DROP TEMPORARY TABLE 语句，但需要使用 autocommit = 1 自动提交。

4. 防止执行不受支持的语句

要防止执行会导致 GTID 复制失败的语句，则需要在启用 GTID 时，在整个复制拓扑的所有实例中启用系统变量 enforce_gtid_consistency。当启用此系统变量之后，上述可能会导致复制出现问题的语句将直接报错，不予执行。

注意：启用系统变量 enforce_gtid_consistency 后，仅对需要写入二进制日志的语句强制执行 GTID 一致性，如果 Server 禁用二进制日志，或者复制过滤选项将可能导致 GTID 复制出问题的语句过滤掉，即最终并没有记录到二进制日志中，则不会对未记录到二进制日志中的语句强制执行 GTID 一致性。

5. 关于跳过事务

使用 GTID 时不支持使用系统变量 sql_slave_skip_counter 来跳过事务。如果需要跳过事务，可以用第 24 章"发生数据误操作之后的处理方案"中提到的注入空事务的方法来跳过事务。

6. 忽略 Server 实例

使用 GTID 时，不推荐在 CHANGE MASTER TO 语句中使用 IGNORE_SERVER_IDS 选项来忽略某个 Server 实例的二进制日志变更，因为在 GTID 复制模式下，已经应用的事务会自动被忽略。在启动 GTID 复制之前，请检查并清除该选项的设置，可通过 SHOW SLAVE STATUS 语句输出中的 Replicate_Ignore_Server_Ids 字段检查，如果未配置该选项，则该字段值为空。

7. GTID 复制模式和 mysql_upgrade

当 Server 启用了 GTID 复制模式时（gtid_mode = ON），如果需要对 Server 使用 mysql_upgrade 进行升级，则不能启用二进制日志记录（--write-binlog 选项）。

第 5 章 半同步复制

除了内置的异步复制，MySQL 5.7 还支持通过插件方式实现半同步复制接口。相对于 MySQL 5.5 和 MySQL 5.6 中的半同步复制，通常我们将 MySQL 5.7 中的半同步复制称为"增强半同步复制"，也称为"无损复制"（MySQL 5.5 和 MySQL 5.6 虽然也支持半同步复制，但不能保证"无损复制"，详见 5.4 节"半同步复制的注意要点"）。本章将详细介绍 MySQL 5.7 中的半同步复制。

5.1 半同步复制的原理

在 MySQL 中配置复制时，如果不额外加载其他任何与复制相关的插件，则默认为异步复制。主库将事件写入二进制日志，但不知道从库是否接收成功，也不知道从库什么时候重放二进制日志。如果主库崩溃，则它已提交的事务可能尚未传输到任何从库。在这种情况下，如果发生从主库到从库的故障转移，可能导致主库中还未来得及传输到从库的那一部分事务丢失。

异步复制的示意图如图 5-1 所示（参考 Oracle MySQL 官方资料）。该图描述了异步复制中主从库数据同步的大致时间线，其中 Master 表示主库，Slave1 和 Slave2 表示主库的两个从库。这里以自动提交的事务为例，因为 MySQL 适合运行一些"短、平、快"的事务，所以通常会启用事务的自动提交，主库发起事务提交，在 execute（执行）阶段执行完对数据的修改操作，然后在 binlog（二进制日志）阶段将修改数据所产生的二进制日志记录写入二进制日志文件中，在 commit（提交）阶段完成存储引擎层的事务提交（事务的状态修改为"提交"）。与此同时，主库会通过 Dump 线程将二进制日志记录发送给两个从库，两个从库收到后会写入 relay log（中继日志）文件中。之后，两个从库各自读取 relay log 文件中的内容进行 apply（应用），即模拟事务在主库中的提交方式来回放 relay log。当这些事务在从库提交时，也可能会写入自己的 binlog 中（从库启用系统变量 log_bin 和 log_slave_updates 时才会写入）。最后，两个从库在 commit 阶段各自修改存储引擎层中的事务状态，结束事务。

由于主库执行提交与发送二进制日志是异步的，也就是说，从库是否成功接收二进制日志不影响主库中的事务执行提交，因此可能会出现"主库发生宕机，但主库中已提交事务的二进制日志并没有被任何从库成功接收"的情况，即发生了数据丢失。

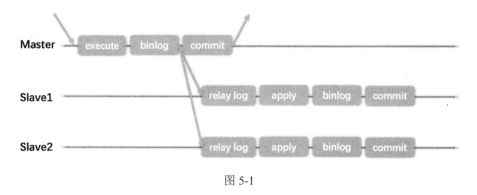

图 5-1

为了避免出现上述问题，MySQL 对异步复制进行改进，引入了半同步复制。但半同步复制不是原生的内置功能模块，要让半同步复制功能正常工作，需要额外加载半同步复制功能模块的插件。主库端和从库端需要安装不同的插件库文件，它们都包含在支持半同步复制的 MySQL 发行版中，不需要单独下载。更多关于安装与配置的信息可参考第 15 章 "搭建半同步复制"。半同步复制处于正常工作状态时，主库提交的事务产生的二进制日志需要至少被一个从库接收并写入中继日志，等返回的 ACK 消息被主库成功接收之后，主库才会确认事务已提交。正常情况下，主库发生故障转移时不会发生数据丢失。

半同步复制的示意图如图 5-2 所示（参考 Oracle MySQL 官方资料）。该图描述了半同步复制中主从库数据同步的大致时间线。从图中我们可以看到，在 binlog 阶段后 commit 阶段前，主库必须等待从库在 relay log 阶段之后回复的 ACK 消息。而从库给主库回复 ACK 消息之前，必须确保已经成功接收主库的二进制日志记录，并写入中继日志。这样，当主库发生故障时，主库已提交的事务如果丢失，可以通过从库的中继日志恢复，避免在主库发生故障时已提交的数据丢失[1]。

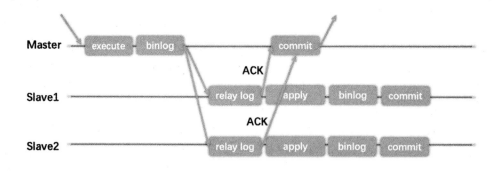

图 5-2

当主库的某个会话提交事务发生阻塞时（在等待从库的 ACK 消息），它不会返回给任何信息给执行事务的会话。当阻塞结束时（已经收到从库的 ACK 消息或者等待 ACK 消息

---

[1] 注：半同步复制从原理上讲，是可以避免数据丢失的，但在实际中还需要看从库落盘相关的系统变量设置、磁盘设备相关的落盘参数设置，以及数据库的生命管理周期设计是否合理等。

超时），主库将事务的提交状态返回给会话，并继续执行其他事务。此时，事务在主库中已经完成提交，而且确保至少有一个从库确认收到这个事务的事件日志。主库可以使用系统变量 rpl_semi_sync_master_wait_for_slave_count（默认值为 1）来设置需要收到多少个从库发回的 ACK 消息，才能执行存储引擎层的事务提交。

当一个修改非事务引擎表的语句发生回滚时，二进制日志仍然会被记录下来，同时主库发起操作的会话也会被阻塞，因为对非事务表的修改实际上是无法回滚的，对非事务表的修改必须发送到从库，而且需要确保有至少一个从库接收了事件日志。

自动提交的每个语句，都会隐式开启一个事务。在半同步复制中，每个自动提交的语句都会被阻塞，直到主库至少接收一个从库的 ACK 消息，阻塞才解除。

与异步复制相比，半同步复制提供了更好的数据完整性，因为当发起事务提交的会话收到提交成功返回的信息时，数据至少已经存在于两个位置（主库和返回 ACK 消息的从库），在主库未收到从库的 ACK 消息之前，主库中发起事务提交的事务处于阻塞状态（事务未提交，处于等待 ACK 消息的状态）。

在**繁忙**的 Server 上，半同步复制确实会对性能产生一些影响，由于需要等待从库成功接收事件日志，所以在主库中的事务提交速度变慢，但这是对提高数据完整性的一种权衡。等待的时长至少为主库将提交事务的二进制日志发送到从库，并等待从库确认收到日志的 TCP/IP 协议通信包（ACK 消息）的整个往返时间。这意味着半同步复制最适合在高速网络中使用，不适合在低速网络中使用。

系统变量 rpl_semi_sync_master_wait_point 控制在半同步复制中主库返回事务提交状态信息给提交事务的客户端之前，等待从库的 ACK 消息的点位。其有效值如下：

- AFTER_SYNC（默认值）：主库将每个事务写入其二进制日志和从库，并将二进制日志同步到磁盘。主库在把二进制日志同步到磁盘之后等待从库确认接收事务。收到从库的 ACK 消息后，主库将在存储引擎层提交事务，并将提交结果返回给发起事务提交的客户端，然后客户端可以继续做其他事情。

- AFTER_COMMIT：主库将每个事务写入其二进制日志和从库，同步二进制日志到磁盘，并继续在存储引擎层执行事务的提交。提交事务后，主库等待接收事务的从库确认。收到从库的 ACK 消息后，主库将提交结果返回给客户端，然后客户端可以继续做其他事情。

系统变量 rpl_semi_sync_master_wait_point 的不同值的特征如下：

- AFTER_SYNC：主库未收到从库的 ACK 消息之前，发起事务提交的会话不会收到事务的提交结果，存储引擎层也不会执行提交。当主库收到从库的 ACK 消息之后，主库在存储引擎层执行该事务的提交，之后返回事务的提交结果给发起事务提交的客户端。因此，所有客户端在主库上看到的数据都相同，如果主库发生故障，则在主库上提交的所有事务都已至少被复制到一个从库中（从库会确保接收的事务日志在中继日志中已落盘）。主库发生故障转移之后数据是无损的。

- AFTER_COMMIT：主库未收到从库的 ACK 消息之前，发起事务提交的会话未收到事务的提交结果，但存储引擎层会先执行提交。所以，在事务执行提交之后、未收到从库的 ACK 消息之前，主库中的其他客户端可以查看到这个事务（只是发起事务提交的会话未收到事务提交的结果信息，但是由于在存储引擎层中该事务已经提交，对于其他会话来说，该事务其实就是正常提交了）。如果在这种情况下主库发生故障，主库中已经提交的事务（可能是最后一个事务）的二进制日志有可能还没来得及被从库成功接收，那么主库发生故障转移之后，原来在主库中能够查询到的事务，在业务切换到从库之后可能在从库中无法查询到（发生数据丢失）。

"半同步复制"中的"半"是什么意思？

- 使用异步复制时，主库将事件写入其二进制日志，而从库在准备就绪之后请求这些日志，并将其写入自身的中继日志。但该过程无法保证任何事件日志都能被从库成功接收并写入中继日志。
- 使用完全同步复制，当主库提交事务时，主库必须等待所有从库中的事务重放完（提交完）之后才能提交，这样会导致在主库中的写事务存在大量的延迟提交，即大量事务被阻塞。
- 半同步复制介于异步复制和完全同步复制之间。主库等待至少收到一个从库的 ACK 消息即可，不需要等到所有从库都收到二进制日志，并且主库只需要等到从库成功接收，不需要等待从库完成重放，即事务提交。

对数据零丢失和数据一致性有一定要求的场景，可以用半同步复制替代异步复制：

- 当一个从库使用半同步复制连接到主库时，从库会告知主库自己是否具有半同步复制的能力，以便主库决定是否需要使用半同步复制方式发送二进制日志给从库。
- 如果主库启用了半同步复制，并且至少有一个保持半同步复制连接的从库，则在主库上执行事务提交时线程会等待，直到收到至少一个半同步从库确认收到事务日志时返回的 ACK 消息（或者主库在等待 ACK 消息时超时）。
- 从库需要确保将收到的主库二进制日志事件写入其中继日志并刷新到磁盘之后，才确认收到主库的事务事件（即此时才会发送 ACK 消息给主库）。
- 如果主库在超时时间内没有收到任何从库的 ACK 消息，则主库将切换为异步复制，直到至少有一个从库恢复与主库的半同步复制连接，主库才重新切换为半同步复制。
- 要正确使用半同步复制，需要主从库两端都启用半同步复制，如果在主库上禁用了半同步复制，或者在从库上禁用了半同步复制，则主库将使用异步复制中的方式传输二进制日志。

提示：半同步复制的搭建步骤详见第 15 章"搭建半同步复制"。更多信息可参考高鹏的"复制"专栏，登录简书网站搜索"第 15 节：MySQL 层事务提交流程简析"。

5.2　半同步复制的管理接口

半同步复制的管理接口指的是下面所述的插件、系统变量和状态变量。两个插件实现半同步功能，主库和从库各自安装一个插件，主库安装 semisync_master.so 插件，从库安装 semisync_slave.so 插件。

系统变量控制插件的行为，如下所述：

- rpl_semi_sync_master_enabled：控制是否在主库上启用半同步复制。要启用或禁用插件，将此变量设置为 1 或 0 即可，默认值为 0（表示禁用插件）。
- rpl_semi_sync_master_timeout：一个以毫秒为单位的值，用于控制主库在超时并切换到异步复制之前，等待从库返回 ACK 消息的时间，默认值为 10000（10 秒）。
- rpl_semi_sync_slave_enabled：与 rpl_semi_sync_master_enabled 类似，但它控制从库端插件 semisync_slave.so 的启用或禁用。

状态变量可以监控半同步复制的状态，如下所述：

- rpl_semi_sync_master_clients：显示半同步从库的数量。
- rpl_semi_sync_master_status：显示当前主库中半同步复制插件是否处于启用状态。如果已启用，则该值为 1；如果未启用，或者由于等待 ACK 消息超时（主库已退回到异步复制），则该值为 0。
- rpl_semi_sync_master_no_tx：从库未成功确认（异步复制模式）事务的提交数。
- rpl_semi_sync_master_yes_tx：从库成功确认（半同步复制模式）事务的提交数。
- rpl_semi_sync_slave_status：显示当前从库中的半同步复制插件是否处于启用状态。如果插件已启用且从库 I/O 线程正在运行，则此值为 1，否则为 0。

提示：只有当主库和从库中使用 INSTALL PLUGIN 语句安装了相应的主库插件或从库插件时，半同步复制相关的系统变量和状态变量在主/从库中才可用。

5.3　半同步复制的监控

半同步复制插件提供了几个系统变量和状态变量，通过检查这些变量，可以确定插件的配置是否正确、操作是否处于半同步复制或者异步复制状态，关键的系统变量和状态变量可参考 5.2 节"半同步复制的管理接口"。

```
# 通过系统变量可以查看半同步复制插件的配置是否正确，使用 SHOW VARIABLES 语句进行查询
mysql> show variables like 'rpl_semi_sync%';

# 通过状态变量可以查看半同步复制插件的工作或操作状态，使用 SHOW STATUS 语句进行查询
mysql> show status like 'Rpl_semi_sync%';
```

当主库提交事务等待从库的 ACK 消息超时之后，半同步复制将回退为异步复制，或者从库重新连接上主库之后切换为半同步复制时，MySQL Server 会相应地设置状态变量 rpl_semi_sync_master_status 的值。当主库从半同步复制自动回退到异步复制时（非人为修改 rpl_semi_sync_master_enabled = 0 导致），此时半同步复制实际上不可操作的，但系统变量 rpl_semi_sync_master_enabled 在主库中的值仍然为 1，即这个值一旦配置好，除非人为修改，始终为 1。可以监控状态变量 rpl_semi_sync_master_status，当它为 ON 时，就表示半同步复制插件为启用状态，为 OFF 就表示为禁用状态。

要在主库中查看其连接了多少个半同步从库，可以通过检查状态变量 rpl_semi_sync_master_clients 值进行确认。

状态变量 rpl_semi_sync_master_yes_tx 和 rpl_semi_sync_master_no_tx 表示通过半同步复制方式已成功或未成功提交的事务数。

在从库端，状态变量 rpl_semi_sync_slave_status 表示半同步复制插件的当前状态。

5.4 半同步复制的注意要点

早期的半同步复制有个缺陷。在正常的半同步复制流程中，当客户端对主库发起一个事务提交之后，主库发送二进制日志给从库，从库收到二进制日志并返回 ACK 消息，然后主库返回事务提交成功的消息给发起提交的客户端。这里对于发起事务提交的客户端看起来没有任何问题，但实际上在早期的半同步复制中，主库在等待 ACK 消息的 InnoDB 存储引擎内部已经提交事务，只是阻塞了返回给发起事务提交的客户端的消息而已。此时，如果有其他会话对该事务修改的数据进行查询，将会查询到最新数据，参见图 5-3（该图来自 my-replication-life blogspot）。

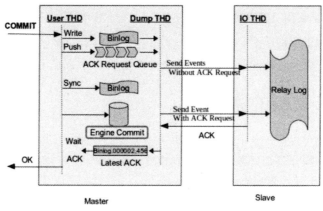

图 5-3

该缺陷可能导致除发起事务提交的客户端会话之外，其他的客户端会话在碰到主库故障转移时发生幻读，参见图 5-4（该图来自 my-replication-life blogspot）。User1 发起一个 INSERT 操作，写入一行数据，正在等待写入成功并返回，此时 User2 就可以在主库上查询到 User1 插入的数据了。当主库发生故障，写业务切换到从库，而从库又没有收到 User1 写入的数据时，User1 会收到事务写入失败的消息，但对于 User2 来说，之前在主库上能查到的数据，切换到从库之后却查不到了，就好像发生幻读一样。

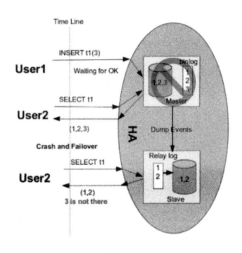

图 5-4

从 MySQL 5.7 开始，Oracle MySQL 官方对半同步复制进行了增强，从字面上来看，增强半同步本质上就是对早期半同步复制中的缺陷进行了修补，而其原理与后者并无差别。那么，增强半同步复制在早期半同步复制的基础上做了什么修改呢？

如图 5-5 所示（该图片来自 my-replication-life blogspot），从左侧底部方框标记的地方可以看到，Engine Commit 步骤下沉到了最后，也就是说，在增强半同步复制中，一个事务在存储引擎内部提交之前，必须先收到从库的 ACK 消息，否则不进行事务最后的提交。这样一来，其他客户端会话在查询数据时，所看到的数据就能够和发起事务提交的客户端会话保持一致，从而解决了主库故障转移之后可能出现的幻读问题。

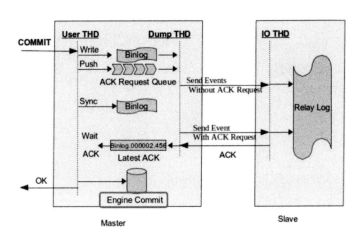

图 5-5

第 6 章 多线程复制

MySQL 5.6 之前的版本不支持从库并行重放主库的二进制日志,所以一旦主库的写压力稍微大一点,从库就容易出现延迟。当然,目前最新的 MySQL 版本已经能够很好地支持多线程复制。为了便于理解复制是如何一步一步演进为多线程复制的,本章将从单线程复制说起。在开始学习本章内容之前,也许你需要回顾一下复制的基本原理,详见第 2 章 "复制的基本原理"。

提示:

- 本章中解析的所有二进制日志示例均为 row 格式。
- 下文中提到的 "单线程复制" 也可称为 "串行复制", "多线程复制" 也可称为 "并行复制"(注意,这里所说的 "串行复制" 与 "并行复制",指的是数据库中数据变更操作的 "串行" 与 "并行",不要和复制拓扑中的主从复制的 "串行" 与 "并行" 混淆)。
- 下文中提到的系统变量详见《千金良方:MySQL 性能优化金字塔法则》附录 C "MySQL 常用配置变量和状态变量详解"。

6.1 单线程复制原理

单线程复制是 MySQL 中最早出现的 Server 之间的数据同步技术,当从库的 I/O 线程将主库二进制日志写入自身的中继日志之后,读取中继日志并进行回放的线程只有一个,也就是 SQL Thread(SQL 线程),参见图 6-1。

下面以单线程复制中主库写入二进制日志的日志解析记录为例,对单线程复制原理进行详细的阐述。假设主库中执行了一个 INSERT 操作,那么在二进制日志中的记录如下(MySQL 5.5):

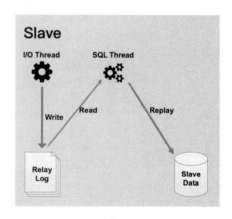

图 6-1

```
# at 605
# Query_log_event 类型的事件用于记录非事务语句的原始内容
```

```
#180830 16:45:37 server id 3306102 end_log_pos 678 Query thread_id=2 exec_time=0 error_code=0
SET TIMESTAMP=1535618737/*!*/;
BEGIN
/*!*/;
# at 678
# at 721
# Table_map_event 类型的事件用于记录 InnoDB 存储引擎层的表映射到 Server 层的 table id
#180830 16:45:37 server id 3306102 end_log_pos 721 Table_map: `test`.`test` mapped to number 33
# Write_rows_log_event 类型的事件用于在二进制日志中记录 INSERT 语句的 Base64 编码，同理，UPDATE 语句使用 Update_rows_log_event 类型的事件来记录，DELETE 语句使用 Delete_rows_log_event 类型的事件来记录
#180830 16:45:37 server id 3306102 end_log_pos 755 Write_rows: table id 33 flags: STMT_END_F

BINLOG '
sa6HWxN2cjIAKwAAANECAAAAACEAAAAAAAEABHRlc3QABHRlc3QAAQMAAQMAAQ==
sa6HWxd2cjIAIgAAAPMCAAAAACEAAAAAAAEAAf/+AQAAAA==
'/*!*/;
### INSERT INTO `test`.`test`
### SET
###   @1=1 /* INT meta=0 nullable=1 is_null=0 */
# at 755
#180830 16:45:37 server id 3306102 end_log_pos 782 Xid = 16
COMMIT/*!*/;
```

如图 6-1 所示，从库的 SQL 线程从中继日志中读取并解析主库二进制日志，由于执行事件的线程只有一个，所以读取的所有事件被串行执行。而二进制日志的写入时机是在事务发起提交之后，也就是说，主库事务先执行，然后产生的二进制日志才会被发送到从库执行。所以，从理论上讲，主从库这一前一后的时差就必然会导致从库复制延迟。如果遇到大事务，则从库延迟会急剧增加，例如，主库执行一个大事务耗费 1 小时，当从库收到这个事务之后开始执行时，就已经落后于主库 1 小时了。也正是因为单线程复制的效率极端低下，倒逼单线程复制向着多线程复制发展。

提示：前面提到，MySQL 5.6 之前的版本，从库不支持多线程复制，但实际上在这之前的版本中，当不启用二进制日志时，InnoDB 存储引擎本身是支持 Group Commit 的（即支持一次提交多个事务），但当启用二进制日志之后，为了保证数据的一致性（也就是必须保证 MySQL Server 层和存储引擎层的提交顺序一致），启用了两阶段提交。而两阶段提交中的 prepare 阶段使用了 prepare_commit_mutex 互斥锁来强制事务串行提交，这也大大降低了数据库的写入效率。

6.2 DATABASE 多线程复制

6.2.1 原理

顾名思义，DATABASE 多线程复制就是在单线程复制的基础之上做了改进，基于库级别的多线程复制。MySQL 从 5.6 版本开始，支持库级别的多线程复制，这也是最早出现的多线程复制机制，在二进制日志中记录类似如下的内容（MySQL 5.6）：

```
# at 861
#180907 18:12:35 server id 3306102 end_log_pos 940 CRC32 0x5b0ad30d Query thread_id=8 exec_time=0 error_code=0
SET TIMESTAMP=1536315155/*!*/;
BEGIN
/*!*/;
......
# at 992
# Table_map_event 类型的事件用于记录 InnoDB 存储引擎层的表映射到 Server 层的 table id。从这里可以看到，该语句操作的是 test_1 库下的表
#180907 18:12:35 server id 3306102 end_log_pos 1043 CRC32 0x04580375 Table_map: `test_1`.`test_1` mapped to number 71
# at 1043
# Write_rows_log_event 类型的事件用于在二进制日志中记录 INSERT 语句的 Base64 编码，同理，UPDATE 语句使用 Update_rows_log_event 类型的事件来记录，DELETE 语句使用 Delete_rows_log_event 类型的事件来记录
#180907 18:12:35 server id 3306102 end_log_pos 1083 CRC32 0xbf0ad4bf Write_rows: table id 71 flags: STMT_END_F
......
# at 1083
#180907 18:12:35 server id 3306102 end_log_pos 1114 CRC32 0xe9e2e9ef Xid = 42
COMMIT/*!*/;
......
# at 1162
#180907 18:12:43 server id 3306102 end_log_pos 1241 CRC32 0xdde19e1c Query thread_id=8 exec_time=0 error_code=0
SET TIMESTAMP=1536315163/*!*/;
BEGIN
/*!*/;
......
# at 1293
# 从这里可以看到，该语句操作的是 test_2 库下的表
#180907 18:12:43 server id 3306102 end_log_pos 1344 CRC32 0xe23b0497 Table_map: `test_2`.`test_2` mapped to number 72
# at 1344
#180907 18:12:43 server id 3306102 end_log_pos 1384 CRC32 0x6c5d94c5 Write_rows: table id 72 flags: STMT_END_F
......
```

```
# at 1384
#180907 18:12:43 server id 3306102 end_log_pos 1415 CRC32 0xefb1bbdd Xid = 45
COMMIT/*!*/;
```

以上对于 INSERT 语句二进制日志的解析内容，与 MySQL 5.5 的解析内容相比，几乎没有什么变化。那么，从 MySQL 5.6 开始，对复制功能的改进主要是什么呢？

对于实例自身的事务而言（这里指的是本地事务，不区分主从库），在原先的两阶段提交中，移除了 prepare_commit_mutex 互斥锁。为保证二进制日志的提交顺序和存储引擎层的一致，引入与 InnoDB 存储引擎层的 Group Commit 类似的机制，并将其称为 Binary Log Group Commit（BLGC）。在 MySQL 数据库上层提交事务时，首先按照顺序将事务放入一个队列中，队列中的第一个事务称为 leader，其他事务称为 follower，leader 控制 follower 的行为。一旦前一个阶段中的队列任务执行完，后一个阶段队列中的 leader 就会带领它的 follower 进入前一个阶段的队列中执行，这样的并行提交可以持续不断地进行。BLGC 的步骤分为图 6-2 所示的 3 个阶段（该图来自 mysqlmusings blogspot）。

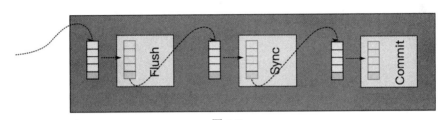

图 6-2

（1）Flush 阶段：将每个事务的二进制日志写入内存。

（2）Sync 阶段：将内存中的二进制日志刷新到磁盘，若队列中有多个事务，那么仅一次 fsync 操作就完成多个事务的二进制日志写入，这就是 BLGC。

（3）Commit 阶段：leader 根据顺序调用存储引擎层事务的提交，InnoDB 存储引擎本就支持 Group Commit，因此修复了原先由 prepare_commit_mutex 互斥锁导致 InnoDB 存储引擎层 Group Commit 失效的问题。这样一来，在启用二进制日志的情况下，就实现了数据库中事务的并行提交。

对于从库事务而言（这里指的是从库通过二进制日志重放的主库事务），主要的改进在于从库复制的 SQL 线程——增加了一个 SQL 协调器线程（Coordinator 线程），真正干活的 SQL 线程被称为工作线程（Worker 线程），当 Worker 线程为 N 个（$N > 1$）以及主库的 DATABASE（Schema）为 N 个时，从库就可以根据多个 DATABASE 之间相互独立（彼此之间无锁冲突）的语句来实现多线程复制；反之，如果 $N = 1$，则多线程复制跟 MySQL 5.6 之前版本中的单线程复制没有太大区别。多线程复制大致的工作流如图 6-3 所示。

提示：对于 DATABASE 多线程复制，如果有跨库事务，并行的 Worker 线程之间可能产生相互等待。

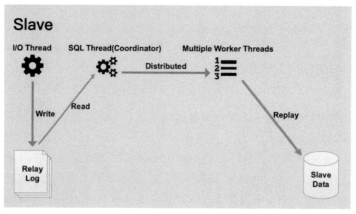

图 6-3

6.2.2 系统变量的配置

1. 主库

```
# 在MySQL 5.7中，DATABASE 多线程复制需要设置如下系统变量，如果是MySQL 5.6，则主库无须做任何设置
slave_parallel_type = DATABASE
```

2. 从库

```
slave_parallel_workers = N
# 在MySQL 5.7中，DATABASE 多线程复制需要设置如下系统变量，如果是MySQL 5.6，则主库无须做任何设置
slave_parallel_type = DATABASE
```

6.3 LOGICAL_CLOCK 多线程复制

6.3.1 原理

MySQL 从 5.7 版本开始，支持 LOGICAL_CLOCK 多线程复制。基于 MySQL 5.6 库级别的 Group Commit 多线程复制做了大幅改进。对于 DATABASE 多线程复制，允许并行回放的粒度为数据库级别，只有在同一时间修改数据且修改操作针对的是不同数据库，才允许并行。而对于 LOGICAL_CLOCK 多线程复制，允许并行回放的粒度为事务级别，即便在同一时间修改数据的操作针对的是同一个数据库，理论上只要事务之间不存在锁冲突，就允许并行，可通过设置系统变量 slave_parallel_type 为 LOGICAL_CLOCK 来启用，如果该变量被设置为 DATABASE，则与 MySQL 5.6 的多线程复制相同。从字面上无法直观地看出 LOGICAL_CLOCK 是基于什么维度来实现多线程复制的。下面我们通过解析二进制日志的内容来进行解读。

```
# at 194
# 二进制日志中新增了一个 Anonymous_gtid_log_event 事件类型来记录 Binlog Group 相关的信息，启用
GTID 时，使用 Gtid_log_event 类型的事件来记录 Binlog Group 相关的信息。同时，为了与 GTID 复制结合，把
未开启 GTID 时的事务称为 ANONYMOUS 事务，last_committed 和 sequence_number 的值在每个二进制日志文件中
```

都重新计数（last_committed 从 0 开始，sequence_number 从 1 开始）。last_committed 表示事务在每个二进制日志文件中的 Binlog Group 编号，sequence_number 为每个二进制日志文件中事务的编号。last_committed 会有重复的值，last_committed 值相同表示事务处于同一个 Binlog Group 中，也表示主库中这些事务在并行提交时没有锁冲突。从库在应用二进制日志时，具有相同 last_committed 值的事务可以并行回放。在每个二进制日志文件中，sequence_number 的值不允许重复

```
#180908 14:49:36 server id 3306102 end_log_pos 259 CRC32 0xe7e2833c Anonymous_GTID
last_committed=0    sequence_number=1    rbr_only=yes
/*!50718 SET TRANSACTION ISOLATION LEVEL READ COMMITTED*//*!*/;
SET @@SESSION.GTID_NEXT= 'ANONYMOUS'/*!*/;  # 当不启用 GTID 复制模式时，这里记录的系统变量
@@SESSION.GTID_NEXT 的值始终为 ANONYMOUS；当启用时，则记录的是具体的 GTID 值
# at 259
# Query_log_event 类型的事件用于记录非事务语句的原始内容
#180908 14:49:36 server id 3306102 end_log_pos 336 CRC32 0x5abf230b Query    thread_id=2
exec_time=0 error_code=0
......
BEGIN
......
# at 386
# Table_map_event 类型的事件用于记录 InnoDB 存储引擎层的表映射到 Server 层的 table id
#180908  14:49:36  server  id  3306102  end_log_pos  433  CRC32  0xf5cc55c6  Table_map:
`test`.`test` mapped to number 108
# at 433
# Write_rows_log_event 类型的事件用于在二进制日志中记录 INSERT 语句的 Base64 编码，同理，UPDATE
语句使用 Update_rows_log_event 类型的事件来记录，DELETE 语句使用 Delete_rows_log_event 类型的事件
来记录
#180908 14:49:36 server id 3306102 end_log_pos 473 CRC32 0xf938ae81 Write_rows: table
id 108 flags: STMT_END_F
......
# at 473
#180908 14:49:36 server id 3306102 end_log_pos 504 CRC32 0x3edc73b0 Xid = 7
COMMIT/*!*/;
# at 504
#180908 14:49:36 server id 3306102 end_log_pos 569 CRC32 0x8716f23e Anonymous_GTID
last_committed=1    sequence_number=2    rbr_only=yes
/*!50718 SET TRANSACTION ISOLATION LEVEL READ COMMITTED*//*!*/;
SET @@SESSION.GTID_NEXT= 'ANONYMOUS'/*!*/;
......
```

从以上对于 INSERT 语句的二进制日志解析内容来看，MySQL 从 5.7 版本开始，新增了两个事件类型，Anonymous_gtid_log_event 类型用于记录未启用 GTID 时的 Binlog Group 信息，Gtid_log_event 类型用于记录启用 GTID 时的 Binlog Group 信息。利用这些信息，从库的 SQL 线程在应用主库的二进制日志时，就可以并行回放，大大提高了从库复制的效率。

从上述代码段的注解中我们可以知道，拥有相同 last_committed 值的事务可以并行回放，但是事务的 last_committed 值是如何确定的呢？

last_committed 值是主库事务在进入 prepare 阶段时获取的已提交事务的最大 sequence_number 值。这个值称为此事务的 commit-parent，被记录在二进制日志中，当从库

回放事务时，如果两个事务有同一个 commit-parent，它们就可以并行执行。

提示：
- 在 LOGICAL_CLOCK 多线程复制中，发生变更的是主库记录二进制日志的算法与格式，以及从库分发事务的算法。至于从库的复制线程，仍然沿用图 6-3 所示的框架。
- 虽然这种方式大幅提高了从库复制的效率，也可以说，允许并行回放的粒度细化到事务级别，甚至可以说细化到行级别（每个事务只修改一行数据）。但是，可以并行回放的事务必须具有相同的 last_committed 值，即使两个事务的数据完全不相关，但如果 last_committed 值不同，也不能并行回放，而有多少事务具有相同的 last_committed 值，则由主库瞬时并发请求的数量而定（系统变量 binlog_group_commit_sync_no_delay_count = 0 时）。如果主库没有什么写压力，写入二进制日志中的每个事务的 last_committed 值都不相同，这时从库的复制实际上仍然是单线程复制。所以，LOGICAL_CLOCK 多线程复制仍然有一定的优化空间。
- 在 MySQL 5.7 较新的版本中，将允许并行回放的算法升级为基于 Lock（锁）的，即通过一个锁的时间范围来确定是否可以并行回放。主库会力求为每一个事务生成尽可能小的 last_committed 值，这样就可以提高从库回放的并行度，因为从库进行回放时判断是否可以并行回放的依据，就是 last_committed 值是否相同。该锁的时间范围设定为：以事务在两阶段提交的 prepare 阶段获取最后一把锁的时间为开始时间点，以 commit 阶段释放第一把锁的时间为结束时间点。如果锁的时间范围有重叠，事务就可以并行回放，无重叠就不能并行回放。在升级为"基于 Lock"的多线程复制之前，主库为事务生成的 last_committed 值是在每个事务执行提交操作时获取的当前已提交完成事务的最大 sequence_number，这种机制称为 Commit-Parent-Based。对于该机制有兴趣的读者可自行研究，这里不展开阐述。

6.3.2 系统变量的配置

1. 主库

```
slave_parallel_type = LOGICAL_CLOCK
```

2. 从库

```
slave_preserve_commit_order = 1
slave_parallel_workers = N
slave_parallel_type = LOGICAL_CLOCK
```

6.4 WRITESET 多线程复制

6.4.1 原理

WRITESET 多线程复制，其实是在 MySQL 5.7 的大于或等于 5.7.22 的版本、MySQL 8.0 的大于或等于 8.0.1 的版本中，对 LOGICAL_CLOCK 多线程复制的优化机制。优化之后，只要不同事务的修改记录不重叠，就可以在从库中并行回放。通过计算每行记录的哈希值（hash）来确定其是否为相同的记录。该哈希值就是 WRITESET 值。

WRITESET 多线程复制允许并行回放的粒度依然为事务级别。严格来说，WRITESET 以及下文中提到 WRITESET_SESSION 都只是在原有的 LOGICAL_CLOCK 多线程复制的基础上优化事务并行度的依赖模式，默认的依赖模式为 COMMIT_ORDER。但由于 WRITESET 和 WRITESET_SESSION 依赖模式在某些场景下，能够显著提高多线程复制的效率，加上这两种依赖模式都是基于全新引入的冲突检测数据库的机制实现的，因此，通常大家都习惯将它们一起称为"WRITESET 多线程复制"，以便和 COMMIT_ORDER 依赖模式的"LOGICAL_CLOCK 多线程复制"区分。

WRITESET 多线程复制的本质就是基于 WRITESET 的值对生成 last_committed 值的依赖模式做了大量优化。下面我们通过解析二进制日志中记录的内容来进行解读。

```
  # at 1817
  # 记录 Binlog Group 信息的 Gtid_log_event 事件。注意，这里事务的 last_committed=2
  #180908 17:12:24 server id 3306102 end_log_pos 1882 CRC32 0x5b468c97 GTID last_
committed=2    sequence_number=5    rbr_only=yes
  /*!50718 SET TRANSACTION ISOLATION LEVEL READ COMMITTED*//*!*/;
  SET @@SESSION.GTID_NEXT= '06188301-b333-11e8-bdfe-0025905b06da:13'/*!*/;
  # at 1882
  # Query_log_event 类型的事件用于记录非事务语句的原始内容
  #180908 17:12:24 server id 3306102 end_log_pos 1959 CRC32 0x6474ec27 Query  thread_id=6
exec_time=0 error_code=0
  SET TIMESTAMP=1536397944/*!*/;
  BEGIN
  ......
  # at 2023
  # Table_map_event 类型的事件用于记录 InnoDB 存储引擎层的表映射到 Server 层的 table id
  #180908  17:12:24  server  id  3306102  end_log_pos  2080  CRC32  0x4b174a04  Table_map:
`test`.`test_writeset` mapped to number 109
  # at 2080
  # Write_rows_log_event 类型的事件用于在二进制日志中记录 INSERT 语句的 Base64 编码，同理，UPDATE
语句使用 Update_rows_log_event 类型的事件来记录，DELETE 语句使用 Delete_rows_log_event 类型的事件
来记录
  #180908 17:12:24 server id 3306102 end_log_pos 2124 CRC32 0x93882793 Write_rows: table
id 109 flags: STMT_END_F
  ......
  # at 2124
  #180908 17:12:24 server id 3306102 end_log_pos 2155 CRC32 0x00afaee3 Xid = 54
  COMMIT/*!*/;
  ......
  # at 3113
  # 记录 Binlog Group 信息的 Gtid_log_event 事件。注意，这里事务的 last_committed=8
  #180908  17:14:38  server  id  3306102  end_log_pos  3178  CRC32  0xd85038b1  GTID
last_committed=8    sequence_number=9    rbr_only=yes
  /*!50718 SET TRANSACTION ISOLATION LEVEL READ COMMITTED*//*!*/;
```

```
  SET @@SESSION.GTID_NEXT= '06188301-b333-11e8-bdfe-0025905b06da:17'/*!*/;
  ......
  # at 3733
  # 注意，这里 Gtid_log_event 事件中的 last_committed 值又回到 2，与之前 last_committed = 2 的事
  务在时间点上相差 2 分钟
    #180908 17:14:57 server id 3306102  end_log_pos 3798 CRC32 0x2373db26 GTID last_committed=2
  sequence_number=11   rbr_only=yes
  /*!50718 SET TRANSACTION ISOLATION LEVEL READ COMMITTED*//*!*/;
  SET @@SESSION.GTID_NEXT= '06188301-b333-11e8-bdfe-0025905b06da:19'/*!*/;
  ......
```

从以上对于 INSERT 语句的二进制日志解析内容来看，WRITESET 多线程复制与 LOGICAL_CLOCK 多线程复制记录的二进制日志相比，格式上没有什么变化，那么它对于复制的改进主要是什么呢？细心的读者可能已经发现，last_committed = 2 的两个事务的时间戳并不是同一个时刻的，并且这两个事务之间，还夹了一个 last_committed = 8 的事务。在以往的 LOGICAL_CLOCK 多线程复制中几乎不可能出现这种情况，这其中发生了什么？

对于主库来说，WRITESET 多线程复制对 LOGICAL_CLOCK 多线程复制的优化，并不是在二进制日志记录的格式上，而是在事务写二进制日志时对 last_committed 值的计算做了大量优化。在 6.3.1 节中，我们提到过 LOGICAL_CLOCK 多线程复制的并行粒度为事务级别，在理论上虽然只要不同事务之间不存在锁冲突即可并行，但实际上采用默认的 COMMIT_ORDER 依赖模式的多线程复制，在从库中的并行分发机制上并不能完全实现。这是因为在默认的 COMMIT_ORDER 依赖模式下，主库生成二进制日志时，只有同一时间提交的事务生成的 last_committed 值才相同，但实际上不同时间提交的事务也有不存在锁冲突的可能。因此，这些事务在主库中生成二进制日志时的 last_committed 值理论上也应该相同。要实现在主库生成中生成二进制日志时，在不同时间提交的且不存在锁冲突的事务生成相同的 last_committed 值，必须引入新机制，而这个新的机制，便是 WRITESET 多线程复制对 LOGICAL_CLOCK 多线程复制优化的关键。

通过唯一索引或主键索引来区分不同的记录，然后和行记录的库表属性以及数据属性一起计算哈希值，计算出的 WRITESET 值存放在一张哈希表中。后面如果新事务的行记录计算出的哈希值在哈希表中无匹配记录，那么此新事务不会产生新的 last_committed 值，就相当于新事务和之前的事务被归并到同一个 Binlog Group，即新旧事务的 last_committed 值相同。如果新事务的行记录计算出的哈希值在哈希表中找到了匹配记录，则表示存在事务冲突，就会产生新的 last_committed 值（即产生了一个新的 Binlog Group）。具体的计算公式如下：

WRITESET = hash(index_name,db_name,db_name_length,table_name,table_name_length,value,value_length)

主库中生成 last_committed 值的依赖模式改进如图 6-4 所示。

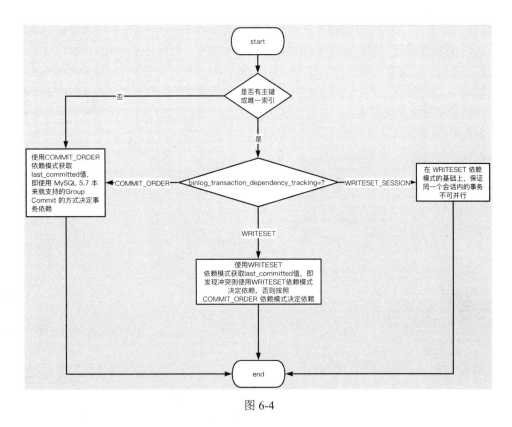

图 6-4

根据系统变量 binlog_transaction_dependency_tracking 的设置，采用不同的依赖模式生成 last_committed 值 。

在 WRITESET 多线程复制中做冲突认证的过程如图 6-5 所示。

对于从库来说，并行应用二进制日志的逻辑几乎没有变化，仍然根据 last_committed 的值判断是否可以并行回放。

提示：

WRITESET 多线程复制在如下场景中不可用：

- DDL 语句。

- 会话当前生效的哈希算法（可以使用系统变量 transaction_write_set_extraction 在会话级别修改）和生成 writeset_history 列表值使用的哈希算法不同（哈希算法被动态修改过之后，碰到不同算法产生的值将无法进行比较）。

- 事务更新了被外键关联的字段。

使用 WRITESET 多线程复制时，如果主库中具有高并发的事务且这些事务具备"短、平、快"的特性，则 WRITESET 多线程复制中的 WRITESET、WRITESET_SESSION 依赖模式与默认的 COMMIT_ORDER 依赖模式相比，对从库的多线程复制效率并没有太大提高，但如果主库中有高并发的事务且这些事务中大事务居多，则使用 COMMIT_ORDER 依赖模式时,二进制日志中的 last_committed 值重复率可能不够高,导致从库多线程复制的效率大大降低。在这种情况下,使用 WRITESET 多线程复制中的 WRITESET、WRITESET_SESSION 依赖模式能够大大提高主库二进制日志中 last_committed 值的重复率,大幅提高从库的多线程复制效率。即便如此,也不建议在 MySQL 中写入大事务,过大的事务在从库的协调器线程做事务分发时极易成为瓶颈,进而使多线程复制的效率难以提高,而且大事务还会带来诸多负面影响。

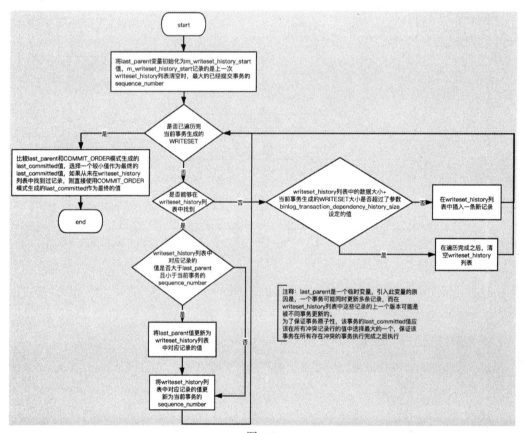

图 6-5

生成 last_committed 值的部分代码如下(仅供参考,有兴趣的读者请自行翻阅源码文件 sql/rpl_trx_tracking.cc,这里不再赘述):

```
......
/**
  Get the sequence_number for a transaction, and get the last_commit based on parallel
committing transactions.
  @param[in] thd Current THD from which to extract trx context.
  @param[in,out] sequence_number Sequence number of current transaction.
  @param[in,out] commit_parent Commit_parent of current transaction,pre-filled with the
commit_parent calculated by the logical clock logic.
*/
```

```cpp
void
Commit_order_trx_dependency_tracker::get_dependency(THD *thd,
                    int64 &sequence_number,
                    int64 &commit_parent)
{
  ......
}
......
void
Writeset_trx_dependency_tracker::get_dependency(THD *thd,
                    int64 &sequence_number,
                    int64 &commit_parent)
{

  Rpl_transaction_write_set_ctx *write_set_ctx=
    thd->get_transaction()->get_transaction_write_set_ctx();
  std::set<uint64> *writeset= write_set_ctx->get_write_set();

#ifndef DBUG_OFF
  /* The writeset of an empty transaction must be empty. */
  if (is_empty_transaction_in_binlog_cache(thd))
    DBUG_ASSERT(writeset->size() == 0);
#endif

  /*
    Check if this transaction has a writeset, if the writeset will overflow the history
size, if the transaction_write_set_extraction is consistent between session and global or
if changes in the tables referenced in this transaction cascade to other tables. If that
happens revert to using the COMMIT_ORDER and clear the history to keep data consistent.
  */

  bool can_use_writesets=
    // empty writeset implies DDL or similar, except if there are missing keys
    (writeset->size() != 0 || write_set_ctx->get_has_missing_keys() ||
     /*
       The empty transactions do not need to clear the writeset history, since they can
be executed in parallel.
     */

     is_empty_transaction_in_binlog_cache(thd)) &&
    // hashing algorithm for the session must be the same as used by other rows in history
    (global_system_variables.transaction_write_set_extraction ==
     thd->variables.transaction_write_set_extraction) &&
    // must not use foreign keys
    !write_set_ctx->get_has_related_foreign_keys();
  bool exceeds_capacity= false;
```

```
    if (can_use_writesets)
    {
      /*
        Check if adding this transaction exceeds the capacity of the writeset history. If
that happens, m_writeset_history will be cleared only after using its information for current
transaction.
      */

      exceeds_capacity=
        m_writeset_history.size() + writeset->size() > m_opt_max_history_size;

      /*
        Compute the greatest sequence_number among all conflicts and add the transaction's
row hashes to the history.
      */

      int64 last_parent= m_writeset_history_start;
      for (std::set<uint64>::iterator it= writeset->begin();
           it != writeset->end(); ++it)
      {
        Writeset_history::iterator hst= m_writeset_history.find(*it);
        if (hst != m_writeset_history.end())
        {
          if (hst->second > last_parent && hst->second < sequence_number)
            last_parent= hst->second;
          hst->second= sequence_number;
        }
        else
        {
          if (!exceeds_capacity)
            m_writeset_history.insert(std::pair<uint64, int64>(*it, sequence_number));
        }
      }

      /*
        If the transaction references tables with missing primary keys revert to COMMIT_ORDER,
update and not reset history, as it is unnecessary because any transaction that refers this
table will also revert to COMMIT_ORDER.
      */

      if (!write_set_ctx->get_has_missing_keys())
      {
        /*
          The WRITESET commit_parent then becomes the minimum of largest parent found using
the hashes of the row touched by the transaction and the commit parent calculated with
COMMIT_ORDER.
        */
```

```
      commit_parent= std::min(last_parent, commit_parent);
    }
  }

  if (exceeds_capacity || !can_use_writesets)
  {
    m_writeset_history_start= sequence_number;
    m_writeset_history.clear();
  }
}
......
/**
  Get the writeset commit parent of transactions using the session dependencies.
  @param[in] thd Current THD from which to extract trx context.
  @param[in,out] sequence_number Sequence number of current transaction.
  @param[in,out] commit_parent Commit_parent of current transaction,
pre-filled with the commit_parent calculated by the Write_set_trx_dependency_tracker as
a fall-back.
*/ void

Writeset_session_trx_dependency_tracker::get_dependency(THD *thd,
                           int64 &sequence_number,
                           int64 &commit_parent)
{
......
}
......
/**
  Get the dependencies in a transaction, the main entry point for the
  dependency tracking work.
*/
void Transaction_dependency_tracker::get_dependency(THD *thd,
                           int64 &sequence_number,
                           int64 &commit_parent) {
  sequence_number = commit_parent = 0;

  switch (m_opt_tracking_mode) {
    case DEPENDENCY_TRACKING_COMMIT_ORDER:

# DEPENDENCY_TRACKING_COMMIT_ORDER 表示系统变量 binlog_transaction_dependency_tracking 设
置为 COMMIT_ORDER 条件为真，此时只调用 m_commit_order.get_dependency()
      m_commit_order.get_dependency(thd, sequence_number, commit_parent);
      break;
    case DEPENDENCY_TRACKING_WRITESET:
      m_commit_order.get_dependency(thd, sequence_number, commit_parent);

# DEPENDENCY_TRACKING_WRITESET 表示系统变量 binlog_transaction_dependency_tracking 设置为
WRITESET 条件为真，此时是在 COMMIT_ORDER 依赖模式的基础上再调用 m_writeset.get_dependency()
```

```
        m_writeset.get_dependency(thd, sequence_number, commit_parent);
        break;
      case DEPENDENCY_TRACKING_WRITESET_SESSION:
        m_commit_order.get_dependency(thd, sequence_number, commit_parent);
        m_writeset.get_dependency(thd, sequence_number, commit_parent);
```

DEPENDENCY_TRACKING_WRITESET_SESSION 表示系统变量 binlog_transaction_dependency_tracking 设置为 WRITESET_SESSION 条件为真，此时是在 WRITESET 依赖模式的基础上再调用 m_writeset_session.get_dependency

```
        m_writeset_session.get_dependency(thd, sequence_number, commit_parent);
        break;
      default:
        DBUG_ASSERT(0); // blow up on debug
        /*
          Fallback to commit order on production builds.
        */
        m_commit_order.get_dependency(thd, sequence_number, commit_parent);
    }
  }
......
```

6.4.2 系统变量的配置

1. 主库

```
slave_parallel_type = LOGICAL_CLOCK
transaction_write_set_extraction = XXHASH64
# WRITESET 和 WRITESET_SESSION 两个值任意设置一个即可。WRITESET_SESSION 依赖模式与 WRITESET 依
赖模式相比，前者表示在后者的基础上，保证同一个会话内的事务不可并行
binlog_transaction_dependency_tracking = WRITESET|WRITESET_SESSION
binlog_transaction_dependency_history_size = 25000
```

2. 从库

```
slave_preserve_commit_order = 1
slave_parallel_workers = N
slave_parallel_type = LOGICAL_CLOCK
```

提示：更多信息可参考高鹏的"复制"专栏，登录简书网站搜索"第 16 节：基于 WRITESET 的并行复制方式"。

第 7 章　多源复制

一些业务数据被打散到多个数据库实例上之后，数据库的备份和恢复就比较烦琐，有没有什么简单的方案能够解决这个问题呢？MySQL 5.7.6 引入了复制通道的概念，使得同一个从库可以同时接收多个主库的数据，一个复制通道逻辑上就对应一个主库。本章将简要介绍如何在复制拓扑中使用通道、通道相关的概念，以及相关系统的设置对单源（单个复制通道）复制的影响。

7.1　复制通道

什么是通道（channel）？通道表示从主库到从库的事务复制路径。在配置多通道时，建议指定一个与主库主机名或者业务用途相关的标识字符作为通道名称，以便在后续维护过程中不会因混淆而导致误操作。

为了提供与先前版本的兼容性，如果在配置复制时没有指定通道名称，MySQL Server 在启动复制时会自动创建一个默认通道，其通道名称为空字符串（""），以便旧版的从库可以正常运行复制。用户不能自行创建或销毁名称为空字符串的复制通道（但在较新的 MySQL 版本中是允许的。例如，MySQL 5.7.23 可以使用"for channel """的形式来创建和删除名称为空字符串的通道）。

复制通道是事务从主库传输到从库时的路径。在多源复制中，从库打开多个通道，每个通道都有自己的中继日志和应用线程（SQL 线程）。一旦某个复制通道的接收器线程（I/O 线程）接收到事务，就会被添加到对应复制通道的中继日志文件中，并传递给对应复制通道的应用线程，使得每个复制通道都可以独立运行。

在 MySQL 5.7 中，多源复制拓扑的从库最多可以添加 256 个复制通道，每个复制通道必须具有唯一的名称，每个复制通道都可以独立配置、独立运行。图 7-1 为从库开启 A、B 两个复制通道的复制流示意图。其中，从库的 A 通道对应着主库 Master-A，从库的 B 通道对应着主库 Master-B。在从库上，Master-A 和 Master-B 两个主库各自有自己的复制线程，也就是说它们的复制过程相互独立，有相互独立的复制进度。

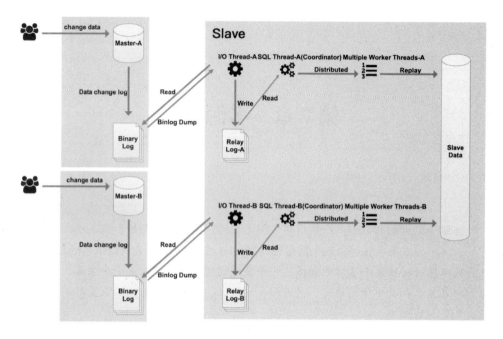

图 7-1

7.2 单通道操作命令

在从库中启用多通道之后,如果想对某个给定的复制通道执行一些操作,与复制相关的某些操作语句就和单通道的有区别了。例如,默认情况下,一些语句针对的是默认通道,一些语句针对的是所有通道,如果需要对某个给定的通道执行操作,则需要结合 FOR CHANNEL 子句一起使用。

```
# 启用多通道之后,如果需要对某个给定的复制通道执行一些操作,如下语句需要结合 FOR CHANNEL 子句一起使用
CHANGE MASTER TO
START SLAVE
STOP SLAVE
SHOW RELAYLOG EVENTS
FLUSH RELAY LOGS
SHOW SLAVE STATUS
RESET SLAVE

# 示例(关于以下语句的详细语法,请使用 HELP START SLAVE 语句来查看,其他语句语法的查看方法与此类似)
START SLAVE FOR CHANNEL 'Master-A';
```

为了更好地支持多通道,以下函数引入了额外的通道参数:

```
# 这里指的是 channel 参数
MASTER_POS_WAIT()
MASTER_POS_WAIT(log_name,log_pos[,timeout][,channel])
WAIT_UNTIL_SQL_THREAD_AFTER_GTIDS()
WAIT_UNTIL_SQL_THREAD_AFTER_GTIDS(gtid_set[, timeout][,channel])
```

7.3 复制语句的向前兼容性

当从库配置了多个通道且未指定 FOR CHANNEL 子句时，一些复制通道操作的命令默认对所有有效的复制通道执行，但有些特殊的通道除外（这里主要指的是组复制通道。关于组复制，本书不赘述，有兴趣的读者可以参考"沃趣技术"微信公众号中的"全方位认识 MySQL 8.0 Group Replication"系列文章），如下所述：

- START SLAVE：启动所有通道的复制线程，但 group_replication_recovery 和 group_replication_applier 通道除外。
- STOP SLAVE：停止所有通道的复制线程，但 group_replication_recovery 和 group_replication_applier 通道除外。
- SHOW SLAVE STATUS：打印所有通道的复制状态信息，但 group_replication_applier 通道除外。
- FLUSH RELAY LOGS：刷新所有通道的中继日志，但 group_replication_applier 通道除外。
- RESET SLAVE：重置所有通道的复制配置信息。注意，执行该命令会清除所有复制通道的中继日志，并重新创建默认的复制通道。

当从库开启多个通道时，以下语句必须使用 FOR CHANNEL channel_name 子句指定操作单个通道，不允许对所有通道执行操作，如果不指定就会报错（ERROR 3079 (HY000): Multiple channels exist on the slave. Please provide channel name as an argument）。

```
SHOW RELAYLOG EVENTS
CHANGE MASTER TO
MASTER_POS_WAIT()
WAIT_UNTIL_SQL_THREAD_AFTER_GTIDS()
```

7.4 启动选项和复制通道选项

必须正确配置以下启动选项才能使用多源复制：

- --relay-log-info-repository：必须设置为 TABLE，如果此选项设置为 FILE，则在从库中配置多源复制时将会失败，显示 ER_SLAVE_NEW_CHANNEL_WRONG_REPOSITORY 错误信息。
- --master-info-repository：必须设置为 TABLE，如果此选项设置为 FILE，则在从库中配置多源复制时将会失败，显示 ER_SLAVE_NEW_CHANNEL_WRONG_REPOSITORY 错误信息。

提示：如果在 MySQL Server 启动之前未配置这两个选项，则可以在 MySQL Server 启动之后，使用系统变量动态配置（系统变量 relay_log_info_repository 和 master_info_repository）。

以下启动选项会影响复制拓扑中的所有通道：

- --log-slave-updates：从库接收的所有事务都写入其自身的二进制日志（多源复制拓扑中所有通道复制的事务都会写入二进制日志）。
- --relay-log-purge：启用之后，每个复制通道各自自动清除自己的中继日志。
- --slave-transaction-retries：所有复制通道的应用线程重试事务的数量。
- --skip-slave-start：启用该选项之后，所有复制通道的线程都关闭自启动，之后如果要启用复制，则需要手工启动复制线程。
- --slave-skip-errors：当遇到复制错误时，所有复制通道按照该选项指定的错误代码来跳过相应的复制错误，并继续后续的复制过程。

提示：如果在 MySQL Server 启动之前未配置这些选项，有的选项可以在 MySQL Server 启动之后，使用系统变量动态配置（系统变量 relay_log_purge 和 slave_transaction_retries）。系统变量 log_slave_updates 和 slave_skip_errors 是只读的，需要在启动 MySQL Server 之前配置好。启动选项--skip-slave-start 没有对应的系统变量，也必须在启动 MySQL Server 之前配置好。

以下启动选项设置的值适用于每个复制通道，但是每个通道都独立使用这些选项设置的值：

- --max-relay-log-size = size：每个复制通道的单个中继日志文件的最大大小，当达到此限制后，文件将旋转滚动生成一个新的中继日志文件。
- --relay-log-space-limit = size：对于每个单独的复制通道，该选项设置所有中继日志的总大小的上限。所以，如果开启了 N 个复制通道，这些中继日志组合总大小的上限为 relay_log_space_limit * N。
- --slave-parallel-workers = value：从库中每个复制通道的 Worker 线程数量（在多线程复制中，原先 SQL 线程的工作由 Worker 线程接替，而原先的 SQL 线程转而负责为 Worker 线程分发二进制日志的工作）。
- --slave-checkpoint-group = N：每个复制源的 I/O 线程等待时间。
- --relay-log-index = filename：每个复制通道的中继日志索引文件的前缀名称。
- --relay-log = filename：每个复制通道的中继日志文件的前缀名称。
- --slave-net-timeout = N：每个复制通道设置此值，之后每个通道等待 N 秒后会检查连接是否断开。
- --slave-skip-counter = N：每个复制通道设置此值，之后每个通道从其所连接的主库跳过 N 个事件。

提示：如果在 MySQL Server 启动之前未配置这些选项，有些可以在 MySQL Server 启动之后，使用系统变量动态配置（系统变量 r max_relay_log_size、slave_parallel_workers、slave_checkpoint_group、slave_net_timeout 和 sql_slave_skip_counter）。系统变量 relay_log_space_limit、relay_log_index 和 relay_log 是只读的，需要在启动 MySQL Server 之前配置好。

7.5 复制通道的命名约定

前面提到，每个复制通道需要有一个唯一的通道名称，但该名称不能过长，最长字符长度为 64，不区分大小写。由于通道名称需要保存在从库表中，因此这些字符串的字符集始终为 UTF-8。另外，虽然可以随意指定复制通道的名称，但以下两个复制通道的名称是保留给组复制使用的，不能由用户指定：

- group_replication_applier
- group_replication_recovery

配置复制时指定的复制通道名称，会影响指定的复制通道的中继日志文件和中继日志索引文件的名称。例如，relay_log_basename-channel_name.xxxxxx，其中 relay_log_basename 是使用选项--relay-log 指定的中继日志文件前缀名称，channel_name 是复制通道的名称。如果未指定选项--relay-log，则使用默认文件名（通常为主机名）作为中继日志文件和中继日志索引文件的前缀名称。

注意：

- 多源复制的从库需要留意总体数据量，一般不建议超过 1 TB，除非有良好的存储设备，例如 PCIe-SSD 或者 SSD 全闪存存储等，这些存储设备具有高带宽、低延迟的特点。即使有良好的存储设备，总体数据量也不建议超过 3 TB，具体情况请自行根据实际的从库复制性能以及备份和恢复时长而定。
- 多源复制与半同步复制混用可能导致问题。在 MySQL 5.7 的官方文档中有一段描述："There must not be multiple replication channels configured. Semisynchronous replication is only compatible with the default replication channel."，大意为"MySQL 5.7 不支持基于半同步复制的多源复制，不能配置多个通道，半同步复制仅与默认复制通道兼容"。如果你有混用的打算，建议关注后面的 MySQL 新版本是否修复了此问题。

提示：多源复制的搭建步骤详见第 21 章"搭建多源复制"，本章中提到的系统变量的详细介绍，可参考《千金良方：MySQL 性能优化金字塔法则》附录 C "MySQL 常用配置变量和状态变量详解"。

第 8 章　从库中继日志和状态日志

从库 I/O 线程从主库读取的二进制日志需要暂时存放在从库的磁盘文件中，这个磁盘文件就是中继日志（relay log）。I/O 线程并不负责解析与重放二进制日志，而是由 SQL 线程负责。当复制线程正常运行时，我们对复制线程的工作位置（这里指的是 I/O 线程读取主库二进制日志的位置和 SQL 线程重放的位置）不是很敏感，但当从库的数据库进程或者主机发生崩溃（crash）时，从库重新启动之后，需要知道上一次复制线程进行到的正确位置，也就是说对复制线程的工作位置需要持久化，否则一旦丢失将无法知道上一次复制进行到哪里了。状态日志被用来对复制线程的工作位置进行持久化（状态日志包括 relay log info 和 master info 两种类型，具体用途详见下文），本章将详细介绍中继日志和状态日志。

8.1　中继日志和状态日志概述

在复制线程正常运行期间，从库会创建多个日志，这些日志包含用于存放同步主库二进制日志记录的中继日志，以及记录有关中继日志当前状态和位置信息的状态日志，一共 3 种类型，如下所示：

- master info：包含从库与主库的连接状态和当前配置信息。此日志包含相关主库的主机名（主库 IP）、登录主库的凭据（复制线程使用的账号和密码）和位置信息（I/O 线程读取的主库二进制日志的文件名和位置信息）。使用 --master-info-repository = TABLE 启动从库复制线程时，可以将此日志信息写入 mysql.slave_master_info 表，而不是保存在磁盘文件 master.info 中（默认 --master-info-repository = FILE）。
- relay log：包括从主库二进制日志中读取的、由从库 I/O 线程写入的事件信息。这些中继日志中保存的事件信息后面会被从库的 SQL 线程读取并在从库中回放，这样就完成了主从库之间的数据同步。
- relay log info：保存有关从库中继日志中执行点的信息，使用 --relay-log-info-repository = TABLE 启动从库复制时，可以将此日志信息写入 mysql.slave_relay_log_info 表，而不是写入磁盘文件 relay_log.info 中（默认 --relay-log-info-repository = FILE）。

当使用表来保存从库的状态日志时，如果 mysqld 无法初始化用于记录复制状态的表，则会发出警告信息，但不影响数据库进程的正常启动。例如，从不支持从库日志状态记录表的 MySQL 版本升级到支持这两张表的 MySQL 版本时，最有可能发生这种情况。

重要提示：请勿尝试手动对 mysql.slave_master_info 或 mysql.slave_relay_log_info 表执行更新和插入操作，这可能会导致未知行为，MySQL 也不支持对这两张表手动执行插入和更新操作。

当复制线程正在运行时，只允许对 mysql.slave_master_info 和 mysql.slave_relay_log_info 表执行只读操作，不允许对其中的任何表执行加锁操作。

8.2 从库中继日志

与二进制日志一样，中继日志也是由一组顺序编号的文件和一个索引文件组成的。其中，顺序编号的文件中包含描述数据库更改的事件信息，索引文件中包含所有顺序编号的中继日志文件名列表。

术语"relay log file"通常指包含数据变更记录的单个中继日志文件，"relay log"通常指所有顺序编号的中继日志文件和中继日志索引文件。

中继日志文件与二进制日志文件格式相同，都可以使用命令行工具 mysqlbinlog 解析并读取其中的内容。默认情况下，中继日志文件存放在系统变量 datadir 指定的目录下，文件名格式为 *host_name*-relay-bin.nnnnnn，其中 host_name 是从库的主机名称，nnnnnn 是数字序列号。连续的中继日志文件使用连续的数字序列号，从 000001 开始。从库使用索引文件来跟踪当前正在使用哪些中继日志文件。中继日志索引文件默认存放在系统变量 datadir 指定的目录下，文件名格式为 *host_name*-relay-bin.index。可以分别使用选项 --relay-log 和 --relay-log-index（或在 my.cnf 中启用系统变量 relay_log 和 relay_log_index）覆盖默认的中继日志文件名前缀和中继日志索引文件名。

如果从库使用了默认的基于主机名的中继日志文件名，则在配置复制后更改从库的主机名可能会导致复制失败（报错：Failed to open the relay log and Could not find target log during relay log initialization），这是一个已知的问题（不是 bug，而是可规避的正常现象）。如果预计将来可能会更改从库主机名，则需要固定中继日志文件和中继日志索引文件的前缀名称（使用系统变量 relay_log 和 relay_log_index），这样就可以避免主机名发生变更带来的问题。

如果在复制线程正常运行一段时间后，不幸发生上述因修改主机名而导致的故障，则解决此问题的一种方法是停止从库数据库进程，然后将旧中继日志索引文件的内容添加到新的索引文件中，再重新启动从库数据库进程。在 UNIX 系统上，可以按照如下命令操作：

```
# 先将新产生的中继日志文件列表追加到旧的中继日志索引文件中
shell> cat new_relay_log_name.index >> old_relay_log_name.index
# 然后为旧的中继日志索引文件重新命名
shell> mv old_relay_log_name.index new_relay_log_name.index
```

从库创建新的中继日志文件有如下几种情况：

- 每次启动 I/O 线程时。

- 刷新日志时。例如，执行 FLUSH LOGS、FLUSH RELAY LOGS 语句或 mysqladmin flush-logs 命令。

- 当前中继日志文件变得"太大"时，按照如下规则滚动创建新的中继日志文件：

    - 如果系统变量 max_relay_log_size 的值大于 0，那么当中继日志文件大小达到这个上限时，就会创建新的中继日志文件。

    - 如果系统变量 max_relay_log_size 的值为 0，则中继日志文件的大小达到系统变量 max_binlog_size 设置的上限时，就会创建新的中继日志文件。

SQL 线程在重放完一个中继日志文件中的所有事件之后就不再需要该文件，会自动定期删除这个文件。中继日志文件的删除由 SQL 线程负责，没有明确的删除机制。但是，执行 FLUSH LOGS 语句会滚动中继日志产生新的文件，这会影响 SQL 线程删除重放过的中继日志文件的时间。

8.3 从库状态日志

从库会创建两个日志，即上文提到的 master info 和 relay log info，用于保存复制状态信息。默认情况下，复制状态信息是保存在磁盘文件中的，日志文件名为 master.info 和 relay-log.info，存放在系统变量 datadir 指定的目录下。可以分别使用选项--master-info-file 和--relay-log-info-file[1]（或在 my.cnf 中启用系统变量 relay_log_info_file）更改这些文件的名称和保存的位置。设置--master-info-repository = TABLE，或在 my.cnf 中启用系统变量 master_info_repository，可以将 master info 状态日志写入 mysql.slave_master_info 表；设置--relay-log-info-repository = TABLE，或在 my.cnf 中启用系统变量 relay_log_info_repository，可以将 relay log info 状态日志写入 mysql.slave_relay_log_info 表。

这两个状态日志中包含 SHOW SLAVE STATUS 语句输出的信息，由于存储在磁盘上（或持久化在表里），因此在从库的数据库进程关闭后，这些状态日志不会丢失。下次重新启动从库时会读取这两个状态日志，以确定从主库读取的二进制日志的位置和从库自己应用的中继日志位置。

master info 状态日志文件或者 mysql.slave_master_info 表中保存着用户连接主库的账号和密码明文信息，需要注意保护这些敏感信息。

如果将系统变量 master_info_repository 和 relay_log_info_repository 设置为 TABLE，位置信息将保存在 InnoDB 引擎创建的表 mysql.slave_master_info 和 mysql.slave_relay_log_info 中。其中，当 relay_log_info_repository = TABLE，且 mysql.slave_relay_log_info 表使用了 InnoDB 引擎，重放 InnoDB 存储引擎表的二进制日志（事务）时，二进制日志中回放事务的操作可以和 relay log info 状态日志的更新操作放在同一个事务中一并执行，这意味着 SQL

---

[1] 选项--master-info-file 没有对应的系统变量，而系统变量 relay_log_info_file 是只读的。

线程的重放位置信息日志始终与二进制日志事务保持一致,当 MySQL Server 发生意外而停止时,就可以利用该特性来进行原子恢复,但必须在从库上启用选项--relay-log-recovery(或在 my.cnf 中启用系统变量 relay_log_recovery)以保证实现一致性恢复。

从库启用多线程复制之后,会创建一个主要供内部使用的附加从库状态日志表,用于保存有关多线程复制的 Worker 线程的状态信息。此日志表中包含每个 Worker 线程对应的中继日志信息以及读取的主库二进制日志的位置。如果将从库的中继日志状态信息保存在表中(系统变量 relay_log_info_repository 的值设置为 TABLE),则会将 Worker 线程日志写入 mysql.slave_worker_info 表。如果将中继日志状态信息保存在文件中(系统变量 relay_log_info_repository 的值设置为 FILE),则会将 Worker 线程日志写入 worker-relay-log.info 文件。如果外部应用需要获取 Worker 线程的状态信息,则可以通过查询 PERFORMANCE SCHEMA 库下的 replication_applier_status_by_worker 表来获取。相对于 mysql.slave_worker_info 表,从 replication_applier_status_by_worker 表中获取的状态信息更实用、更易读。

master info 日志信息由从库的 I/O 线程负责更新。表 8-1 显示了 master.info 文件中的行编号、mysql.slave_master_info 表中的列,以及 SHOW SLAVE STATUS 展示的列之间的对应关系。

表 8-1

master.info 文件中的行编号	mysql.slave_master_info 表的列	SHOW SLAVE STATUS 展示的列	描述信息
1	Number_of_lines	无	表示文件中的行编号或表中的列编号
2	Master_log_name	Master_Log_File	表示当前正在从主库读取的主二进制日志文件名称
3	Master_log_pos	Read_Master_Log_Pos	表示当前正在从主库读取的二进制日志文件的当前位置
4	Host	Master_Host	主库的主机名(IP 地址)
5	User_name	Master_User	表示从库用于连接到主库的用户名
6	User_password	无	表示用于连接到主库的用户密码
7	Port	Master_Port	表示要连接的主库数据库端口
8	Connect_retry	Connect_Retry	表示从库在尝试重新连接到主库之前将等待的时间间隔(以秒为单位)
9	Enabled_ssl	Master_SSL_Allowed	表示 Server 是否支持 SSL 连接
10	Ssl_ca	Master_SSL_CA_File	表示证书颁发机构(CA)的证书文件名称
11	Ssl_capath	Master_SSL_CA_Path	表示 CA 的证书路径
12	Ssl_cert	Master_SSL_Cert	表示 CA 颁发的 Server 证书文件的名称

续表

master.info 文件中的行编号	mysql.slave_master_info 表的列	SHOW SLAVE STATUS 展示的列	描述信息
13	Ssl_cipher	Master_SSL_Cipher	表示 SSL 连接握手中可能使用的加密算法列表
14	Ssl_key	Master_SSL_Key	表示 SSL 密钥文件的名称
15	Ssl_verify_server_cert	Master_SSL_Verify_Server_Cert	表示是否检查 Server 证书
16	Heartbeat	无	表示复制机制的心跳检测间隔时间，以秒为单位。注：心跳检测间隔时间由从库配置复制时指定，从库向主库发起复制连接时将配置注入主库，主库按照这个间隔时间定期发送心跳包给从库。但要注意的是，从库会根据主库发送的心跳包判断主库是否存活，如果从库未收到心跳包的次数超过一定阈值（3 次左右），则从库尝试重连主库（按照重试间隔时间与重试次数持续进行重试）。心跳间隔时间可以单独使用 CHANGE MASTER TO 语句的 MASTER_HEARTBEAT_PERIOD 选项指定，如果不指定，则默认为系统变量 slave_net_timeout 值的一半。如果指定 MASTER_HEARTBEAT_PERIOD 选项值为 0，则表示关闭心跳检测。另外，需要注意，指定的 MASTER_HEARTBEAT_PERIOD 的值不能大于系统变量 slave_net_timeout 的值，否则主库会因为在超时时间内没有探测到从库的心跳回包而频繁断开与从库的连接
17	Bind	Master_Bind	表示从库使用哪个网络接口连接到主库
18	Ignored_server_ids	Replicate_Ignore_Server_Ids	表示从库需要忽略的主库 Server ID 列表。注意，对于 Ignored_server_ids，其 Server ID 值列表中最前面的数字表示要忽略的 Server ID 总数，而不是具体的 Server ID 值
19	Uuid	Master_UUID	表示主库的 UUID
20	Retry_count	Master_Retry_Count	表示允许从库重连主库的最大次数
21	Ssl_crl	无	表示 SSL 证书吊销列表文件的路径
22	Ssl_crlpath	无	表示包含 SSL 证书吊销列表文件的目录路径

续表

master.info 文件中的行编号	mysql.slave_master_info 表的列	SHOW SLAVE STATUS 展示的列	描述信息
23	Enabled_auto_position	Auto_position	表示是否正在使用自动定位功能
24	Channel_name	Channel_name	表示复制通道的名称
25	Tls_Version	Master_TLS_Version	表示在 Master 上的 TLS 版本

relay log info 状态日志由从库的 SQL 线程负责更新。relay-log.info 文件包括行计数和复制延迟值等信息。表 8-2 显示了 relay-log.info 文件中的行编号、mysql.slave_relay_log_info 表中的列，以及 SHOW SLAVE STATUS 展示的列之间的对应关系。

表 8-2

relay-log.info 文件中的行编号	mysql.slave_relay_log_info 表中的列	SHOW SLAVE STATUS 展示的列	描述信息
1	Number_of_lines	无	表示文件中的行编号或表中的列编号
2	Relay_log_name	Relay_Log_File	表示当前中继日志文件的名称
3	Relay_log_pos	Relay_Log_Pos	表示当前中继日志文件中的当前位置，从中继日志中读取的每个中继日志事件被成功执行之后，此位置信息会被更新
4	Master_log_name	Relay_Master_Log_File	表示从库 SQL 线程读取的中继日志文件中的事件对应的主库二进制日志文件的名称
5	Master_log_pos	Exec_Master_Log_Pos	表示从库 SQL 线程已执行的中继日志文件中的事件对应的主库二进制日志文件中的位置
6	Sql_delay	SQL_Delay	表示从库必须滞后于主库的秒数
7	Number_of_workers	无	表示并行执行事务的 Worker 线程数
8	Id	无	内部使用，目前的值总是 1
9	Channel_name	Channel_name	表示复制通道的名称

在 MySQL 5.6 之前的版本中，relay-log.info 文件不包括行编号和延迟值（并且 mysql.slave_relay_log_info 表不可用），其包含的内容如表 8-3 所示。

表 8-3

状态列名	描述信息
Relay_Log_File	表示当前中继日志文件的名称
Relay_Log_Pos	表示当前中继日志文件中的当前位置，从中继日志中读取的每个中继日志事件被成功执行之后，此位置信息会更新
Relay_Master_Log_File	表示从库 SQL 线程读取的中继日志文件中的事件对应的主库二进制日志文件的名称
Exec_Master_Log_Pos	表示从库 SQL 线程已执行的中继日志文件中的事件对应的主库二进制日志文件中的位置

如果 relay-log.info 文件尚未被刷新到磁盘，则该文件中的内容和 SHOW SLAVE STATUS 语句显示的状态可能不匹配。所以，正常情况下，建议使用 SHOW SLAVE STATUS 语句查看从库状态日志信息，或查询 mysql.slave_master_info 和 mysql.slave_relay_log_info 表。

备份从库数据时，应备份这两个状态日志以及中继日志文件。因为使用从库备份来做数据恢复时，需要使用状态日志与对应的中继日志文件，才能正确恢复备份数据对应的主库二进制日志位置，以便将复制恢复到正常状态。如果丢失了中继日志（这里指的是 relay log file 中还未来得及重放的事件），但 relay log info 状态日志保存完好，则可以检查 relay log info 状态日志以确定 SQL 线程对应的主库二进制日志位置。然后，使用 CHANGE MASTER TO 语句结合 MASTER_LOG_FILE 和 MASTER_LOG_POS 选项，指定从主库的哪个二进制日志的哪个位置开始重新读取，但这要求指定的二进制日志文件和二进制日志位置在主库上没有被清理。

注意：如果将从库降级到早于 MySQL 5.6 的版本，则此旧版本无法正确读取新版本的 relay-log.info 文件。可以通过删除 relay-log.info 文件中多余的行来解决这个问题。

以下是状态日志的磁盘文件、表数据、SHOW SLAVE STATUS 语句输出内容的示例：

```
# 状态日志的磁盘文件
## master.info 文件的内容
[root@localhost mydata]# cat master.info
25
mysql-bin.000072
74459488
10.10.30.162
repl
password
3306
10
0
......
0
```

```
5.000

0
938c7a23-27af-11ea-9599-5254002a54f2
86400
......
1
```

relay-log.info 文件的内容

```
[root@localhost mydata]# cat relay-log.info
7
/home/mysql/data/mysqldata1/relaylog/mysql-relay-bin.000002
924
mysql-bin.000072
74459998
0
0
1
```

状态日志表

mysql.slave_master_info 表的内容

```
mysql> select * from mysql.slave_master_info\G
*************************** 1. row ***************************
       Number_of_lines: 25
       Master_log_name: mysql-bin.000072
        Master_log_pos: 74459488
                  Host: 10.10.30.162
             User_name: repl
         User_password: password
                  Port: 3306
         Connect_retry: 10
           Enabled_ssl: 0
                Ssl_ca:
            Ssl_capath:
              Ssl_cert:
            Ssl_cipher:
               Ssl_key:
 Ssl_verify_server_cert: 0
             Heartbeat: 5
                  Bind:
     Ignored_server_ids: 0
                  Uuid: 938c7a23-27af-11ea-9599-5254002a54f2
           Retry_count: 86400
               Ssl_crl:
           Ssl_crlpath:
  Enabled_auto_position: 1
          Channel_name:
           Tls_version:
```

```
1 row in set (0.00 sec)

## mysql.slave_relay_log_info 表的内容
mysql> select * from mysql.slave_relay_log_info\G
*************************** 1. row ***************************
  Number_of_lines: 7
   Relay_log_name: /home/mysql/data/mysqldata1/relaylog/mysql-relay-bin.000215
    Relay_log_pos: 74459701
  Master_log_name: mysql-bin.000072
   Master_log_pos: 74459488
        Sql_delay: 0
Number_of_workers: 16
               Id: 1
     Channel_name:
1 row in set (0.00 sec)

# SHOW SLAVE STATUS 语句输出的内容
mysql> show slave status\G
*************************** 1. row ***************************
               Slave_IO_State: Waiting for master to send event
                  Master_Host: 10.10.30.162
                  Master_User: repl
                  Master_Port: 3306
                Connect_Retry: 10
              Master_Log_File: mysql-bin.000072
          Read_Master_Log_Pos: 74459488
               Relay_Log_File: mysql-relay-bin.000215
                Relay_Log_Pos: 74459701
        Relay_Master_Log_File: mysql-bin.000072
             Slave_IO_Running: Yes
            Slave_SQL_Running: Yes
              Replicate_Do_DB:
          Replicate_Ignore_DB:
           Replicate_Do_Table:
       Replicate_Ignore_Table:
      Replicate_Wild_Do_Table:
  Replicate_Wild_Ignore_Table:
                   Last_Errno: 0
                   Last_Error:
                 Skip_Counter: 0
          Exec_Master_Log_Pos: 74459488
              Relay_Log_Space: 74459995
              Until_Condition: None
               Until_Log_File:
                Until_Log_Pos: 0
           Master_SSL_Allowed: No
           Master_SSL_CA_File:
           Master_SSL_CA_Path:
```

```
                  Master_SSL_Cert: 
                Master_SSL_Cipher: 
                   Master_SSL_Key: 
            Seconds_Behind_Master: 0
Master_SSL_Verify_Server_Cert: No
                    Last_IO_Errno: 0
                    Last_IO_Error: 
                   Last_SQL_Errno: 0
                   Last_SQL_Error: 
      Replicate_Ignore_Server_Ids: 
                 Master_Server_Id: 33061
                      Master_UUID: 938c7a23-27af-11ea-9599-5254002a54f2
                 Master_Info_File: mysql.slave_master_info
                        SQL_Delay: 0
              SQL_Remaining_Delay: NULL
          Slave_SQL_Running_State: Slave has read all relay log; waiting for more updates
               Master_Retry_Count: 86400
                      Master_Bind: 
          Last_IO_Error_Timestamp: 
         Last_SQL_Error_Timestamp: 
                   Master_SSL_Crl: 
               Master_SSL_Crlpath: 
               Retrieved_Gtid_Set: 938c7a23-27af-11ea-9599-5254002a54f2:1-7374239
                Executed_Gtid_Set: 938c7a23-27af-11ea-9599-5254002a54f2:1-7374239
                    Auto_Position: 1
             Replicate_Rewrite_DB: 
                     Channel_Name: 
               Master_TLS_Version: 
1 row in set (0.00 sec)
```

第 9 章 通过 PERFORMANCE_SCHEMA 库检查复制信息

通常，我们在查看与复制相关的状态信息时，都习惯性地使用 SHOW SLAVE STATUS 语句。也许你会说，"我也会用 PERFORMANCE_SCHEMA 库的表查看一些在复制时的报错信息"，但是，你知道 SHOW SLAVE STATUS 语句输出的信息和 PERFORMANCE_SCHEMA 库中的复制信息记录表之间有什么异同吗？本章将详细介绍二者的异同。

注意：这些复制信息表也支持组复制，但限于篇幅，对于组复制技术，本书不做过多介绍，有兴趣的读者可自行研究。

9.1 PERFORMANCE_SCHEMA 库中的复制信息记录表概述

在 PERFORMANCE_SCHEMA 库中，以 replication_开头的几个表提供了复制机制的配置与状态的详细信息，这与我们用 SHOW SLAVE STATUS 语句查到的一些信息类似。虽然 SHOW SLAVE STATUS 语句的输出更为直观（事实上，大多数 DBA 通常都用该语句来查看复制信息），但如果用程序来获取这些值就比较麻烦。两者的差别大致有如下三点：

- 对于 PERFORMANCE_SCHEMA 库中的表保存的复制信息，可以使用 SELECT 语句来查询（还可以使用一些复杂的 WHERE 条件及 JOIN 连接查询），而使用 SHOW SLAVE STATUS 语句做不到这一点。

- 对于多线程从库，SHOW SLAVE STATUS 语句的 Last_SQL_Errno 和 Last_SQL_Error 字段分别显示的是所有 Coordinator 线程（协调器线程）与所有 Worker 线程（工作线程）的错误汇总信息，这意味着除最后一次错误信息外，之前的错误信息都可能丢失。而采用表存储，则每个 Coordinator 线程和 Worker 线程的错误都会各自保存一行数据，不会出现使用 SHOW SLAVE STATUS 语句时，前面的错误信息被后来的错误信息覆盖的尴尬情况。

- 在 replication_applier_status_by_worker 表中，所有线程都有一个字段 Last_Seen_Transaction 来记录从库中最后一个事务的 GTID，即所有的 Worker 线程都可以看到自己执行的最后一个事务，而 SHOW SLAVE STATUS 语句不提供这样的信息。

PERFORMANCE_SCHEMA 库提供了如下几个与复制信息相关的表（这里简单介绍这些表的作用，其具体的含义详见下文）：

- replication_applier_configuration：记录从库延迟复制时的配置参数值。
- replication_applier_status：记录从库当前的普通事务执行状态（该表也记录组复制中的复制状态信息）。
- replication_applier_status_by_coordinator：当使用多线程复制时，该表中记录 Coordinator 线程工作状态；当使用单线程复制时，该表为空。
- replication_applier_status_by_worker：如果是以单线程复制运行的（slave_parallel_workers = 0），则该表记录一条 WORKER_ID = 0 的状态；如果以多线程复制运行（slave_parallel_workers > 0），则该表记录所有 Worker 线程的状态。
- replication_connection_configuration：记录从库用于连接到主库的配置参数值。
- replication_connection_status：记录从库 I/O 线程的连接状态信息。
- replication_group_member_stats：记录组复制成员的事务状态统计信息。
- replication_group_members：记录组复制成员的网络和状态信息。

这些表记录的信息的生命周期如下（生命周期指的是这些表中的信息什么时候写入，什么时候会被修改，什么时候会被清理等）：

- 在执行 CHANGE MASTER TO 语句之前，这些表是空的。
- 执行 CHANGE MASTER TO 语句之后，在配置参数表 replication_applier_configuration 和 replication_connection_configuration 中就可以查看配置信息。此时，由于并没有启动复制线程，所以表中 THREAD_ID 字段为 NULL，SERVICE_STATE 字段的值为 OFF（这两个字段也存在于 replication_applier_status、replication_applier_status_by_coordinator、replication_applier_status_by_worker、replication_connection_status 等状态信息表中）。
- 执行 START SLAVE 语句后，如果相应的复制线程正常运行，则会更新这些表中相应线程的状态。例如，THREAD_ID 字段被分配了一个数值（THREAD_ID 字段值与在 SHOW PROCESSLIST 语句的输出中看到的线程 id 值不同，后者与前者的关联关系可以通过 performance_schema.threads 表查看），而且 SERVICE_STATE 字段被修改为 ON。
- 执行 STOP SLAVE 语句之后，所有的复制 I/O 线程、Coordinator 线程、Worker 线程状态表中，THREAD_ID 字段的值为 NULL，SERVICE_STATE 字段的值变为 OFF。

注意：停止与复制相关的线程之后，这些记录并不会被清理，因为如果发生意外而导致复制中止或者临时停止复制线程，在重新启动复制时，可能需要获取一些状态信息用于排错或者其他用途。

- 执行 RESET SLAVE 语句之后，所有记录复制的配置和状态信息的表中的内容都会被清除。但是用 SHOW SLAVE STATUS 语句还是能查看一些复制的状态和配置信息，因为该语句是从内存中获取这些信息的，RESET SLAVE 语句并没有清理内存，而是清理了磁盘文件、表中记录的信息（还包括 mysql.slave_master_info 和 mysql.slave_relay_log_info 两个表）。如果需要清理内存里保存的与复制相关的信息，需要使用 RESET SLAVE ALL 语句。

PERFORMANCE_SCHEMA 库中的这些复制信息表是基于 GTID 记录的，而不是基于二进制日志的位置记录的，所以这些表记录的是 Server UUID 值，而不是 Server ID 值。所以，PERFORMANCE_SCHEMA 库中记录的一些复制信息可能是 SHOW SLAVE STATUS 语句输出的信息中没有的，但 SHOW SLAVE STATUS 语句输出的信息中也有一些是 PERFORMANCE_SCHEMA 库中没有的。具体的差异如下：

（1）在 SHOW SLAVE STATUS 语句的输出中，以下的字段用于引用二进制日志文件、二进制日志文件位置、中继日志文件和中继日志文件位置等信息，它们在 PERFORMANCE_SCHEMA 库的表中不记录：

- Master_Log_File
- Read_Master_Log_Pos
- Relay_Log_File
- Relay_Log_Pos
- Relay_Master_Log_File
- Exec_Master_Log_Pos
- Until_Condition
- Until_Log_File
- Until_Log_Pos

（2）在 SHOW SLAVE STATUS 语句的输出中，Master_Info_File 字段记录主/从库的配置保存方式等信息，在 PERFORMANCE_SCHEMA 库的表中不记录。默认情况下，Master_Info_File 字段表示 master.info 文件的路径，如果 master_info_repository 设置为 TABLE，则该文件中的信息被保存在 mysql.slave_master_info 表中，Master_Info_File 字段显示 mysql.slave_master_info 字符串的值。

（3）在 SHOW SLAVE STATUS 语句的输出中，以下字段是基于 Server ID 而不是基于 Server UUID 使用的，在 PERFORMANCE_SCHEMA 库的表中不记录：

- Master_Server_Id
- Replicate_Ignore_Server_Ids

（4）在 SHOW SLAVE STATUS 语句的输出中，Skip_Counter 字段是基于事件而不是基于 GTID 计数的，在 PERFORMANCE_SCHEMA 库的表中不记录。

（5）在 SHOW SLAVE STATUS 语句的输出中，以下字段是 Last_SQL_Errno 和 Last_SQL_Error 字段的别名，在 PERFORMANCE_SCHEMA 库的表中不记录：

- Last_Errno
- Last_Error

（7）在 PERFORMANCE_SCHEMA 库中，replication_applier_status_by_coordinator 表的 LAST_ERROR_NUMBER 和 LAST_ERROR_MESSAGE 字段的值对应 SHOW SLAVE STATUS 语句输出的 Last_SQL_Errno 和 Last_SQL_Error 字段值（如果从库为多线程复制，则具体的 Worker 线程报错信息还需要查询表 replication_applier_status_by_worker，这些表提供比 Last_SQL_Errno 和 Last_SQL_Error 字段更具体的信息）。

（8）在 SHOW SLAVE STATUS 语句的输出中，以下的字段用于复制过滤，在 PERFORMANCE_SCHEMA 库的表中不记录：

- Replicate_Do_DB
- Replicate_Ignore_DB
- Replicate_Do_Table
- Replicate_Ignore_Table
- Replicate_Wild_Do_Table
- Replicate_Wild_Ignore_Table

（9）在 SHOW SLAVE STATUS 语句的输出中，Slave_IO_State 和 Slave_SQL_Running_State 字段在 PERFORMANCE_SCHEMA 库的表中不记录。如果需要使用 SELECT 语句查询，可以用对应的复制状态表中的 THREAD_ID 字段与 information_schema.processlist 表中的 ID 字段相关联，然后使用 information_schema.processlist 表的 STATE 字段作为 SELECT 字段执行 JOIN 查询，得到两个线程的状态值。

- 查询 I/O 线程的状态值可以使用语句：

```
select STATE as IO_THREAD_STATE, ID as PROCESS_ID,pr.THREAD_ID from performance_schema.replication_ connection_status as pr join information_schema.processlist as ip on pr.THREAD_ID=sys.ps_thread_id(ip.ID);
```

- 查询 SQL 线程的状态值，可以使用语句：

```
select STATE as SQL_THREAD_STATE,ID as PROCESS_ID, pr.THREAD_ID from performance_schema.replication_applier_status_by_coordinator as pr join information_schema.processlist as ip on pr.THREAD_ID=sys.ps_thread_id(ip.ID);
```[1]

---

[1] 为 sys.ps_thread_id()函数传入一个 processlist id，会返回对应会话的 thread id。

（8）在 SHOW SLAVE STATUS 语句的输出中，Seconds_Behind_Master 和 Relay_Log_Space 字段在 PERFORMANCE_SCHEMA 库的表中不记录。

提示：如下状态变量在 MySQL 5.7.5 之前可用，但在之后的版本中，其对应的状态值被移到复制状态表中进行记录（performance_schema.replication_connection_status 和 performance_schema.replication_applier_status 表）：

- Slave_retried_transactions
- Slave_last_heartbeat
- Slave_received_heartbeats
- Slave_heartbeat_period
- Slave_running

对于组复制的集群拓扑，复制信息散布在如下几张表中：

- replication_group_member_stats
- replication_group_members
- replication_applier_status
- replication_connection_status
- threads

系统变量 slave_parallel_workers 的值决定是否启用单线程或多线程复制：当该变量为 0 时，表示关闭多线程复制（使用单线程复制）；为 1 时，表示启用多线程复制，但只有一个 Coordinator 线程和一个 Worker 线程；大于 1 时，表示启用多线程复制，且有一个 Coordinator 线程和多个 Worker 线程。

9.2　PERFORMANCE_SCHEMA 库中的复制信息记录表详解

9.2.1　replication_applier_configuration 表

replication_applier_configuration 表记录从库延迟复制的配置参数，延迟复制的线程被称为普通线程。通过表中的 CHANNEL_NAME 和 DESIRED_DELAY 字段可以查看某个复制通道是否配置了延迟复制，该表中的记录在 Server 运行时可以使用 CHANGE MASTER TO 语句更改。

查看 replication_applier_configuration 表中记录的内容：

```
# 该表会为每个复制通道记录一条类似如下的信息（多源复制时存在多个通道，因此会有多条记录）
mysql> select * from replication_applier_configuration;
+--------------+---------------+
| CHANNEL_NAME | DESIRED_DELAY |
```

```
+---------------+---------------+
|               |      0        |
+---------------+---------------+
1 row in set (0.00 sec)

# 如果是 MGR 集群，则表中会记录类似如下 MGR 集群信息
mysql> select * from replication_applier_configuration;
+----------------------------+---------------+
| CHANNEL_NAME               | DESIRED_DELAY |
+----------------------------+---------------+
| group_replication_applier  |       0       |
| group_replication_recovery |       0       |
+----------------------------+---------------+
2 rows in set (0.00 sec)
```

replication_applier_configuration 表中各字段的含义及其与 SHOW SLAVE STATUS 语句输出的字段的对应关系如表 9-1 所示。

表 9-1

replication_applier_configuration 表中的字段名	含　义	对应 SHOW SLAVE STATUS 语句输出的字段名
CHANNEL_NAME	显示复制通道名称	Channel_Name
DESIRED_DELAY	在该通道下，从库复制 SQL 线程必须滞后于主库的秒数	SQL_Delay

对 replication_applier_configuration 表不允许执行 TRUNCATE TABLE 语句。

9.2.2　replication_applier_status 表

replication_applier_status 表记录从库当前的普通事务执行状态（也记录组复制架构中的复制状态信息）。此表提供了所有线程执行二进制日志重放事务时的普通状态信息。线程重放事务时特定的状态信息保存在 replication_applier_status_by_coordinator 和 replication_applier_status_by_worker 表中。

查看 replication_applier_status 表中记录的内容：

```
# 该表中的记录行数与复制通道数量相关，与是否使用多线程复制无关，即如果是多源复制，则每个复制通道记录一行信息；反之，只记录一行记录
mysql> select * from replication_applier_status;
+--------------+---------------+-----------------+---------------------------+
| CHANNEL_NAME | SERVICE_STATE | REMAINING_DELAY | COUNT_TRANSACTIONS_RETRIES |
+--------------+---------------+-----------------+---------------------------+
|              |      ON       |      NULL       |             0             |
+--------------+---------------+-----------------+---------------------------+
1 row in set (0.00 sec)
```

```
# 如果是MGR集群，则表中会记录MGR复制专用通道的状态信息
mysql> select * from replication_applier_status;
+----------------------------+---------------+-----------------+---------------------------+
|CHANNEL_NAME                |SERVICE_STATE  |REMAINING_DELAY  |COUNT_TRANSACTIONS_RETRIES |
+----------------------------+---------------+-----------------+---------------------------+
| group_replication_applier  | ON            | NULL            | 0                         |
| group_replication_recovery | OFF           | NULL            | 0                         |
+----------------------------+---------------+-----------------+---------------------------+
2 rows in set (0.00 sec)
```

replication_applier_status 表中各字段的含义及其与 SHOW SLAVE STATUS 语句输出的字段的对应关系如表 9-2 所示。

表 9-2

replication_applier_status 表中的字段名	含 义	对应 SHOW SLAVE STATUS 语句输出的字段名
CHANNEL_NAME	显示复制通道名称	Channel_Name
SERVICE_STATE	在该复制通道下，从库应用线程的状态。其有效值有：ON（如果从库应用线程处于活跃状态或空闲状态时，显示 ON），OFF（如果从库应用线程未处于活动状态则为 OFF，可能没有启动复制）	无
REMAINING_DELAY	在该复制通道下，如果从库应用一个事件后，由于 DESIRED_DELAY 指定了延迟的秒数而执行等待，则该字段显示需要等待的剩余延迟秒数。如果不需要等待，此字段显示为 NULL。DESIRED_DELAY 的值保存在 replication_applier_configuration 表中	SQL_Remaining_Delay
COUNT_TRANSACTIONS_RETRIES	显示该通道下由于从库 SQL 线程无法应用事务而进行的重试次数	无

对 replication_applier_status 表不允许执行 TRUNCATE TABLE 语句。

9.2.3 replication_applier_status_by_coordinator 表

replication_applier_status_by_coordinator 表中记录的是从库使用多线程复制时，从库的 Coordinator 线程工作状态，每个通道将创建一个 Coordinator 线程，即有多少个复制通道，该表中就会有多少行状态记录。如果从库使用单线程复制，则此表为空，对应的记录转移到 replication_applier_status_by_worker 表中。

查看 replication_applier_status_by_coordinator 表中记录的内容：

```
# 若为单线程主从复制，该表为空；若为多线程主从复制，表中记录 Coordinator 线程状态信息，多源复制时每个复制通道记录一行信息
mysql> select * from replication_applier_status_by_coordinator;
```

```
+-----------+-----------+---------------+-------------------+-------------------+
| CHANNEL_NAME | THREAD_ID | SERVICE_STATE | LAST_ERROR_NUMBER | LAST_ERROR_MESSAGE |
LAST_ERROR_TIMESTAMP |
+-----------+-----------+---------------+-------------------+-------------------+
|           | 43        | ON            | 0                 |                   | 0000-00-00 00:00:00 |
+-----------+-----------+---------------+-------------------+-------------------+
1 row in set (0.00 sec)

# 如果是 MGR 集群，则表中会记录类似如下的 MGR 集群信息
mysql> select * from replication_applier_status_by_coordinator;
+--------------------------+-----------+---------------+-------------------+-------------------+
| CHANNEL_NAME             | THREAD_ID | SERVICE_STATE | LAST_ERROR_NUMBER | LAST_ERROR_MESSAGE |
LAST_ERROR_TIMESTAMP |
+--------------------------+-----------+---------------+-------------------+-------------------+
| group_replication_applier | 91       | ON            | 0                 |                   | 0000-00-00 00:00:00 |
+--------------------------+-----------+---------------+-------------------+-------------------+
1 row in set (0.00 sec)
```

replication_applier_status_by_coordinator 表中各字段的含义及其与 SHOW SLAVE STATUS 语句输出中字段的对应关系如表 9-3 所示。

表 9-3

replication_applier_status_by_coordinator 表中的字段名	含 义	对应 SHOW SLAVE STATUS 语句输出中的字段名
CHANNEL_NAME	显示复制通道名称	Channel_Name
THREAD_ID	在该通道下，从库 Coordinator 线程的 ID	无
SERVICE_STATE	在该通道下，从库 Coordinator 线程的状态。其有效值有：ON（Coordinator 线程存在且处于活跃状态或空闲状态），OFF（Coordinator 线程不再存在，可能没有启动）	Slave_SQL_Running
LAST_ERROR_NUMBER LAST_ERROR_MESSAGE	在该通道下，从库 Coordinator 线程发生错误而停止的最新错误编号和错误消息。如果错误编号为 0，错误信息字段为空串，则表示"无错误"。如果 LAST_ERROR_MESSAGE 字段值不为空，则错误值也会被输出到从库的错误日志中。注意，在执行 RESET MASTER 或 RESET SLAVE 语句时，这两个字段的值会被重置	Last_SQL_Errno Last_SQL_Error
LAST_ERROR_TIMESTAMP	在该通道下，从库 Coordinator 线程发生错误的时间，时间格式为：YYMMDD HH: MM: SS	Last_SQL_Error_Timestamp

对 replication_applier_status_by_coordinator 表不允许执行 TRUNCATE TABLE 语句。

9.2.4 replication_applier_status_by_worker 表

replication_applier_status_by_worker 表记录 SQL/Worker 线程的工作状态。当从库使用单线程复制时，则该表只记录一条 WORKER_ID = 0 的 SQL 线程的状态；当从库使用多线程复制时，则该表记录由系统变量 slave_parallel_workers = N 指定的 N 个数量的 Worker 线程状态（WORKER_ID 从 1 开始编号），每一个通道都有自己独立的 N 个 Worker 线程。如果是 MGR 集群，则该表中记录的是 N 个 group_replication_applier 线程及一个 group_replication_recovery 线程的工作状态。

查看 replication_applier_status_by_worker 表中记录的内容：

```
# 单线程主从复制，表中记录的内容如下
mysql> select * from replication_applier_status_by_worker;
+------+------+------+--------+---------+----------+----------+---------+
| CHANNEL_NAME | WORKER_ID | THREAD_ID | SERVICE_STATE | LAST_SEEN_TRANSACTION | LAST_ERROR_NUMBER | LAST_ERROR_MESSAGE | LAST_ERROR_TIMESTAMP |
+------+------+------+--------+---------+----------+----------+---------+
|  | 0 | 82 | ON |  | 0 |  | 0000-00-00 00:00:00 |
+------+------+------+--------+---------+----------+----------+---------+
1 row in set (0.00 sec)

# 多线程主从复制，表中的记录内容如下（如果是多源复制，则每个复制通道记录系统变量 slave_parallel_
workers 指定个数的 Worker 线程信息）
mysql> select * from replication_applier_status_by_worker;
+------+------+------+--------+--------+--------+------------+
| CHANNEL_NAME | WORKER_ID | THREAD_ID | SERVICE_STATE | LAST_SEEN_TRANSACTION | LAST_ERROR_NUMBER | LAST_ERROR_MESSAGE | LAST_ERROR_TIMESTAMP |
+------+------+------+--------+--------+--------+------------+
|  | 1 | 44 | ON |  | 0 |  | 0000-00-00 00:00:00 |
|  | 2 | 45 | ON |  | 0 |  | 0000-00-00 00:00:00 |
|  | 3 | 46 | ON |  | 0 |  | 0000-00-00 00:00:00 |
|  | 4 | 47 | ON |  | 0 |  | 0000-00-00 00:00:00 |
+------+------+------+--------+--------+--------+------------+
4 rows in set (0.00 sec)

# 如果是 MGR 集群，则表中会记录类似如下的信息
mysql> select * from replication_applier_status_by_worker;
+--------+------+------+--------+------+----------+--------+
| CHANNEL_NAME | WORKER_ID | THREAD_ID | SERVICE_STATE | LAST_SEEN_TRANSACTION | LAST_ERROR_NUMBER | LAST_ERROR_MESSAGE | LAST_ERROR_TIMESTAMP |
+--------+------+------+--------+------+----------+--------+
|group_replication_recovery|0|NULL|OFF|  | 0 |  | 0000-00-00 00:00:00 |
|group_replication_applier|1| 92 | ON | aaaaaaaa-aaaa-aaaa-aaaa-aaaaaaaaaaaa:104099082 | 0 |  | 0000-00-00 00:00:00 |
|group_replication_applier|2| 93 | ON |  | 0 |  | 0000-00-00 00:00:00 |
```

```
......
+-----+-------+------+------+---------+---------+----------+-------+
17 rows in set (0.00 sec)
```

replication_applier_status_by_worker 表中各字段的含义及其与 SHOW SLAVE STATUS 语句输出的字段的对应关系如表 9-4 所示。

表 9-4

replication_applier_status_by_worker 表中的字段名	含 义	对应 SHOW SLAVE STATUS 语句输出的字段名
CHANNEL_NAME	显示复制通道名称	Channel_Name
WORKER_ID	该通道下在该表中的 Worker 线程标识符（与 mysql.slave_worker_info 表中的 Worker 线程 ID 字段值相同），按照从 1 到系统变量 slave_parallel_workers 定义的值的数值范围依次编号（如 slave_parallel_workers 定义为 4，就会产生 1、2、3、4 这几个 WORKER_ID）。当执行 STOP SLAVE 语句之后，THREAD_ID 字段变为 NULL，但 WORKER_ID 值不会发生变化	无
THREAD_ID	在该通道下 Worker 线程的 ID	无
SERVICE_STATE	在该通道下 Worker 线程的状态。有效值有：ON（Worker 线程存在且处于活跃状态或空闲状态），OFF（Worker 线程不再存在，可能没有启动）	无
LAST_SEEN_TRANSACTION	该通道下 Worker 线程最后看到的事务，即 Worker 线程正在执行的事务号。如果系统变量 gtid_mode 值为 OFF，则此字段为 ANONYMOUS，表示事务不具有 GTID。如果 gtid_mode 为 ON，则此字段值可能有几种情况：如果 Worker 线程从启动复制以来一直处于空闲状态，则该字段为空串；如果 Worker 线程正在执行事务，则此字段值总是显示正在执行的事务的 GTID。 如果 Worker 线程执行过事务，后续再处于空闲状态时，则一直保留最后一个执行的事务的 GTID，直到下一个事务被分配到该 Worker 线程时进行更新 GTID；如果 Worker 线程在应用事务时发生错误，则此字段值是发生错误时[1]该 Worker 线程正确执行完的最后一个事务的 GTID	无

---

[1] 注：任何一个 Worker 线程发生错误，都会被 Coordinator 线程捕获，然后该通道下的所有 Worker 线程都会被停止。

replication_applier_status_by_worker 表中的字段名	含 义	对应 SHOW SLAVE STATUS 语句输出的字段名
LAST_ERROR_NUMBER LAST_ERROR_MESSAGE	该通道下从库 SQL/Worker 线程发生错误而停止时的最新错误号和错误消息。如果错误编号为 0，错误信息字段为空串，则表示"无错误"。如果 LAST_ERROR_MESSAGE 字段值不为空，则错误值也会打印在从库的错误日志中。注意，在执行 RESET MASTER 或 RESET SLAVE 语句时这两个字段值会被重置	从库使用单线程复制时，分别对应 Last_SQL_Errno 和 Last_SQL_Error。从库使用多线程复制时，在 SHOW SLAVE STATUS 语句输出中不显示，这两个字段显示 Coordinator 线程的错误编号和错误信息
LAST_ERROR_TIMESTAMP	该通道下从库 SQL/Worker 线程发生错误的时间，时间格式为：YYMMDD HH: MM: SS	Last_SQL_Error_Timestamp

对 replication_applier_status_by_worker 表不允许执行 TRUNCATE TABLE 语句。

9.2.5　replication_connection_configuration 表

replication_connection_configuration 表记录从库用于连接主库的配置参数，该表中存储的配置信息在执行 CHANGE MASTER TO 语句时会被更新。

与 replication_connection_status 表相比，replication_connection_configuration 表更改的频率更低，因为它只包含从库连接主库的配置参数，在连接（I/O 线程）正常工作期间，这些配置参数的值保持不变；而只要 I/O 线程的状态发生变化，replication_connection_status 表中包含的连接状态信息就会更新。如果是多源复制，则从库指向多少个主库就会记录多少行记录。在组复制中，每个节点有两条记录，但这两条记录并未记录完整的连接配置参数，例如，host 等信息并不在该表中，而是被记录在 replication_group_members 表中。

查看 replication_connection_configuration 表中记录的内容：

```
# 单线程和多线程的主从复制，表中记录的内容相同。如果是多源复制，则每个复制通道各自有一行记录
mysql> select * from replication_connection_configuration\G;
*************************** 1. row ***************************
                 CHANNEL_NAME: 
                         HOST: 10.10.20.14
                         PORT: 3306
                         USER: qfsys
            NETWORK_INTERFACE: 
                AUTO_POSITION: 1
                  SSL_ALLOWED: NO
......
    CONNECTION_RETRY_INTERVAL: 60
```

```
            CONNECTION_RETRY_COUNT: 86400
                HEARTBEAT_INTERVAL: 5.000
                       TLS_VERSION:
1 row in set (0.00 sec)

# 如果是 MGR 集群，则表中会记录类似如下的信息
mysql> select * from replication_connection_configuration\G
*************************** 1. row ***************************
              CHANNEL_NAME: group_replication_applier
                      HOST: <NULL>
......
*************************** 2. row ***************************
              CHANNEL_NAME: group_replication_recovery
                      HOST: <NULL>
......
2 rows in set (0.00 sec)
```

replication_connection_configuration 表中各字段的含义及其与 CHANGE MASTER TO 语句选项的对应关系如表 9-5 所示。

表 9-5

replication_connection_ configuration 表中的字段名	含 义	对应 CHANGE MASTER TO 语句的选项名
CHANNEL_NAME	显示复制通道名称	使用 FOR CHANNEL *channel* 子句指定的通道（*channel*）值
HOST	在该通道下，从库连接的主库的数据库实例 IP 地址或主机名	MASTER_HOST
PORT	在该通道下，从库连接的主库的数据实例端口	MASTER_PORT
USER	在该通道下，从库连接主库的数据库实例账号的密码	MASTER_USER
NETWORK_INTERFACE	在该通道下，从库连接主库的数据库实例网卡名称	MASTER_BIND
AUTO_POSITION	在该通道下，如果从库使用自动定位，则值为 1，否则为 0	MASTER_AUTO_POSITION
SSL_ALLOWED	是否开启 SSL 连接。如果主库允许 SSL 连接且从库支持，则该字段为 Yes；如果主库不允许 SSL 连接，则为 No；如果主库允许 SSL 连接而从库不支持，则为 Ignored	MASTER_SSL

续表

replication_connection_configuration 表中的字段名	含 义	对应 CHANGE MASTER TO 语句的选项名
SSL_CA_FILE SSL_CA_PATH SSL_CERTIFICATE SSL_CIPHER SSL_KEY SSL_VERIFY_SERVER_CERTIFICATE SSL_CRL_FILE SSL_CRL_PATH	主库使用 SSL 连接的 CA 证书、密钥（key），以及是否校验 SSL 相关证书的字段	MASTER_SSL_CA MASTER_SSL_CAPATH MASTER_SSL_CERT MASTER_SSL_CRL MASTER_SSL_CRLPATH MASTER_SSL_KEY MASTER_SSL_CIPHER MASTER_SSL_VERIFY_SERVER_CERT
CONNECTION_RETRY_INTERVAL	在该通道下，从库与主库的连接丢失后，从库每隔多少秒重试一次连接主库	MASTER_CONNECT_RETRY
CONNECTION_RETRY_COUNT	在该通道下，允许从库重试连接主库的次数	MASTER_RETRY_COUNT
HEARTBEAT_INTERVAL	在该通道下，从库连接主库的 I/O 线程的保活心跳包的间隔时间，以秒为单位。注意，心跳配置信息由从库推送给主库，但心跳包由主库发起，用于探测从库的存活状况	MASTER_HEARTBEAT_PERIOD

对 replication_connection_configuration 表不允许执行 TRUNCATE TABLE 语句。

9.2.6　replication_connection_status 表

replication_connection_status 表中记录的是从库 I/O 线程的连接状态信息，也记录组复制中其他节点的连接信息。在组复制中，一个节点加入集群之前的数据需要使用异步复制通道进行同步，组复制的异步复制通道信息在 SHOW SLAVE STATUS 语句的输出中不可见。

如果是多源复制，每个复制通道的 I/O 线程都会在表中记录一行状态信息。

查看 replication_connection_status 表中记录的内容：

```
# 多线程和单线程的主从复制，表中的记录相同。如果是多源复制，则每个复制通道在表中各记录一行信息
mysql> select * from replication_connection_status\G
*************************** 1. row ***************************
            CHANNEL_NAME: 
              GROUP_NAME: 
             SOURCE_UUID: ec123678-5e26-11e7-9d38-000c295e08a0
               THREAD_ID: 101
           SERVICE_STATE: ON
COUNT_RECEIVED_HEARTBEATS: 136
```

```
      LAST_HEARTBEAT_TIMESTAMP: 2018-06-12 00:55:22
      RECEIVED_TRANSACTION_SET:
           LAST_ERROR_NUMBER: 0
          LAST_ERROR_MESSAGE:
        LAST_ERROR_TIMESTAMP: 0000-00-00 00:00:00
1 row in set (0.00 sec)

# 如果是 MGR 集群，则表中会记录类似如下的信息
mysql> select * from replication_connection_status\G
*************************** 1. row ***************************
             CHANNEL_NAME: group_replication_applier
               GROUP_NAME: aaaaaaaa-aaaa-aaaa-aaaa-aaaaaaaaaaaa
              SOURCE_UUID: aaaaaaaa-aaaa-aaaa-aaaa-aaaaaaaaaaaa
                THREAD_ID: NULL
            SERVICE_STATE: ON
COUNT_RECEIVED_HEARTBEATS: 0
  LAST_HEARTBEAT_TIMESTAMP: 0000-00-00 00:00:00
  RECEIVED_TRANSACTION_SET: aaaaaaaa-aaaa-aaaa-aaaa-aaaaaaaaaaaa:104099082
        LAST_ERROR_NUMBER: 0
       LAST_ERROR_MESSAGE:
     LAST_ERROR_TIMESTAMP: 0000-00-00 00:00:00
*************************** 2. row ***************************
             CHANNEL_NAME: group_replication_recovery
......
2 rows in set (0.00 sec)
```

replication_connection_status 表中各字段的含义及其与 SHOW SLAVE STATUS 语句输出的字段的对应关系如表 9-6 所示。

表 9-6

replication_connection_status 表中的字段名	含义	对应 SHOW SLAVE STATUS 语句输出的字段名
CHANNEL_NAME	显示复制通道名称	Channel_Name
GROUP_NAME	如果该行记录的是组复制中的组成员，则此字段显示该成员所属的组名称	
SOURCE_UUID	在该通道下，从库连接的主库的 server_uuid 值	Master_UUID
THREAD_ID	在该通道下，从库连接的主库的 I/O 线程 ID	无
SERVICE_STATE	在该通道下，从库连接的主库的 I/O 线程状态。其有效值有：ON（I/O 线程已创建且处于活跃状态或空闲状态）、OFF（I/O 线程不再存在，可能没有启动）和 CONNECTING（I/O 线程存在并正在尝试连接主库，还未成功建立连接）	Slave_IO_Running

续表

replication_connection_ status 表中的字段名	含 义	对应 SHOW SLAVE STATUS 语句输出的字段名
RECEIVED_ TRANSACTION_SET	在该通道下，从库接收的主库事务对应的 GTID SET。如果未启用 GTID，则该字段为空	Retrieved_Gtid_Set
LAST_ERROR_NUMBER LAST_ERROR_MESSAGE	在该通道下，从库 I/O 线程发生错误而停止时的最新错误编号和错误消息。如果错误编号为 0，错误消息字段为空字符串，则表示"无错误"；如果 LAST_ERROR_MESSAGE 字段值不为空，则错误值也会输出到从库的错误日志中。注意，在执行 RESET MASTER 或 RESET SLAVE 语句时，这两个字段的值会被重置	Last_IO_Errno Last_IO_Error
LAST_ERROR_ TIMESTAMP	在该通道下，从库 I/O 线程发生错误的时间，时间格式为：YYMMDD HH: MM: SS	Last_IO_Error_Timestamp
LAST_HEARTBEAT_ TIMESTAMP	在该通道下，从库 I/O 线程最近一次收到主库心跳信号的时间，时间格式为：YYMMDD HH: MM: SS	无
COUNT_RECEIVED_ HEARTBEATS	在该通道下，自上一次重新启动数据库进程，或最近一次执行 RESET 语句，或最近一次执行 CHANGE MASTER TO 语句以来，从库收到主库心跳信号的总次数	无

对 replication_connection_status 表不允许执行 TRUNCATE TABLE 语句。

9.2.7 replication_group_member_stats 表

replication_group_member_stats 表记录组复制成员的事务状态统计信息。仅在组复制组件运行时，此表中才会有记录。

查看 replication_group_member_stats 表中记录的内容：

```
mysql> select * from replication_group_member_stats\G
*************************** 1. row ***************************
                          CHANNEL_NAME: group_replication_applier
                               VIEW_ID: 15287289928409067:1
                             MEMBER_ID: 5d78a458-30d2-11e8-a66f-5254002a54f2
           COUNT_TRANSACTIONS_IN_QUEUE: 0
            COUNT_TRANSACTIONS_CHECKED: 0
              COUNT_CONFLICTS_DETECTED: 0
    COUNT_TRANSACTIONS_ROWS_VALIDATING: 0
    TRANSACTIONS_COMMITTED_ALL_MEMBERS: 0a1e8349-2e87-11e8-8c9f-525400bdd1f2:1-148826,
2d623f55-2111-11e8-9cc3-0025905b06da:1-2,
aaaaaaaa-aaaa-aaaa-aaaa-aaaaaaaaaaaa:1-104099082
```

```
    LAST_CONFLICT_FREE_TRANSACTION:
1 row in set (0.00 sec)
```

表中各字段的含义如下:

- CHANNEL_NAME:组成员所在的组使用的复制通道名称,通道名称为 group_replication_applier。

- VIEW_ID:组成员所在组的当前视图标识符。

- MEMBER_ID:显示当前组成员 Server 的 UUID,组中每个节点具有不同的值,而且是唯一的(因为使用的是组成员实例的 UUID,该 UUID 是随机生成的,保证全局唯一)。

- COUNT_TRANSACTIONS_IN_QUEUE:表示当前队列中等待冲突检查的事务数(等待全局事务认证的事务数),一旦通过冲突检测,它们将排队等待应用。

- COUNT_TRANSACTIONS_CHECKED:表示已通过冲突检查机制检查的事务数(已通过全局事务认证的事务数,从节点加入组复制时开始计算)。

- COUNT_CONFLICTS_DETECTED:表示未通过冲突检测机制检查的事务数(在全局事务认证时未通过的事务数)。

- COUNT_TRANSACTIONS_ROWS_VALIDATING:表示冲突检测数据库当前的大小(冲突检测数据库存放了每个通过了认证但还未完成提交的事务相关的检测数据),冲突检测数据库可用于认证新事务(通过检测新事务是否和之前的事务相冲突)。

- TRANSACTIONS_COMMITTED_ALL_MEMBERS:显示已在当前视图中所有成员上成功提交的事务(类似于所有成员实例的 gtid_executed 集合的交集),该值在固定时间间隔中更新(所以并不实时)。

- LAST_CONFLICT_FREE_TRANSACTION:显示最后一次无冲突校验检查的事务标识符(最后一个没有冲突的事务的 GTID)。

对 replication_group_member_stats 表不允许执行 TRUNCATE TABLE 语句。

9.2.8　replication_group_members 表

replication_group_members 表记录组复制成员的网络和状态信息,仅在组复制组件运行时表中才会有记录。

查看 replication_group_members 表中记录的内容:

```
mysql> select * from replication_group_members;
+------------+------------+-------------+-------------+--------------+
| CHANNEL_NAME|MEMBER_ID | MEMBER_HOST | MEMBER_PORT | MEMBER_STATE |
+------------+------------+-------------+-------------+--------------+
```

```
| group_replication_applier | 5d78a458-30d2-11e8-a66f-5254002a54f2 | node1 | 3306 | ONLINE |
+---------------------------+--------------------------------------+-------+------+---------+
1 row in set (0.00 sec)
```

表中各字段的含义如下：

- CHANNEL_NAME：组复制中使用的通道名称，通道名称为 group_replication_applier。
- MEMBER_ID：显示当前组成员 Server 的 UUID，组中每个节点具有不同的值。
- MEMBER_HOST：组复制中组成员的网络地址（主机名或 IP 地址，与成员实例的系统变量 hostname 或 report_host 的值相同）。
- MEMBER_PORT：组复制中组成员的侦听端口，与组成员实例的系统变量 port 或 report_port 的值相同。
- MEMBER_STATE：组复制中组成员的状态。

  其有效的状态值如下：

  - OFFLINE：组复制成员已经安装组复制插件，但未启动。
  - RECOVERING：组复制成员已经加入复制组，正在从组中接收数据，即正在加入集群。
  - ONLINE：组复制成员处于正常运行状态。

提示：在组复制中，如果组成员的组复制状态发生错误，无法正常从组中接收数据，可能会变成 ERROR 状态。如果网络发生故障或者其他成员宕机，那么剩余存活的孤立节点的状态可能会变为 UNREACHABLE。

对 replication_group_members 表不允许执行 TRUNCATE TABLE 语句。

第 10 章　通过其他方式检查复制信息

第 8 章详细介绍了如何通过 mysql.slave_master_info 和 mysql.slave_relay_log_info 表来检查复制信息，第 9 章详细介绍了如何通过 PERFORMANCE_SCHEMA 库下的复制记录表来检查复制信息，本章将对前面章节中未提及的与复制相关的小细节进行补充说明。例如，通过 SHOW PROCESSLIST 语句来查看 I/O 线程和 SQL 线程的状态信息、通过 PERFORMANCE_SCHEMA 库中的 user_variables_by_thread 表来查看 I/O 线程向主库注册的自定义变量信息等。

注意：mysql.slave_worker_info 表中记录了多线程复制的一些状态信息，但是该表中的内容主要在 MySQL 内部使用，可读性较差，本章不做介绍，有兴趣的读者可自行研究。

10.1　复制状态变量

在第 9 章中，我们提到一些状态变量的值被转移到 PERFORMANCE_SCHEMA 库下的复制信息表中保存，从 MySQL 5.7.5 开始，不推荐使用 SHOW STATUS 语句来查看这些状态变量的值。但是，如果你的数据库刚刚升级到 MySQL 5.7，而且某些依赖于这些状态变量值的应用程序还来不及改造，你可能仍然希望继续通过这些状态变量来获取一些复制状态信息。那么，可以通过启用系统变量 show_compatibility_56（设置为 ON）来回到 MySQL 5.6 及其之前版本的查询方式。这样，你就可以继续使用 SHOW STATUS 语句来查看下面这些复制状态变量的值了：

```
Slave_retried_transactions
Slave_last_heartbeat
Slave_received_heartbeats
Slave_heartbeat_period
Slave_running
```

10.2　复制心跳信息

1. 心跳间隔配置信息

（1）从库端的记录位置

performance_schema.replication_connection_configuration 表提供了复制心跳间隔的配置

信息（详见 9.2.5 节），该心跳间隔的值是从库在使用 CHANGE MASTER TO 语句配置复制时指定并用于向主库注册的，默认情况下为从库系统变量 slave_net_timeout 值的一半。

mysql.slave_master_info 表的 Heartbeat 字段提供了复制心跳间隔的配置信息。

（2）主库端的记录位置

performance_schema.user_variables_by_thread 表记录了从库 I/O 线程向主库注册的所有用户会话变量，其中也包括心跳间隔时间。

2．心跳包的状态信息

心跳包的状态信息仅在从库中才会有专用的表进行记录。

performance_schema.replication_connection_status 表提供了复制心跳包的状态信息（详见 9.2.6 节），可以检查复制连接是否处于活跃状态，以及最近收到的心跳包时间和心跳信号数量。如果主库最近没有产生任何新的二进制日志并发送给从库，主库会定期向从库发送心跳信号。

下面分别对前面提到的 4 张表中的信息做简单的说明。

```
# 从库端 PERFORMANCE_SCHEMA 库中的心跳配置信息
mysql> select * from performance_schema.replication_connection_configuration\G
*************************** 1. row ***************************
        CHANNEL_NAME:
                HOST: 10.10.30.161
                PORT: 3306
                USER: qbench
   NETWORK_INTERFACE:
       AUTO_POSITION: 1
......
  CONNECTION_RETRY_INTERVAL: 60
     CONNECTION_RETRY_COUNT: 86400
         HEARTBEAT_INTERVAL: 5.000   # 在这里可以看到，复制心跳间隔被配置为 5s
                TLS_VERSION:

# 从库 mysql schema 中的心跳间隔配置信息
mysql> select * from mysql.slave_master_info\G
*************************** 1. row ***************************
        Number_of_lines: 25
         Master_log_name: mysql-bin.000008
          Master_log_pos: 194
......
          Connect_retry: 60
......
              Heartbeat: 5   # 在这里可以看到，复制心跳间隔被配置为 5 秒
......
            Retry_count: 86400
......
    Enabled_auto_position: 1
```

```
        Channel_name:
        Tls_version:
1 row in set (0.00 sec)

# 主库端复制心跳间隔的配置信息
mysql> select * from performance_schema.user_variables_by_thread;
+-----------+------------------------+--------------------------------------+
| THREAD_ID | VARIABLE_NAME          | VARIABLE_VALUE                       |
+-----------+------------------------+--------------------------------------+
| 44        | slave_uuid             | 2d623f55-2111-11e8-9cc3-0025905b06da |
| 44        | master_heartbeat_period | 5000000000 |    # 在这里可以看到，从库向主库注册的复制心跳间
隔被配置为 5s（这里的数值单位为 ns）
| 44        | master_binlog_checksum | CRC32                                |
+-----------+------------------------+--------------------------------------+
3 rows in set (0.00 sec)

# 从库端 PERFORMANCE_SCHEMA 库中 replication_connection_status 记录的心跳包状态信息
mysql> select * from performance_schema.replication_connection_status\G
*************************** 1. row ***************************
          CHANNEL_NAME:
            GROUP_NAME:
           SOURCE_UUID: db40f850-1226-11e9-afa1-6c92bf672fea
             THREAD_ID: 83
         SERVICE_STATE: ON
COUNT_RECEIVED_HEARTBEATS: 0   # 自上一次重新启动数据库进程，或最近一次执行 RESET 语句，或最
近一次执行 CHANGE MASTER TO 语句以来，从库收到的主库心跳信号的总次数
LAST_HEARTBEAT_TIMESTAMP: 0000-00-00 00:00:00   # 从库 I/O 线程最近一次收到主库心跳信号的
时间
RECEIVED_TRANSACTION_SET: db40f850-1226-11e9-afa1-6c92bf672fea:1-54887820
......
1 row in set (0.02 sec)
```

10.3　SHOW SLAVE STATUS 语句输出信息详解

关于 SHOW SLAVE STATUS 语句的输出信息，第 8 章和第 9 章都或多或少提到过一些，但在第 8 章中主要是与 mysql.slave_master_info 和 mysql.slave_relay_log_info 表做比较，在第 9 章中主要是和 PERFORMANCE_SCHEMA 库中的表做比较，并不完整。本节将完整列出该语句的输出信息。

我们先列出该语句的一个输出示例：

```
mysql> show slave status\G
        Slave_IO_State: Waiting for master to send event
         Master_Host: 10.10.30.161
         Master_User: qbench
         Master_Port: 3306
        Connect_Retry: 60
      Master_Log_File: mysql-bin.000008
```

```
                    Read_Master_Log_Pos: 194
                       Relay_Log_File: mysql-relay-bin.000006
                        Relay_Log_Pos: 367
                Relay_Master_Log_File: mysql-bin.000008
                     Slave_IO_Running: Yes
                    Slave_SQL_Running: Yes
                      Replicate_Do_DB:
                  Replicate_Ignore_DB:
                   Replicate_Do_Table:
               Replicate_Ignore_Table:
              Replicate_Wild_Do_Table:
          Replicate_Wild_Ignore_Table:
                           Last_Errno: 0
                           Last_Error:
                         Skip_Counter: 0
                  Exec_Master_Log_Pos: 194
                      Relay_Log_Space: 574
                      Until_Condition: None
                       Until_Log_File:
                        Until_Log_Pos: 0
                   Master_SSL_Allowed: No
                   Master_SSL_CA_File:
                   Master_SSL_CA_Path:
                      Master_SSL_Cert:
                    Master_SSL_Cipher:
                       Master_SSL_Key:
                Seconds_Behind_Master: 0
        Master_SSL_Verify_Server_Cert: No
                        Last_IO_Errno: 0
                        Last_IO_Error:
                       Last_SQL_Errno: 0
                       Last_SQL_Error:
          Replicate_Ignore_Server_Ids:
                     Master_Server_Id: 3306102
                          Master_UUID: f3372787-0719-11e8-af1f-0025905b06da
                     Master_Info_File: mysql.slave_master_info
                            SQL_Delay: 0
                  SQL_Remaining_Delay: NULL
              Slave_SQL_Running_State: Slave has read all relay log; waiting for more updates
                   Master_Retry_Count: 86400
                          Master_Bind:
              Last_IO_Error_Timestamp:
             Last_SQL_Error_Timestamp:
                       Master_SSL_Crl:
                   Master_SSL_Crlpath:
                   Retrieved_Gtid_Set:
                    Executed_Gtid_Set: 06188301-b333-11e8-bdfe-0025905b06da:1-270852,
2d623f55-2111-11e8-9cc3-0025905b06da:1,
```

```
f3372787-0719-11e8-af1f-0025905b06da:1-34
              Auto_Position: 1
      Replicate_Rewrite_DB:
              Channel_Name:
        Master_TLS_Version:
```

对 SHOW SLAVE STATUS 语句输出信息的完整解释如下:

- Slave_IO_State:SHOW PROCESSLIST 语句输出的从库 I/O 线程状态字段的副本,显示 I/O 线程正在做什么。

- Master_Host:连接的主库 IP,用 CHANGE MASTER TO 语句的 MASTER_HOST 选项来指定。

- Master_User:连接的主库用户名,用 CHANGE MASTER TO 语句的 MASTER_USER 选项来指定。

- Master_Port:连接的主库实例端口,用 CHANGE MASTER TO 语句的 MASTER_PORT 选项来指定。

- Connect_Retry:连接主库的重试间隔(默认值为 60,单位为 s),可以用 CHANGE MASTER TO MASTER_CONNECT_RETRY 语句来设置。当从库没有正常收到主库发送的心跳包时(系统变量 slave_net_timeout = 60 控制超时,在这个超时时间内,如果与主库连接时没有响应,从库就认为已经丢失与主库的连接),从库开始尝试重连主库,如果无法连接,则使用该值作为下一次重试的间隔时间。

- Master_Log_File:复制 I/O 线程当前读取的主库二进制日志的文件名,可以用 CHANGE MASTER TO 语句的 MASTER_LOG_FILE 选项来指定。

- Read_Master_Log_Pos:复制 I/O 线程当前读取的主库二进制日志文件的位置,可以用 CHANGE MASTER TO 语句的 MASTER_LOG_POS 选项来指定。

- Relay_Log_File:从库 SQL 线程当前正在读取的中继日志文件名,SQL 线程正在从这个文件中读取二进制日志事件并重放。可以用 CHANGE MASTER TO 语句的 RELAY_LOG_FILE 选项指定此文件名。

- Relay_Log_Pos:从库 SQL 线程当前正在读取与回放的中继日志文件的位置。可以用 CHANGE MASTER TO 语句的 RELAY_LOG_POS 选项来指定。

- Relay_Master_Log_File:从库 SQL 线程当前正在重放的、对应主库的二进制日志文件名。

- Slave_IO_Running:从库 I/O 线程的运行状态。

  在 MySQL 5.1.46 及其之后的版本中有 3 个值,对应的状态如下:

  - No:从库 I/O 线程没有运行,对应着内部的 MYSQL_SLAVE_NOT_RUN 状态。

  - Connecting:从库 I/O 线程正在运行,但还未成功连接主库,对应着内部的

MYSQL_SLAVE_RUN_NOT_CONNECT 状态，对于这种状态，MySQL 4.1.14 ~ 5.1.46 版本中 Slave_IO_Running 显示为 No，在 MySQL 4.1.14 之前的版本中 Slave_IO_Running 显示为 Yes。

- Yes：从库 I/O 线程正在运行且已成功连接主库，对应的内部状态为 MYSQL_SLAVE_RUN_CONNECT。

● Slave_SQL_Running：从库 SQL 线程的运行状态，有 Yes 和 No 两个值。

● Replicate_Do_DB、Replicate_Ignore_DB：用 mysqld 程序的选项--replicate-do-db 和 --replicate-ignore-db 来指定。如果已指定，则显示指定的内容，否则为空。

● Replicate_Do_Table、Replicate_Ignore_Table、Replicate_Wild_Do_Table、Replicate_Wild_Ignore_Table：用 mysqld 程序的选项--replicate-do-table、--replicate-ignore-table、--replicate-wild-do-table 和--replicate-wild-ignore-table 来指定。如果已指定，则显示指定的内容，否则为空。

● Last_Errno、Last_Error：这两个列是 Last_SQL_Errno 和 Last_SQL_Error 列的别名。执行 RESET MASTER 或 RESET SLAVE 语句会重置这两列的值。

● Skip_Counter：系统变量 sql_slave_skip_counter 当前的设置值。

● Exec_Master_Log_Pos：从库 SQL 线程当前正在重放的、对应主库的二进制日志文件位置。

● Relay_Log_Space：当前存在的所有中继日志的总大小。

● Until_Condition、Until_Log_File、Until_Log_Pos：START SLAVE 语句的 UNTIL 子句中指定的值。

其中，Until_Condition 有如下可能的值：

- None：如果没有指定 UNTIL 语句，就为 None 值。
- Master：如果使用 UNTIL 语句指定了主库的二进制日志文件位置，且从库正在读取主库的二进制日志，就直至到达给定的主库二进制日志文件位置为止。
- Relay：如果使用 UNTIL 语句指定了中继日志的文件和位置，且从库正在读取中继日志文件的位置，就直至到达给定的中继日志文件位置为止。
- SQL_BEFORE_GTIDS：如果从库 SQL 线程正在处理事务，就直至它到达其 GTID 在 GTID SET 中列出的第一个事务为止。
- SQL_AFTER_GTIDS：如果从库 SQL 线程正在处理所有事务，就直到 GTID SET 中的最后一个事务被线程处理为止。
- SQL_AFTER_MTS_GAPS：如果从库使用多线程复制，且 SQL 线程都正在运行，则直到在中继日志中找不到更多的日志组间隙为止。

- Master_SSL_Allowed、Master_SSL_CA_File、Master_SSL_CA_Path、Master_SSL_Cert、Master_SSL_Cipher、Master_SSL_CRL_File、Master_SSL_CRL_Path、Master_SSL_Key、Master_SSL_Verify_Server_Cert：这些字段显示了从库如何使用 SSL 连接主库（对应着 CHANGE MASTER TO 语句的选项：MASTER_SSL_CA、MASTER_SSL_CAPATH、MASTER_SSL_CERT、MASTER_SSL_CIPHER、MASTER_SSL_CRL、MASTER_SSL_CRLPATH、MASTER_SSL_KEY、MASTER_SSL_VERIFY_SERVER_CERT）。

  其中 Master_SSL_Allowed 有如下可能的值：

  - Yes：主库允许使用 SSL 进行连接。
  - No：主库不允许使用 SSL 进行连接。
  - Ignored：主库允许使用 SSL 进行连接，但是从库没有开启对 SSL 的支持。

- Seconds_Behind_Master：显示从库的复制延迟时间。关于该值的计算方法，详见第 11 章 "MySQL 复制延迟 Seconds_Behind_Master 究竟是如何计算的"。

- Last_IO_Errno、Last_IO_Error：导致 I/O 线程停止的最新错误的编号和错误信息。Last_IO_Errno 为 0，Last_IO_Error 为空字符串时，表示 "无错误" 发生。如果 Last_IO_Error 值不为空，则错误信息也会被同时写入从库的错误日志。

  - I/O 错误信息中包含一个时间戳，表示最近的 I/O 线程错误发生的时间。此时间戳的格式 YYMMDD HH:MM:SS，也显示在 Last_SQL_Error_Timestamp 列中。
  - 执行 RESET MASTER 或 RESET SLAVE 语句会复位这些列中显示的值。

- Last_SQL_Errno、Last_SQL_Error：导致 SQL 线程中止的最新错误的编号和错误信息。Last_SQL_Errno 为 0，Last_SQL_Error 为空字符串时，表示 "无错误" 发生，如果 Last_SQL_Error 值不为空，则错误信息也会被写入从库的错误日志中。

  - 如果从库是多线程的，则 SQL 线程是 Worker 线程的协调器。在这种情况下，从 MySQL 5.7.2 开始，Last_SQL_Error 字段显示的值对应着 MySQL 5.7 中新增的 performance_schema.replication_applier_status_by_coordinator 表中 Last_Error_Message 列的值。在这个表中可以看到每个 Coordinator 线程的状态，以及更详细的错误信息。如果该表不可用，则可以使用从库的错误日志来查看错误信息。performance_schema.replication_applier_status_by_worker 表详细记录了 SHOW SLAVE STATUS 输出的 SQL 线程具体的故障信息。在多线程复制时，Last_SQL_Error 字段可能输出的是某一类错误信息而不是 Worker 线程的具体错误信息，例如，某表、某记录不存在之类。在这个表中可以查看在 Worker 线程上发生的具体错误信息。
  - SQL 线程错误信息包含一个时间戳，表示最近发生的 SQL 线程错误的时间。此时间戳的格式为 YYMMDD HH: MM: SS，也会显示在 Last_SQL_Error_

Timestamp 列中。

- 执行 RESET MASTER 或 RESET SLAVE 语句会复位这些列中显示的值。

● Replicate_Ignore_Server_Ids：在 MySQL 5.7 中，可以使用 CHANGE MASTER TO 语句的 IGNORE_SERVER_IDS 选项，设置从库忽略哪些 Server ID 的主库二进制日志。

- 默认情况下，该选项为空值。通常，仅在使用环形拓扑或多源复制拓扑时设置该选项。当其不为空时，如果需要忽略多个主库，则用逗号隔开每个主库的 Server ID。例如：

```
Replicate_Ignore_Server_Ids: 2, 6, 9
```

注意：使用 Ignored_server_ids 忽略多个主库时，第一个数字代表需要忽略的主库数量，在实际过滤时第一个数字会被忽略。从第二个数字开始，其才被识别为主库的 Server ID。

例如，CHANGE MASTER TO 语句的选项 IGNORE_SERVER_IDS 设置了忽略 Server ID 值为 2、6、9 的主库，而实际上 SHOW SLAVE STATUS 查到该字段的值是：

```
Ignored_server_ids: 3, 2, 6, 9
```

第一个数字（在此例中为 3）被忽略，它在这里代表的是 Server ID 的数量。

- Replicate_Ignore_Server_Ids 设置的过滤主库二进制日志的行为由 I/O 线程执行，而不是由 SQL 线程执行的，这意味着被过滤的事件不会写入中继日志。这与 Server 选项（如--replicate-do-table）所采用的过滤操作不同，那只适用于 SQL 线程。

● Master_Server_Id：主库的 Server ID 值。

● Master_UUID：主库的 UUID 值。

● Master_Info_File：本地用于保存 I/O 线程信息的 master.info 文件位置。如果 master_info_repository = FILE，则这是一个文件系统路径；如果 master_info_repository = TABLE，则这是 mysql.slave_master_info，表示记录的位置是在 mysql 库下的 slave_master_info 表中。

● SQL_Delay：显示从库 SQL 线程必须滞后于主库的二进制日志时间戳多长时间，用 CHANGE MASTER TO 语句的 MASTER_DELAY 选项来指定。

● SQL_Remaining_Delay：显示的是从库正在执行相对于主库必须延迟的那些事件时，剩余的延迟秒数。如果从库没有配置复制延迟，该字段值为 NULL。

● Slave_SQL_Running_State：与 Slave_IO_State 字段类似，显示的是 SHOW PROCESSLIST 语句中 SQL 线程状态的副本信息。

● Master_Retry_Count：设置在从库 I/O 线程丢失与主库的连接之后，能够重试连接的次数，默认为 24 × 3600 = 86 400 次，可以使用 CHANGE MASTER TO 语句的

MASTER_RETRY_COUNT 选项来指定。

- Master_Bind：显示从库复制线程绑定在哪个网络接口上（用于在从库有多个网卡的场景下指定使用哪个网卡）。如果需要绑定，可以使用 CHANGE MASTER TO 语句的 MASTER_BIND 选项指定网卡名称。

- Last_IO_Error_Timestamp：显示最近一次 I/O 线程发生错误的时间，格式为 YYMMDD HH: MM: SS。

- Last_SQL_Error_Timestamp：显示最近一次 SQL 线程发生错误的时间，格式为 YYMMDD HH: MM: SS。

- Retrieved_Gtid_Set：从库收到的所有事务对应的 GTID SET，如果未启用 GTID 复制，则该字段为空。

    - 该字段表示 GTID SET 中最大的 GTID 值与中继日志中最大的 GTID 值对应，一旦收到主库的 Gtid_log_event，就会增加 GTID 的值。所以，当发生意外时，有可能事务的二进制日志事件仅传输了一部分，这个事务的 GTID 也被记录在这个 GTID SET 中。

    - 执行 RESET SLAVE 或 CHANGE MASTER TO 会导致所有中继日志被清除（包括 SQL 线程重放完的和没有重放的）。选项--relay-log-recovery 被设置为 1 时（或在 my.cnf 文件中设置系统变量 relay_log_recovery = 1），重启从库会导致没有重放完的中继日志被清除（包括 SQL 线程重放完的和没有重放的），以 SQL 线程重放的中继日志位置为准向主库重新发送请求，当 relay_log_purge = 1 时，中继日志文件中始终保留最新的中继日志，SQL 线程重放完的中继日志才会被清理。

    提示：在 MySQL 5.7.1 之前，此字段的值以大写字母输出。在 MySQL 5.7.1 及更高版本中，它始终以小写字母输出。

- Executed_Gtid_Set：从库在自身二进制日志中写入的 GTID SET。它与此 Server 上的全局系统变量 gtid_executed、SHOW MASTER STATUS 语句的 Executed_Gtid_Set 字段值相同。如果未启用 GTID，则该字段值为空。

    提示：在 MySQL 5.7.1 之前，此字段值以大写字母输出。在 MySQL 5.7.1 及更高版本中，它始终以小写字母输出。

- Auto_Position：如果启用了自动定位功能，则该值为 1，否则为 0，可以使用 CHANGE MASTER TO 语句的 MASTER_AUTO_POSITION 选项来指定。

- Replicate_Rewrite_DB：主库上的数据库名在从库重放时被重新指定到另一个数据库名下。

    例如，使用 CHANGE REPLICATION FILTER REPLICATE_REWRITE_DB =((db1, db2),(db3,db4))语句指定主库的 db1 中的数据写到从库的 db2 里，主库的 db3 中的数据写到从库的 db4 里。SHOW SLAVE STATUS 语句的 Replicate_Rewrite_DB 字段

值显示为：

```
Replicate_Rewrite_DB: (db1, db2), (db3, db4)
```

从 MySQL 5.7.3 开始，可以使用 CHANGE REPLICATION FILTER 语句在从库上指定复制过滤规则，这些过滤子句与原来的复制过滤参数的作用相同，并且可以动态设置而不需要重启 Server。

- Channel_name：总是显示默认的复制通道名称，如果没有使用多源复制，则该字段为空。
- Master_TLS_Version：主库上使用的 TLS 版本号，这是 MySQL 5.7.10 版本中新增的选项。

10.4 通过 SHOW PROCESSLIST 语句查看复制线程状态

在主库上可以使用 SHOW PROCESSLIST 语句查看已连接的从库的状态。从库连接主库的线程，在 Command 字段中显示的值为"Binlog Dump"，如下所示：

```
mysql> show processlist\G
......
*************************** 4. row ***************************
     Id: 2
   User: qbench
   Host: 10.10.30.162:53941
     db: NULL
Command: Binlog Dump GTID
   Time: 1388472
  State: Master has sent all binlog to slave; waiting for more updates
   Info: NULL
......
```

在从库上可以使用 SHOW PROCESSLIST 语句查看 I/O 线程和 SQL 线程的状态，如下所示：

```
+----+-------------+------+------+---------+---------+-----------+------+
| Id | User        | Host | db   | Command | Time    | State     | Info |
+----+-------------+------+------+---------+---------+-----------+------+
| 1  | system user |      | NULL | Connect | 1388608 | Waiting for master to send event | NULL |  # 这里就是 I/O 线程
| 2  | system user |      | NULL | Connect | 1388548 | Slave has read all relay log; waiting for more updates | NULL |  # 这里就是 SQL 线程
| 3  | system user |      | NULL | Connect | 1388608 | Waiting for an event from Coordinator | NULL |  # 如果启用了多线程复制，则还会有 Worker 线程
......
| 19 | system user |      | NULL | Connect | 1388608 | Waiting for an event from Coordinator | NULL |
......
+----+-------------+------+------+---------+---------+-----------+------+
19 rows in set (0.00 sec)
```

提示：关于线程状态的详细介绍，可参考《千金良方：MySQL 性能调优金字塔法则》的附录 A。

10.5　SHOW MASTER STATUS 语句输出详解

SHOW MASTER STATUS 语句输出的内容如下：

```
mysql> show master status\G
*************************** 1. row ***************************
             File: mysql-bin.000014
         Position: 161081896
     Binlog_Do_DB:
 Binlog_Ignore_DB:
Executed_Gtid_Set: 2016f827-2d98-11e7-bb1e-00163e407cfb:1-2629355,
402872e0-33bd-11e7-8e8d-00163e4fde29:1-592
1 row in set (0.00 sec)
```

对输出内容的详细说明如下（执行 SHOW MASTER STATUS 语句需要有 SUPER 或者 REPLICATION CLIENT 权限）：

- File：主库当前正在使用的二进制日志文件名。
- Position：主库当前正在使用的二进制日志文件的位置。
- Binlog_Do_DB：主库当前生效的复制过滤参数，只有这个选项指定的库列表中的库发生的数据写入，才能被记录到二进制日志文件中。
- Binlog_Ignore_DB：主库当前生效的复制过滤参数，这个选项指定的库列表中的库发生数据写入，不会被记录到二进制日志文件中。
- Executed_Gtid_Set：当启用 GTID 时，Executed_Gtid_Set 显示主库上数据当前的 GTID SET，这与此 Server 上的系统变量 gtid_executed 的值以及此 Server 上 SHOW SLAVE STATUS 语句输出的 Executed_Gtid_Set 字段值相同。

10.6　SHOW SLAVE HOSTS 语句

在主库上执行 SHOW SLAVE HOSTS 语句，可以显示当前主库所连接的从库列表，一个从库显示一行信息，如下所示：

```
mysql> show slave hosts;
+-----------+------+------+-----------+--------------------------------------+
| Server_id | Host | Port | Master_id | Slave_UUID                           |
+-----------+------+------+-----------+--------------------------------------+
| 3306241   |      | 3306 | 3306217   | 799ef59c-4126-11e7-83ce-00163e407cfb |
| 3306250   |      | 3306 | 3306217   | f9b1a9b6-46b7-11e7-9e8b-00163e4fde29 |
+-----------+------+------+-----------+--------------------------------------+
2 rows in set (0.00 sec)
```

```
# 在较旧的版本中，输出的字段可能有所不同，如下
mysql> show slave hosts;
+-----------+--------+------+------------------+-----------+
| Server_id | Host   | Port | Rpl_recovery_rank | Master_id |
+-----------+--------+------+------------------+-----------+
| 10        | slave1 | 3306 | 0                | 1         |
+-----------+--------+------+------------------+-----------+
1 row in set (0.00 sec)
```

以上输出结果中，Host、Port 等字段，需要依赖一些以 report 开头的选项（如 report-host、report-port）来设置从库自己的 IP 地址和端口信息，以便当从库连接主库时，在主库中可以使用该语句来查询从库的 IP 地址和端口信息。对于 Server ID、UUID，从库 I/O 线程向主库注册时自己会带上这些信息。输出字段的含义如下：

- Server_id：从库的 Server ID，在复制架构中应该是全局唯一的，可以使用系统变量 server_id 进行设置。如果人为设置两个以上的从库为相同的 Server ID，或者不设置，当两个 Server ID 都使用默认值时，会造成一个复制架构中出现相同的 Server ID。

- Host：从库的主机名，可以使用 IP 地址。从库需要使用选项--report-host = host 来显式指定该值（或在 my.cnf 文件中用系统变量 report_host 来指定，该变量为只读的），否则主库上执行 SHOW SLAVE HOSTS 语句时，这个字段的值显示为空字符串。

  注意：从库可以使用选项--report-host 随意指定一个主机名或者 IP 地址，该 Host 信息仅仅用于从库向主库注册，注册信息可以用于在查看主库上的从库列表时生成 Host 字段值。

- User：从库连接主库使用的用户名。从库需要使用选项--report-user=user_name 来显式指定该值（或在 my.cnf 文件中使用系统变量 report_user 来指定，该变量为只读的），并且在主库使用选项--show-slave-auth-info 启动时，SHOW SLAVE STATUS 语句的输出信息中才会包含此字段。

  注意：从库可以使用选项--report-user 随意指定一个账号名称，因此该账号名称仅用于向主库注册，并不接入 MySQL 的权限系统进行验证，与复制 I/O 线程连接主库时使用的账号名称并没有关联，但其注册信息可以用于在查看主库上的从库列表时生成 User 字段值。

- Password：从库连接主库时使用的用户密码，从库需要使用选项--report-password 来显式指定该值（或在 my.cnf 文件中使用系统变量 report_password 来指定，该变量为只读的），并且在主库使用选项--show-slave-auth-info 启动时，SHOW SLAVE STATUS 语句的输出信息中才会包含此列。

  注意：从库可以使用选项--report-password 随意指定一个用户密码，与 User 值类似，该用户名称仅用于向主库注册，注册信息可以用于在查看主库上的从库列表时生成 Password 字段值。

- Port：从库可以使用选项--report-port 随意指定一个端口号，默认情况下使用从库系统变量 port 的值。如果从库未显式指定选项--report-port 的值，则在主库上 SHOW SLAVE STATUS 语句显示的 Port 字段为系统变量 port 的值。该信息仅用于从库向主库注册，注册信息可以用于在查看主库上的从库列表时生成 Port 字段值。
- Master_id：表示当前从库正在从哪个主库进行复制。所显示的对应主库的 Server ID，在复制架构中应该是全局唯一的，当一个主库有多个从库时，该列的所有行显示为相同的值。
- Slave_UUID：在复制架构中，从库唯一的全局 UUID，该值默认由数据库实例启动时自动生成，保存在系统变量 datadir 指定目录下的 auto.cnf 文件中，以便下次启动时继续使用。如果 auto.cnf 文件被删除，则启动时会重新生成一个新的 UUID。

第 11 章　MySQL 复制延迟 Seconds_Behind_Master 究竟是如何计算的

在主从复制拓扑中，监控复制延迟是必不可少的工作。如果应用场景对复制延迟并不敏感，那么大多数时候通过采集 SHOW SLAVE STATUS 语句输出信息中 Seconds_Behind_Master 字段的值监控复制延迟就已经足够了。相信有 MySQL 使用经验的人对这种方法并不陌生，我们都知道 Seconds_Behind_Master 的值在某些场景下并不是那么可靠，也或多或少都知道一些计算这个值的方法。但这些计算方法真的正确吗？接下来，本章将对此进行讨论并确认正确的计算方法。

11.1　"口口相传"的计算方法

方法一：计算从库 I/O 线程读取的主库二进制日志事件的时间戳，与 SQL 线程正在执行的二进制日志事件的时间戳之间的差值，单位为 s。

该方法其实就是在计算两个线程处理日志的时间差，也是目前最"流行"的一种算法。基于这个算法，如果主从库之间的网络存在很大延迟，主库中就可能存在大量二进制日志还没来得及发送给从库，那么这时使用该方法计算出来的延迟，跟主从库之间数据真正的延迟就没有太大关系了。

方法二：计算从库的系统（主机）时间，与 I/O 线程读取的主库二进制日志事件的时间之间的差值，单位为 s。

这种算法没那么"流行"，基于这个算法，如果从库的操作系统时间被更改，或者主库的操作系统时间被修改，即主从库主机的时间差本身就比较大，那么计算出来的结果也毫无参考意义。

看起来以上两种算法似乎都不太靠谱，为了一探究竟，我们需要找到可靠的信息源进行确认，那么从哪里可以获得可靠的信息源呢？

11.2 探寻"正确"的计算方法

信息源一：MySQL 官方手册

在 Oracle MySQL 官方手册中搜索 "Seconds_Behind_Master"，即可查到关于如何计算 Seconds_Behind_Master 字段值的原文，其中文含义大致如下：

- 当从库在不断处理更新时——持续不断地有事件（event）被 SQL 线程或者 I/O 线程处理，此字段显示的是从库主机当前时间戳与主库（原始）二进制日志中记录的时间戳之间的差。

- 当从库没有任何需要处理的更新时，如果 I/O 线程和 SQL 线程状态都为 Yes，则此字段显示为 0。如果任何线程的状态不为 Yes，则此字段显示为 NULL。

- 实际上，这个字段度量的是从库 SQL 线程和 I/O 线程之间的时间差，单位为秒。如果主从库之间的网络非常快，那么从库 I/O 线程读取的主库二进制日志会与主库中最新的二进制日志非常接近，这样计算出来的值就可以作为主从库之间的数据延迟时间。但是如果主从库之间的网络非常慢，可能导致从库 SQL 线程正在重放的主库二进制日志非常接近从库 I/O 线程读取的主库二进制日志，而 I/O 线程因为网络慢的原因，读取的主库二进制日志可能远远落后于主库最新的二进制日志，此时计算出来的值是不可靠的。尽管这个时候该字段可能显示为 0，但实际上从库已经落后于主库非常多了。所以，对于网络比较慢的情况，该值并不可靠。

- 如果主库与从库 Server 自身的时间不一致，只要从库的复制线程启动之后，没有做过任何时间变更，也可以正常计算这个字段的值，但是如果修改过 Server 的时间，则可能导致时钟偏移，计算出来的这个值就不可靠。

- 如果从库的 SQL 线程没有运行，或者 SQL 线程正在运行，已经消费完所有中继日志，而且 I/O 线程没有运行，则该字段显示为 NULL（如果 I/O 线程已经停止，但还有中继日志未重放完，该字段仍然会显示为复制延迟时间，等到所有中继日志重放完之后，其显示为 NULL）。如果 SQL 线程和 I/O 线程都在运行，但处于空闲状态（SQL 线程已经重放完 I/O 线程产生的中继日志），则该字段显示为 0。

- 该字段的值是通过存储在主库二进制日志事件中的时间戳与从库当前时间戳之差计算出来的（这个时间戳会通过复制拓扑同步到从库，如果主从库的操作系统时间戳存在差异，则还需要减去此差值）。这就意味着在正常的复制下（排除人为在从库写入数据的情况），主库与从库上的二进制日志事件的时间戳都来自主库。目前的计算该字段值的算法中有一个问题：在单线程复制场景下，如果在从库上通过客户端连接进入，并直接更新数据，可能导致该字段的值随机波动，因为有时候事件来源于主库，有时候事件是由从库直接更新产生的，而这个字段的值会受后者的影响。但如果是多线程复制，则此值是基于 Exec_Master_Log_Pos 的事件时间戳来计算的，因此可能不会反映从库最近提交的事务的位置。

信息源二：源码

以下是对源码中关于延迟时间计算方法的注释的说明：

```
# 位于 rpl_mi.h 中对 clock_diff_with_master 的定义附近（笔者翻阅了 MySQL 5.6.34 和 MySQL 5.7.22
两个版本的源码，两者对于复制延迟的计算公式一致）
# 从源码注释上来看，复制延迟的计算公式为 clock_of_slave - last_timestamp_executed_by_
SQL_thread - clock_diff_with_master
# 该公式的含义为：从库的当前系统（主机）时间 - 从库 SQL 线程正在执行的事件的时间戳 - 主从库的系统（主
机）之间的时间差
    /*
        The difference in seconds between the clock of the master and the clock of
        the slave (second - first). It must be signed as it may be <0 or >0.
        clock_diff_with_master is computed when the I/O thread starts; for this the
        I/O thread does a SELECT UNIX_TIMESTAMP() on the master.
        "how late the slave is compared to the master" is computed like this:
        clock_of_slave - last_timestamp_executed_by_SQL_thread - clock_diff_with_master

    */
    # clock_diff_with_master 值为主从 Server 的主机时间差，该值只在 I/O 线程启动时计算一次，后面每次
计算 Seconds_Behind_Master 字段值时，直接复用这个计算结果，每次重启 I/O 线程时会重新计算该值
    long clock_diff_with_master;

    # master_row[0] 为从库在主库上执行 SELECT UNIX_TIMESTAMP() 的操作
    mi->clock_diff_with_master=
        (long) (time((time_t*) 0) - strtoul(master_row[0], 0, 10));

    # 从 rpl_slave.cc 文件中启动 I/O 线程时可以看出：
        start_slave_thread->
            handle_slave_io->
                get_master_version_and_clock   # 获取当前从库和主库的系统（主机）之间的时间差
(clock_diff_with_master)
```

以下是源码中关于 Seconds_Behind_Master 计算结果的一些判定值：

```
    /*
        The pseudo code to compute Seconds_Behind_Master:   # 阐明这是一段关于如何计算
Seconds_Behind_Master 的伪代码
        if (SQL thread is running)   # 如果 SQL 线程正在运行，则进入这个 if 判断语句内，假设这里标记
为 if one
        {
            if (SQL thread processed all the available relay log)   # 如果 SQL 线程应用完所有可用
的中继日志，则进入这个 if 判断语句内，假设这里标记为 if two
            {
                if (IO thread is running)   # 如果 I/O 线程正在运行，则进入这个 if 判断语句内，假设这里标
记为 if three
                    print 0;   # 如果 if one/two/three 三个条件都为真，则延迟值判定为 0
                else
                    print NULL;   # 如果 if one/two 为真，if three 为假，则延迟值判定为 NULL
            }
```

```
          else
            compute Seconds_Behind_Master;    # 如果 if one 为真，if two 为假，则执行公式，计算延
迟值
        }
        else
          print NULL;    # 如果 if one 为假，则延迟值判定为 NULL
*/

    if (mi->rli->slave_running)
    {
      /*
        Check if SQL thread is at the end of relay log
        Checking should be done using two conditions
        condition1: compare the log positions and
        condition2: compare the file names (to handle rotation case)
      */
      if ((mi->get_master_log_pos() == mi->rli->get_group_master_log_pos()) &&
          (!strcmp(mi->get_master_log_name(), mi->rli->get_group_master_log_name())))
      {
        if (mi->slave_running == MYSQL_SLAVE_RUN_CONNECT)
          protocol->store(0LL);
        else
          protocol->store_null();
      }
      else
      {
        long time_diff= ((long)(time(0) - mi->rli->last_master_timestamp)
                         - mi->clock_diff_with_master);
        /*
          Apparently on some systems time_diff can be <0. Here are possible
          reasons related to MySQL:
          - the master is itself a slave of another master whose time is ahead.
          - somebody used an explicit SET TIMESTAMP on the master.
          Possible reason related to granularity-to-second of time functions
          (nothing to do with MySQL), which can explain a value of -1:
          assume the master's and slave's time are perfectly synchronized, and
          that at slave's connection time, when the master's timestamp is read,
          it is at the very end of second 1, and (a very short time later) when
          the slave's timestamp is read it is at the very beginning of second
          2. Then the recorded value for master is 1 and the recorded value for
          slave is 2. At SHOW SLAVE STATUS time, assume that the difference
          between timestamp of slave and rli->last_master_timestamp is 0
          (i.e. they are in the same second), then we get 0-(2-1)=-1 as a result.
          This confuses users, so we don't go below 0: hence the max().

          last_master_timestamp == 0 (an "impossible" timestamp 1970) is a
          special marker to say "consider we have caught up".
        */
```

```
                    protocol->store((longlong)(mi->rli->last_master_timestamp ?
                                    max(0L, time_diff) : 0));   # 这里 time_diff 其实就是最终
计算的 Seconds_Behind_Master 值,如果该值为负数,则直接归零
    }
}
```

11.3 验证

11.3.1 我们想确认什么

所谓"尽信书不如无书",不能书上说是这样的我们就信了,所以这里我们对官方手册和源码中提及的计算公式、场景进行简单的验证,看看实际的结果。

根据 11.2 节源码中的注释,我们找到了计算复制延迟的"正确"方法(公式)为:

clock_of_slave - last_timestamp_executed_by_SQL_thread - clock_diff_with_master

此公式的含义为"从库的当前系统(主机)时间 - 从库 SQL 线程正在执行的事件的时间戳 - 主从库的系统(主机)之间的时间差"。公式中的 clock_diff_with_master 值,也就是主从库的主机时间差只在 I/O 线程启动的时候计算一次(后面会复用该计算结果来计算复制延迟,直到下次重启 I/O 线程时才会重新计算)。

根据官方手册中的描述,主从库的系统时间不一致,也能计算出正确的复制延迟:如果主从库的系统时间不一致,在复制线程(I/O 线程)启动之后,如果没有对主库或者从库的系统时间再次进行修改,那么根据公式是可以正确计算出复制延迟的。如果在复制线程启动之后再次修改了主从库的系统时间,就会导致计算出的复制延迟不可靠(因为计算公式中的 clock_diff_with_master 只在 I/O 线程启动时才会计算,后面计算复制延迟时会复用该计算结果)。

根据 11.2 节源码中的描述,当 Seconds_Behind_Master 计算结果为负数时,该值直接归零。

11.3.2 提前确认一些信息

首先,我们需要确认当 I/O 线程启动之后,会在主库上执行一些什么操作。然后,我们需要确认直接修改主机时间之后,在数据库中执行一些时间函数时返回的时间信息是否会跟着改变,写入二进制日志的时间信息是否会跟着改变。

先来看第一个问题,当 I/O 线程启动之后,会在主库上执行一些什么操作?

在主库打开 general_log:

```
mysql> set global general_log=1;
Query OK, 0 rows affected (0.01 sec)
```

从库启动 I/O 线程:

```
mysql> start slave io_thread;
Query OK, 0 rows affected (0.00 sec)
```

查看主库 general_log 中记录的内容:

```
2019-04-18T10:16:36.414222+08:00 8 Connect    qbench@10.10.30.162 on using TCP/IP
2019-04-18T10:16:36.414632+08:00 8 Query    SELECT UNIX_TIMESTAMP() # 在主库查询系统时间的
```
语句。这里可以证实 I/O 线程启动时会获取主库的系统时间。当从库获知主库的系统时间之后,就可以计算主从库的系统时间之差

```
2019-04-18T10:16:36.415401+08:00 8 Query    SELECT @@GLOBAL.SERVER_ID # 查询主库的 SERVER_ID
2019-04-18T10:16:36.415638+08:00 8 Query    SET @master_heartbeat_period= 5000000000 # 对
```
Dump 线程在会话级别设置心跳间隔时间,I/O 线程向主库注册的会话级别的自定义变量可以在 performance_schema. user_variables_by_thread 表中查询

```
2019-04-18T10:16:36.415814+08:00 8 Query    SET @master_binlog_checksum= @@global.binlog_
checksum
2019-04-18T10:16:36.416129+08:00 8 Query    SELECT @master_binlog_checksum
2019-04-18T10:16:36.416335+08:00 8 Query    SELECT @@GLOBAL.GTID_MODE # 查询主库的 GTID 模
式值
2019-04-18T10:16:36.416527+08:00 8 Query    SELECT @@GLOBAL.SERVER_UUID # 查询主库的 UUID
2019-04-18T10:16:36.416693+08:00 8 Query    SET @slave_uuid= '2d623f55-2111-11e8-9cc3-
0025905b06da' # 在会话级别设置从库自己的 UUID
2019-04-18T10:16:36.417224+08:00 8 Binlog Dump GTID Log: '' Pos: 4 GTIDs: '06188301-
b333-11e8-bdfe-0025905b06da:1-270852,
    2d623f55-2111-11e8-9cc3-0025905b06da:1,
    f3372787-0719-11e8-af1f-0025905b06da:1-21'
```

现在,我们来看第二个问题:修改主机时间之后(数据库进程不重启),在数据库中执行一些时间函数时返回的时间信息是否会跟着改变,写入二进制日志的时间信息是否会跟着改变?

在主库的主机中修改时间:

```
# 在修改之前先查询一下主机和数据库中的时间,可以看到它们都在 2019-04-19 16:14 附近
[root@localhost ~]# date
2019 年 04 月 19 日 星期五 16:14:19 CST

# 在修改之前先查看数据库中的时间
mysql> select now(),unix_timestamp,from_unixtime(unix_timestamp());
+---------------------+----------------+---------------------------------+
| now()               | unix_timestamp | from_unixtime(unix_timestamp()) |
+---------------------+----------------+---------------------------------+
| 2019-04-19 16:14:20 |     1555661660 | 2019-04-19 16:14:20             |
+---------------------+----------------+---------------------------------+

# 将主机时间修改为 2020-04-19 17:15:00
[root@localhost ~]# date -s '2020-04-19 17:15:00'
2020 年 04 月 19 日 星期日 17:15:00 CST

# 发现在数据库中使用 unix_timestamp() 函数查到的时间随着主机时间变了,也变为 2020-04-19
```

```
mysql> select now(),unix_timestamp(),from_unixtime(unix_timestamp());
+---------------------+------------------+---------------------------------+
| now()               | unix_timestamp() | from_unixtime(unix_timestamp()) |
+---------------------+------------------+---------------------------------+
| 2020-04-19 17:15:03 | 1587287703       | 2020-04-19 17:15:03             |
+---------------------+------------------+---------------------------------+

# 插入一点测试数据，解析二进制日志，发现二进制日志中的时间戳也跟着发生了变化，变为 2020-04-19。这就
可以证实，修改主机时间将直接影响数据库中时间函数获取的时间以及二进制日志的时间戳
......
# at 468
#190419 16:14:29 server id 3306154 end_log_pos 499 CRC32 0x80c44c50 Xid = 74
COMMIT/*!*/;
# at 499
#200419 17:15:08 server id 3306154 end_log_pos 564 CRC32 0xc2794ccb GTID last_committed=1
sequence_number=2 rbr_only=yes
/*!50718 SET TRANSACTION ISOLATION LEVEL READ COMMITTED*//*!*/;
SET @@SESSION.GTID_NEXT= 'c413c893-56b0-11e9-a705-000c29c10fa5:35'/*!*/;
......
```

11.3.3 执行验证

场景一：在主库每秒一直有数据写入的情况下（这里我们使用脚本来持续执行每隔 1 秒插入 10 行数据的操作），验证主从库各自向前和向后修改主机时间、主从库同时向前和向后修改主机时间时，从库的复制延迟如何变化（脚本持续每秒获取一次 Seconds_Behind_Master 字段值）。

场景二：利用 sysbench 持续对主库加压，直到从库出现复制延迟，然后，立即停止 sysbench 对主库的施压或立即停止从库 I/O 线程，持续观察从库的复制延迟如何变化。

由于验证过程较为烦琐，有兴趣的读者请自行验证。这里仅直接展示验证结果供大家参考，如表 11-1 和表 11-2 所示。

表 11-1

场景一的验证结果			
角色	修改操作	复制延迟状态	重启复制线程
主库	时间向前改早一年	频繁出现延迟 29 002 339 s	无延迟
主库	时间向后改晚两年	无延迟（仅根据公式计算应该出现负数值）	无延迟
从库	时间向前改早一年	无延迟（仅根据公式计算应该出现负数值）	无延迟
从库	时间向后改晚两年	频繁出现延迟 63 098 955 s	无延迟

表 11-2

角色	场景二的验证结果	
	修改操作	复制延迟状态
主库	利用 sysbench 加压，直到从库开始延迟时停止加压	复制延迟逐渐增加，达到某个高峰值后瞬间变为 0
从库	利用 sysbench 加压，直到从库开始延迟时停止 I/O 线程	复制延迟逐渐增加，达到某个高峰值瞬间变为 NULL

11.4 小结

根据以上可靠的信息来源以及验证过程可以简单得出如下结论：

- 对于主从库主机时间不一致的情况，当 I/O 线程第一次启动时，会计算主从库之间的主机时间差，在后面计算复制延迟时，减去这个时间差，就可以保证获取的是正确的复制延迟。但是，只有 I/O 线程启动时才会计算该时间差，所以 I/O 线程启动之后，如果修改主从库的主机时间，会导致根据公式计算的复制延迟不可靠，但是当 I/O 线程重启之后就可以恢复正常（因为 I/O 线程重启时，会重新计算主从库主机之间的时间差）。

- 在计算复制延迟时（执行 SHOW SLAVE STATUS 语句就会进行计算），对 Seconds_Behind_Master 的计算结果做一些判定（11.2 节介绍源码的伪代码注释时讲过，这里再总结一下）：

  - 如果 I/O 线程和 SQL 线程同时为 Yes（表示两个线程处于运行状态），且 SQL 线程没有做任何事情（没有需要执行的事件），此时直接判定复制延迟为 0，不会用公式来计算复制延迟，否则会用公式进行计算（所以，在该前置条件下，不会出现当主库没有写任何二进制日志事件时，从库延迟不断增加的情况）。

  - 如果 SQL 线程为 Yes，而且 I/O 线程已经读取的中继日志未应用完，则会用公式计算延迟时间，不管 I/O 线程是否正在运行。但当 SQL 线程重放完所有中继日志时，如果 I/O 线程不为 Yes，则直接判定复制延迟的结果为 NULL。

  - 任何时候，如果 SQL 线程不为 Yes，则直接判定复制延迟的结果为 NULL。

- 当计算的复制延迟为负数时，将该值直接归零。

提示：当 SQL 线程重放大事务时，SQL 线程的时间戳更新相当于被暂停了（因为一个大事务的事件在重放时需要很长时间才能完成，虽然这个大事务也可能会有很多事件，但是这些事件的时间戳可能全部相同），此时根据计算公式可以得出，无论主库是否有新的数据写入，从库的复制延迟仍然会持续增加（也就是说，此时的复制延迟值是不可靠的），所以就会出现主库停止写入之后，从库的复制延迟逐渐增加到某个最大值之后突然变为 0 的情况。

更多信息可参考高鹏的"复制"专栏，可登录简书网站搜索"第 27 节：从库 Seconds_Behind_Master 的计算方式"和"第 28 节：从库 Seconds_Behind_Master 延迟总结"。

第 12 章　如何保证从库在意外中止后安全恢复

为了保证复制线程能够在从库意外中止之后正确恢复到中止之前的状态（有时称为 crash-safe），需要对从库中的某些系统变量与复制选项设置合适的值。本章将详细介绍如何设置这些系统变量与复制选项及其对应的不同影响。

12.1　从库的崩溃与恢复概述

从库在意外中止之后重新启动时，其 SQL 线程需要恢复意外中止时已执行完的事务，恢复过程中所需的信息存储在从库的中继日志状态信息中。早期的 MySQL 版本，此信息只能保存在数据目录下的磁盘文件中，存在丢失 SQL 线程正确的重放位置的风险（具体情况要由从库重放的事务所处的阶段而定），甚至还有中继日志状态信息文件本身发生损坏的风险。MySQL 5.6 中新增了 mysql.slave_relay_log_info 表，该表是 InnoDB 存储引擎，用来存储中继日志状态信息，它可以使用事务的特性来支持在 MySQL Server 意外中止时的安全恢复。同时，使用表存储中继日志状态信息时，该表中的位置信息可以随着业务数据的事务一并提交。这就意味着从库意外中止时，从库的数据重放进度（位置）信息能够和业务数据的事务保持一致。

要将中继日志状态信息保存到 InnoDB 表中，需要将系统变量 relay_log_info_repository 值设置为 TABLE，然后 mysql.slave_relay_log_info 表就会存储从库 SQL 线程所需的位置信息。

12.2　从库的崩溃与恢复详解

从库意外中止后的恢复方式会受到所选的复制方法、单线程与多线程、系统变量的设置（relay_log_recovery）、是否使用了自动定位复制选项（MASTER_AUTO_POSITION = 1）等因素的影响。下面将从单线程与多线程的维度，说明不同系统变量与复制选项的组合对从库恢复的影响。

12.2.1　单线程复制的安全恢复

图 12-1（参考 Oracle MySQL 官方资料）中列出了在从库意外中止之后，单线程复制中不同因素的组合对恢复结果的影响。

gtid_mode	MASTER_AUTO_POSITION	relay_log_recovery	relay_log_info_repository	Crash type	Recovery guaranteed	Relay log impact
OFF	Any	1	TABLE	Server	Yes	Lost
OFF	Any	1	Any	OS	No	Lost
OFF	Any	0	TABLE	Server	Yes	Remains
OFF	Any	0	TABLE	OS	No	Remains
ON	ON	Any	Any	Any	Yes	Lost
ON	OFF	0	TABLE	Server	Yes	Remains
ON	OFF	0	Any	OS	No	Remains

图 12-1

如图 12-1 中的粗线灰框标记处所示，使用单线程复制时，这样的配置对于从库意外中止之后的安全恢复来说，弹性最好。

- 同时启用 GTID 和复制选项 MASTER_AUTO_POSITION = 1 时，是否设置系统变量 relay_log_recovery = 1、是否设置系统变量 relay_log_info_repository = TABLE，以及其他系统变量的设置，都不会影响从库意外中止之后的恢复。请注意，为了保证能顺利恢复，还必须在从库上设置"双一"：系统变量 sync_binlog = 1 和 innodb_flush_log_at_trx_commit = 1，以便从库的二进制日志和事务的重做日志（redo log）能够及时同步到磁盘。否则，已提交事务的二进制日志可能会在从库意外宕机时丢失。

- 采用基于二进制文件和位置的传统复制时，必须设置系统变量 relay_log_recovery = 1 和 relay_log_info_repository = TABLE，否则在操作系统崩溃（OS Crash）之后无法保证复制的安全恢复。

- 当设置系统变量 relay_log_recovery = 1 时，在复制的恢复期间，已经从主库拉取的且还未在从库中重放的中继日志会丢失（被清理掉并重新从主库获取）。

12.2.2　多线程复制的安全恢复

图 12-2（参考 Oracle MySQL 官方资料）中列出了从库意外中止之后，多线程复制中不同因素的组合对从库恢复的影响。

gtid_mode	sync_relay_log	MASTER_AUTO_POSITION	relay_log_recovery	relay_log_info_repository	Crash type	Recovery guaranteed	Relay log impact
OFF	1	Any	1	TABLE	Any	Yes	Lost
OFF	>1	Any	1	TABLE	Server	Yes	Lost
OFF	>1	Any	1	Any	OS	No	Lost
OFF	1	Any	0	TABLE	Server	Yes	Remains
OFF	1	Any	0	TABLE	OS	No	Remains
ON	Any	ON	Any	Any	Any	Yes	Lost
ON	1	OFF	0	TABLE	Server	Yes	Remains
ON	1	OFF	0	Any	OS	No	Remains

图 12-2

如图 12-2 中粗线灰框标记处所示，使用多线程复制时，这样的配置对于从库意外中止之后的安全恢复来说，弹性最好：

- 同时启用 GTID 和复制选项 MASTER_AUTO_POSITION = 1 时，是否设置系统变量 relay_log_recovery = 1、是否设置系统变量 relay_log_info_repository = TABLE，以及其他系统变量的设置，都不会影响从库意外中止之后的恢复。
- 采用传统复制时，必须设置系统变量 relay_log_recovery = 1、sync_relay_log = 1 和 relay_log_info_repository = TABLE，否则在操作系统崩溃之后不能保证复制的安全恢复。
- 当设置系统变量 relay_log_recovery = 1 时，在从库的恢复期间，已经从主库拉取的且还未在从库中重放的中继日志会丢失（被清理掉）。

当设置 sync_relay_log = 1 时，每个事务的中继日志都需要落盘，尽管此时对于从库来说意外中止之后的恢复能力最强（最多丢失最后一个事务的中继日志），但该设置也会大大增加存储的负载。如果 sync_relay_log 被设置为除 1 以外的值，则从库在意外中止之后的恢复能力，取决于操作系统如何处理中继日志（系统变量 relay_log_recovery = 1 时，从库意外中止之后会清除未重放的中继日志，在 SQL 线程应用的位置重新从主库拉取二进制日志生成新的中继日志。系统变量 relay_log_recovery = 0 时，从库意外中止之后恢复时，其恢复过程将包含对中继日志的处理（中继日志中未来得及回放的数据），等待中继日志中的数据处理完之后，清除中继日志。

注意： 当系统变量 relay_log_recovery = 0 时，如果从库崩溃时 I/O 线程正在接收大事务，则可能造成该大事务在从库中继日志中的记录不完整。从库从崩溃中恢复之后，会重新生成一个全新的中继日志，之前未写入完的中继日志中的事务会在新的中继日志中重新从主库拉取整个事务的二进制日志。当 SQL 线程读取到事务不完整的中继日志时，会在这个地方挂（hang）死（重放过程无法继续，因为正常的中继日志的切换必须保证一个事务的二进制日志事件是完整的，这一点与二进制日志的切换原则是相同的，所以当 SQL 线程碰到中继日志中不完整的二进制日志事件时，会死等 I/O 线程将后续的二进制日志事件补充完整，从而导致重放过程挂死）。

要解决这个问题，只需要在从库上执行 RESET SLAVE 语句（执行该语句时会清理掉还未应用的中继日志），然后执行 START SLAVE 语句（执行该语句时，会重新从主库拉取全新的二进制日志）。

采用基于二进制文件和位置的传统复制时，如果从库意外中止，则可能导致从库中继日志信息日志与事务不一致（事务序列中间产生了间隙）。在 MySQL 5.7.13 及之前的版本中，需要手工设置系统变量 relay_log_recovery = 0, slave_parallel_workers = 0，启动 Server，然后使用 START SLAVE UNTIL SQL_AFTER_MTS_GAPS 语句启动复制以修复事务不一致的情况（修改为单线程复制，串行地逐个处理不一致的事务），接着再修改系统变量 relay_log_recovery = 1, slave_parallel_workers = N（$N > 1$），重新启动从库 Server。在 MySQL

5.7.13 及更高的版本中，如果恢复过程中碰到中继日志信息日志与事务不一致，会自动填充间隙，处理不一致的事务，并继续后续的恢复过程。

如果使用多源复制，同时设置系统变量 relay_log_recovery = 1，则从库在意外中止之后的恢复过程中，所有复制通道都将会经历中继日志恢复过程，如果在任何复制通道中发现了任何中继日志信息与事务不一致的情况，都会自动修复。

第 13 章　MySQL Server 复制过滤

MySQL Server 的复制过滤功能，虽然通常在标准的生产环境中不建议使用（因为如果使用不当，可能导致各种各样的主从库数据不一致的问题），但在某些场合下，为了区分业务数据或保证数据安全性（例如，对业务数据所有的写操作都统一写入主库，但是对业务数据的读取则需要按照业务模块来划分，因为不同业务系统之间有数据安全隔离的需求），MySQL Server 的复制过滤功能可能是一个简单、快捷的实现方案。在有足够的了解，而且通过缜密的规划来规避其缺点的前提下，也可以将其列为能解决此类需求的可选方案之一。本章将简单梳理一遍 MySQL Server 复制过滤功能的实现逻辑。

13.1　MySQL Server 复制过滤规则概述

如果主库在执行事务提交时，未将语句写入其二进制日志，则该语句不会被复制到从库。如果主库把该语句记录到其二进制日志中，则该语句将被发送给所有从库，但在每个从库上该语句是否会被执行，则取决于每个从库自身的配置是否忽略了该语句。

在主库上，可以使用选项--binlog-do-db 和--binlog-ignore-db 来控制哪些数据库的数据变更需要记录到二进制日志中，哪些数据变更需要被忽略。正常情况下，不建议在主库中使用复制过滤选项，建议在从库上使用复制过滤选项来控制哪些数据库、哪些表需要被执行或需要被忽略。

在从库方面，是否执行或是否忽略从主库接收到的语句，是由从库在启动数据库进程时设置的选项（以"--replicate-"为前缀的启动选项）决定的，也可以使用 CHANGE REPLICATION FILTER 语句动态设置复制过滤规则。这两种方式设置的过滤器规则都相同。但要注意，不要在组复制（MGR，MySQL Group Replication）中使用复制过滤器，因为在不同的节点上配置不同的复制过滤规则，可能会导致整个集群中的数据一致性被破坏。

在最简单的情况下，如果从库未设置复制过滤选项（以"--replicate-"为前缀的启动选项），从库会执行从主库接收到的所有语句；如果从库设置了复制过滤选项，则具体如何过滤取决于从库指定了哪些选项。

从库复制过滤选项的检查过程大致如下：当从库在执行接收到的某个语句时，首先检查数据库级别的选项（启动选项--replicate-do-db 和--replicate-ignore-db）。如果未使用任何数据库级别的过滤选项，则继续检查表级别的过滤选项（启动选项--replicate-do-table 或--replicate-wild-do-table、--replicate-ignore-table 或--replicate-wild-ignore-table）。具体过程详见 13.2 节。

对于仅对数据库产生影响的一些语句（如 CREATE DATABASE、DROP DATABASE、ALTER DATABASE），数据库级别的过滤选项始终优先于任何表级别的过滤选项生效。换句话说，对于此类语句，当且仅当没有适用的数据库级别过滤选项时，才会检查表级别的过滤选项。这是对早期 MySQL 版本（MySQL 5.6 之前的版本）行为的修正。在早期的版本中，如果从库同时启动库级别的过滤选项--replicate-do-db = dbx 和表级别的过滤选项--replicate-wild-do-table = db%.t1，则不会复制 CREATE DATABASE dbx 语句，因为这些 MySQL 版本中无论有没有合适的库级别过滤选项，都会检查表级别过滤选项，而表级别过滤选项只是需要复制 t1 表，而不是整个库，所以 dbx 库相关的操作语句被忽略。这是个 Bug。

为了更容易确定复制过滤选项产生的影响，也为了更容易确定什么内容需要复制，什么内容需要被过滤,不建议混合使用包含 do 和 ignore 字符前缀或包含 wild-do 和 wild-ignore 字符前缀的选项。

注意：
- 如果指定了任何--replicate-rewrite-db 选项，则会在 rewrite 之前检测复制过滤规则。
- 所有复制过滤选项都遵循与 MySQL Server 中生效的相同的区分大小写规则，即复制过滤时匹配的库名与表名的大小写敏感度由系统变量 lower_case_table_names 控制。

13.2 库级别复制过滤选项的评估

13.1 节提到，从库在评估复制过滤选项时，首先检查是否存在适用的--replicate-do-db 或--replicate-ignore-db 选项（当主库使用--binlog-do-db 或--binlog-ignore-db 复制过滤选项时，其评估过程与此类似，但主库不支持表级别的选项）。

检查复制过滤选项时，如何匹配数据库还取决于主库的二进制日志格式。如果使用 row 格式记录语句，则只有在需要检查的数据库中（这里指数据库配置文件中的复制过滤选项指定的数据库）的数据发生变更时才会执行检查；如果使用 statement 格式记录语句，则默认数据库就是需要执行检查的数据库。使用 USE 语句指定的数据库被称为默认数据库，只有使用 USE 语句指定的数据库中的数据发生变更时，才会执行检查。所以，如果有多个数据库中的数据需要执行复制过滤检查，则需要在写数据前，先使用 USE 语句切换默认数据库。

注意：row 格式的二进制日志只能记录 DML 语句，但 DDL 语句只能记录为 statement 格式。因此，对于 DDL 语句，始终按照 statement 格式的复制过滤规则进行筛选。也就是说，必须显式使用 USE 语句切换到正确的数据库，DDL 语句才能够正确地按照复制过滤规则进行筛选。

对于以"--replication_"开头的复制过滤选项（这里指的是从库），下面列出了对语句执行筛选的步骤。

1. 根据主库使用的二进制日志格式选择对应的检查方法。

（1）statement 格式：检查由复制过滤选项设置的需要检查的库，是否与默认数据库匹配（用 USE 语句切换的数据库，如果不使用 USE 语句切换到默认数据库，则不会进行检查）。

（2）row 格式：复制过滤选项设置数据库中的数据发生变更时执行检查（不需要使用 USE 语句切换数据库）。

2. 是否有--replicate-do-db 选项？

（1）是：执行的语句所涉及的数据库是否与复制过滤选项值中的任何一个匹配？

- 是：继续执行第 4 步。
- 否：忽略应用更新并退出判断。

（2）否：继续执行第 3 步。

3. 是否有--replicate-ignore-db 选项？

（1）是：执行的语句所涉及的数据库是否与复制过滤选项值中的任何一个匹配？

- 是：忽略应用更新并退出判断。
- 否：继续执行第 4 步。

（2）否：继续执行第 4 步。

4. 继续检查表级别的复制选项（如果有的话）。

重要提示：在此判断阶段，这些语句还未执行，需要等到检查完所有表级别复制选项之后，才会真正执行。

从库中库级别复制过滤选项的评估示意图如图 13-1 所示（非标准流程图）。

对于以"--binlog-"为前缀的复制过滤选项（这里指的是主库），下面列出了对语句执行筛选的步骤：

1. 是否有--binlog-do-db 或--binlog-ignore-db 选项？

（1）是：继续执行第 2 步。

（2）否：将语句记录到二进制日志中，并退出判断。

2. 是否切换了默认数据库（使用 USE 语句选择了某个数据库）？

（1）是：继续执行第 3 步。

（2）否：忽略语句（不记录到二进制日志中），并退出判断。

3. 已经切换了默认数据库，检查是否有--binlog-do-db 选项？

（1）是：检查默认数据库是否与库级别的复制过滤选项值匹配。

- 是：把该语句记录到二进制日志中，并退出判断。
- 否：忽略语句（不记录到二进制日志中），并退出判断。

（2）否：继续执行第 4 步。

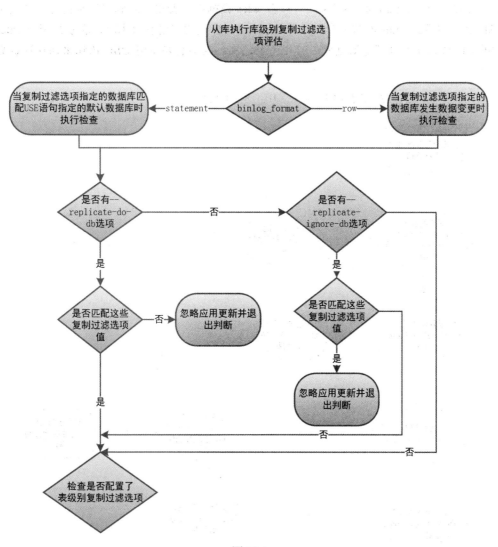

图 13-1

4. 是否有任何--binlog-ignore-db 选项与数据库匹配？

（1）是：忽略语句（不记录到二进制日志中），并退出判断。

（2）否：将语句记录到二进制日志中，并退出判断。

重要提示：对于 statement 格式的二进制日志，判断 CREATE DATABASE、ALTER DATABASE 和 DROP DATABASE 等语句时会出现异常。在这些情况下，在判断是记录语句还是忽略更新时，正在执行创建、更改或删除操作的语句中所涉及的数据库将成为默认数据库。

选项--binlog-do-db 可以理解为"忽略其他数据库"。例如，使用 statement 格式的二进制日志时，--binlog-do-db = sales 表示只将 sales 数据库的变更记录到二进制日志中（使用 USE 语句切换到 sales 库时才会生效）；使用 row 格式的二进制日志时，表示只记录 sales 数据库的数据变更（其他数据库中的变更不记录，且不需要使用 USE 语句来切换默认数据库）。

主库的复制过滤选项的评估示意图如图 13-2 所示。

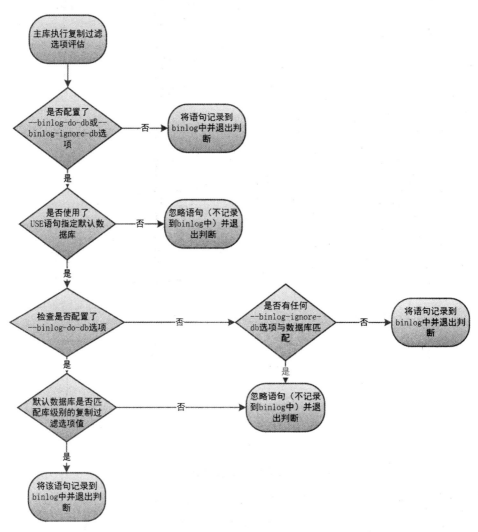

图 13-2

13.3　表级别复制过滤选项的评估

仅当满足以下两个条件之一时，从库才会检查并评估表级别的复制过滤选项：

- 找不到匹配的数据库级别复制过滤选项。
- 找到了一个或多个数据库级别复制过滤选项，并根据 13.2 节中描述的匹配规则进行评估，达到"执行"判断表级别选项的条件（13.2 节中的执行流程到达第 4 步）。

首先，检查是否使用了基于 statement 的复制。如果是，并且语句又是在存储函数内执行的，则从库执行语句并退出判断。如果使用的是基于 row 的复制，从库不知道在存储函数内执行了什么语句（因为二进制日志中记录的是 row 格式的事件信息，并没有记录原始的语句文本，也就无法进行复制过滤），则此时无法进行过滤。

注意：对于基于 statement 的复制，复制事件记录的就是原始的 SQL 语句文本（即二进制日志中的事件数据就是主库中执行的原始 SQL 语句）；对于基于 row 的复制，一个语句修改多行时，可能会产生多个基于行的事件（即对于大事务产生的多个用于记录数据变更的事件，会用特殊的标记表示其是否是最后一个事件，详见第 26 章"二进制日志文件的基本组成"）。从事件的角度来看，检查表级别复制过滤选项的过程，对于基于 row 和基于 statement 的复制都是相同的。

如果从库中未配置表级别的复制过滤选项，则从库会执行所有事件。如果有任何 --replicate-do-table 或 --replicate-wild-do-table 选项，则事件必须匹配其中一个才会被执行，否则，就会被忽略。如果有任何 --replicate-ignore-table 或 --replicate-wild-ignore-table 选项，则除与这些 ignore 选项匹配的表外，其他表的所有事件都会被执行。

以下步骤详细描述了此评估过程。

1. 是否有表级别复制过滤选项？

（1）是：继续执行第 2 步。

（2）否：执行更新并退出判断。

2. 使用哪种二进制日志格式？

（1）statement：对将要执行更新的每个语句执行后续判断步骤。

（2）row：对将要对表的行执行更新的每个语句执行后续判断步骤。

3. 是否有 --replicate-do-table 选项？

（1）是：将要执行更新的表是否匹配选项中的任意一个值？

- 是：执行更新并退出判断。
- 否：继续执行第 4 步。

（2）否：继续执行第 4 步。

4. 是否有--replicate-ignore-table 选项？

（1）是：将要执行更新的表是否匹配选项中的任意一个值？

- 是：忽略更新并退出判断。
- 否：继续执行第 5 步。

（2）否：继续执行第 5 步。

5. 是否有--replicate-wild-do-table 选项？

（1）是：将要执行更新的表是否匹配选项中的任意一个值？

- 是：执行更新并退出判断。
- 否：继续执行第 6 步。

（2）否：继续执行第 6 步。

6. 是否有--replicate-wild-ignore-table 选项？

（1）是：将要执行更新的表是否匹配选项中的任意一个值。

- 是：忽略更新并退出判断。
- 否：继续执行第 7 步。

（2）否：继续执行第 7 步。

7. 是否还有另一张表要检测复制过滤？

（1）是：回到第 3 步，继续执行检测。

（2）否：继续执行第 8 步。

8. 是否有--replicate-do-table 或--replicate-wild-do-table 选项？

（1）是：忽略更新并退出判断。

（2）否：执行更新并退出判断。

从库的表级别复制过滤选项的评估示意图如图 13-3 所示。

注意：如果查询语句是一个多表连接的 SQL 语句，且语句中的某个表使用了 --replicate-do-table 选项或--replicate-wild-do-table 选项指定需要被复制，语句中另外的表使用了 --replicate-ignore-table 或 --replicate-wild-ignore-table 选项指定需要被忽略，则基于 statement 的复制将中止，因为从库对事件的执行或忽略操作必须是完整的语句，逻辑上无法执行此操作（但如果二进制日志设置为 row 格式且执行的语句为 DML 语句，则可以正常执行）。这个限制逻辑也适用于基于 row 的复制中的 DDL 语句，因为 DDL 语句始终被记录为 statement 格式（DDL 语句的记录格式与二进制日志格式无关），而不考虑有效的二进制日志格式。

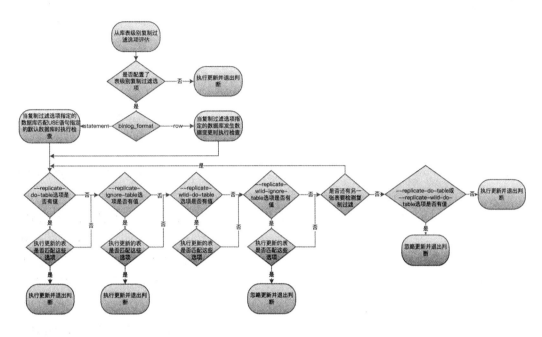

图 13-3

13.4 复制过滤规则的应用

本节提供有关复制过滤选项的不同组合的说明和用法示例,表 13-1 中给出了复制过滤规则选项类型的一些典型组合。

表 13-1

条件(选项类型)	最终结果
无任何复制过滤选项	从库执行从主库接收到的所有事件
--replicate-*-db 库级别选项,但没有表级别选项	从库使用数据库级别选项来决定是接受还是忽略某个数据库中的事件
--replicate-*-table 表级别选项,但没有数据库级别选项	由于没有数据库级别的选项,因此在数据库检查阶段接收所有事件。从库仅根据表选项来决定是否执行或忽略某个表的事件
数据库级别和表级别选项的组合	从库使用数据库级别选项来决定是接受还是忽略某个数据库中的事件。然后,它根据表级别选项评估这些选项允许的所有事件。这样做有时会导致结果看起来违反直觉,并且可能会有所不同,具体取决于你使用的是基于 statement 的复制还是基于 row 的复制

下面是一个复制过滤选项的使用示例,在基于 statement 的复制和基于 row 的复制下的效果完全不同。

```
# 假设我们在主库中有 2 个表,db1 中有 mytbl1 表,db2 中有 mytbl2 表,且从库中配置了如下复制过滤选项(并且没有其他复制过滤选项)
```

```
replicate-ignore-db = db1
replicate-do-table = db2.tbl2

# 现在我们在主库上执行以下语句：
use db1;
INSERT INTO db2.tbl2 VALUES(1);
```

以上示例中，根据二进制日志格式不同，从库的执行结果有很大差异，可能与初始期望不匹配。

- 基于 statement 的复制：USE 语句使 db1 成为默认数据库，因此 --replicate-ignore-db 选项匹配成功。在默认数据库内，INSERT 语句被忽略，不检查表级别的选项。
- 基于 row 的复制：使用基于 row 的复制时，默认数据库对从库复制过滤配置选项没有影响。因此，是否使用 USE 语句切换默认数据库，对如何处理 --replicate-ignore-db 选项指定的库没有区别，此选项指定的数据库与 INSERT 语句更改数据的数据库不匹配，因此从库继续检查表级别选项。--replicate-do-table 指定的表与要更新的表匹配，执行插入行操作。

提示：本章重点介绍复制过滤相关的选项，实际上 MySQL 5.7 独有的 CHANGE REPLICATION FILTER 语法可以支持在从库中动态配置复制过滤规则（但需要重启 SQL 线程才会生效），可以使用如下帮助命令或官方手册查看相关详细信息。

```
# 直接登录 MySQL 5.7 版本的数据库查看帮助信息
mysql> help change replication filter
Name: 'CHANGE REPLICATION FILTER'
Description:
Syntax:
CHANGE REPLICATION FILTER filter[, filter][, ...]

filter:
    REPLICATE_DO_DB = (db_list)
  | REPLICATE_IGNORE_DB = (db_list)
  | REPLICATE_DO_TABLE = (tbl_list)
  | REPLICATE_IGNORE_TABLE = (tbl_list)
  | REPLICATE_WILD_DO_TABLE = (wild_tbl_list)
  | REPLICATE_WILD_IGNORE_TABLE = (wild_tbl_list)
  | REPLICATE_REWRITE_DB = (db_pair_list)

db_list:
    db_name[, db_name][, ...]

tbl_list:
    db_name.table_name[, db_table_name][, ...]
wild_tbl_list:
    'db_pattern.table_pattern'[, 'db_pattern.table_pattern'][, ...]

db_pair_list:
```

```
    (db_pair)[, (db_pair)][, ...]

db_pair:
    from_db, to_db
......
```

关于复制过滤的配置示例详见第 23 章 "将不同数据库的数据复制到不同实例"。

方案篇

第 14 章　搭建异步复制

异步复制是相对于同步复制和半同步复制而言的，这三者之间的区别，可参考第 1 章 "复制的概述"和第 5 章 "半同步复制"。相比于其他两种复制，异步复制的速度最快，且主库性能不受从库复制性能的影响，它也是 MySQL 中最早出现的复制技术。在 MySQL 5.6 之前，无论是同步复制、半同步复制，还是异步复制，都是传统复制（基于二进制日志文件和位置），维护复制拓扑时比较麻烦。MySQL 5.6 及其之后的版本支持 GTID 复制模式，使得对复制拓扑的维护变得非常方便快捷。本章将详细介绍在传统复制和 GTID 复制的两种复制模式下，搭建异步复制的过程，关于传统复制和 GTID 复制的原理，可参考第 4 章。

14.1　操作环境信息

1. 操作系统版本

   CentOS Linux release 7.2.1511（Core）

2. 数据库版本

   MySQL 5.7.26

3. XtraBackup 版本

   percona-xtrabackup-24-2.4.4-1.el7.x86_64

4. sysbench 版本

   sysbench 1.0.7

5. 服务器列表

   - 主库：10.10.30.161
   - 从库：10.10.30.162

6. 复制拓扑规划

   两个数据库实例，配置一主一从

   提示：为方便读者更直观地看到传统复制和 GTID 复制的搭建过程的差异，本章将介绍全新初始化和已有数据两种场景，在每一种场景中分别介绍两种复制模式的搭建步骤。

- 全新初始化：这里指的是未上线数据库实例（或者在即将上线的系统中，只分配了用于安装数据库程序的服务器，但还未安装任何数据库程序），其中没有任何历史数据的情况，所以在搭建过程中就不需要考虑业务是否会受影响。相对于已有数据的场景，这种场景下的复制模式搭建起来方便、快捷、轻量，甚至无须备份步骤。
- 已有数据：这里指的是已上线数据库实例，且当前承载业务的是一个单实例数据库的情况。在搭建过程中需要考虑到业务不能中断，以及对数据库服务器的性能影响的问题。

14.2 全新初始化场景

14.2.1 传统复制

在全新初始化场景下搭建传统复制的复制拓扑，其过程中的关键步骤如图 14-1 所示。

图 14-1

接下来，我们将介绍在全新初始化场景下搭建传统复制的复制拓扑的整个过程。

14.2.1.1 单实例的安装

这里省略初始化安装数据库实例的步骤，详情可参考《千金良方：MySQL 性能优化金字塔法则》的第 1 章 "MySQL 初始化安装、简单安全加固"。

对于全新初始化的场景，可事先分别对主从数据库实例执行初始化安装，并分别启动数据库进程，然后配置复制关系。但是在初始化之前，主从数据库实例需要先在各自的配置文件中配置好系统变量 server_id 和 log_bin，详情可参考 14.4.1 节 "传统复制模式的变量模板"。

对于系统变量 server_id 和 log_bin 的要求，在所有复制场景下都相同，下文不再赘述。

14.2.1.2 配置主从复制关系

假设即将进行复制关系配置的两个数据库实例，都已完成初始化安装且已启动，主库 server_id 为 3306102，从库 server_id 为 3306162。现在我们就可以配置复制关系了。

因为主从库实例各自初始化时，都会产生自己的二进制日志位置信息，尤其是一些自动化管理系统，还可能在数

据库中初始化创建一些心跳表或者管理用户等。所以，建议主从库各自先清理二进制日志（需要使用具有 SUPER 权限的用户来操作，这里使用的是 root 用户。本章中统一使用 root 用户操作所有的复制配置过程，下文不再赘述）：

```
mysql> reset master;
Query OK, 0 rows affected (0.08 sec)
```

在主库中创建复制用户[1]（如果所有实例中都已存在复制用户，则跳过该步骤）：

```
mysql> grant replication slave on *.* to repl@'%' identified by 'replpass';
Query OK, 0 rows affected, 1 warning (0.00 sec)
```

在从库中使用 CHANGE MASTER TO 语句将复制的配置指向主库：

```
# 使用 CHANGE MASTER TO 语句将复制的配置指向主库 10.10.30.161
mysql> change master to master_host='10.10.30.161',master_user='repl',master_password=
'replpass',master_port=3306,master_log_file='mysql-bin.000001',master_log_pos=0;
Query OK, 0 rows affected, 2 warnings (0.03 sec)
......
```

在从库中启动复制线程：

```
mysql> start slave;
Query OK, 0 rows affected (0.05 sec)
```

在从库中查看复制线程的工作状态。关于复制线程工作状态的详情，可参考第 9 章 "通过 PERFORMANCE_SCHEMA 库检查复制信息" 和第 10 章 "通过其他方式检查复制信息"，下文不再赘述。

```
# 这里使用 SHOW SLAVE STATUS 语句查看复制线程的工作状态
mysql> show slave status\G
*************************** 1. row ***************************
               Slave_IO_State: Waiting for master to send event
                  Master_Host: 10.10.30.161
                  Master_User: repl
                  Master_Port: 3306
                Connect_Retry: 60
              Master_Log_File: mysql-bin.000001
          Read_Master_Log_Pos: 442
               Relay_Log_File: mysql-relay-bin.000002
                Relay_Log_Pos: 655
        Relay_Master_Log_File: mysql-bin.000001
# 当看到两个 Yes 时，则表示复制线程启动正常
             Slave_IO_Running: Yes
            Slave_SQL_Running: Yes
......
           Retrieved_Gtid_Set:
            Executed_Gtid_Set:
......
```

现在，采用传统复制模式的一个简单的一主一从复制拓扑就搭建完成了，但还需要执

---

[1] 是指提供给从库与自身建立复制连接，并接收数据变更日志的用户。

行一个重要的验证步骤。下面我们将在主库中写入一些测试数据，在从库中验证数据是否能正常同步。

```
# 在主库中创建测试库、测试表，并插入测试数据
mysql> create database test_repl;
Query OK, 1 row affected (0.01 sec)

mysql> use test_repl
Database changed
mysql> create table test_table(id int);
Query OK, 0 rows affected (0.02 sec)

mysql> insert into test_table values(1),(2);
Query OK, 2 rows affected (0.01 sec)
Records: 2  Duplicates: 0  Warnings: 0

mysql> select * from test_table;select @@server_id;
+------+
| id   |
+------+
|  1   |
|  2   |
+------+
2 rows in set (0.00 sec)

# 确认 server_id = 3306102，角色为主库
+-------------+
| @@server_id |
+-------------+
|   3306102   |
+-------------+
1 row in set (0.00 sec)

# 在从库中查询 test_table 表中是否存在 id = 1 和 id = 2 这两行数据
mysql> use test_repl
Database changed
mysql> show tables;
+--------------------+
| Tables_in_test_repl |
+--------------------+
| test_table         |
+--------------------+
1 row in set (0.00 sec)

# 发现数据被成功同步到从库
mysql> select * from test_table;select @@server_id;
+------+
| id   |
```

```
+------+
| 1    |
| 2    |
+------+
2 rows in set (0.00 sec)

# 确认 server_id = 3306162，角色为从库
+-------------+
| @@server_id |
+-------------+
| 3306162     |
+-------------+
1 row in set (0.00 sec)
```

如果需要搭建双主复制，只需要将主从角色互换，重复上述搭建步骤即可。但在双主复制拓扑中，需要留意一些特定变量的设定，详见 14.4.1 节 "传统复制模式的变量模板"。

注意：

- 如果不配置系统变量 server_id（或将其显式设置为默认值 0），则主库会拒绝来自从库通过复制线程发起的连接请求。
- 为了在使用支持事务的 InnoDB 存储引擎进行复制时，尽可能保证数据的持久性和一致性，应该在配置文件中使用系统变量 innodb_flush_log_at_trx_commit = 1 和 sync_binlog = 1，但启用这样的设置后会影响性能。
- 确保主库未启用系统变量 skip-networking。否则，从库无法与主库通信，造成复制失败。

以上搭建步骤仅适用于数据库中无任何业务数据，且整个搭建过程中数据库无任何写请求的情形。

复制用户的账号和密码以明文的形式保存在复制配置表中，为了安全起见，建议单独创建一个仅具有 REPLICATION SLAVE 权限的用户以最大限度地降低安全风险。

要创建复制用户，除了在主库上执行 GRANT 授权语句时一并创建用户之外，还可以在主库上先使用 CREATE USER 语句创建用户和密码，然后使用 GRANT 语句来授予该用户账号 REPLICATION SLAVE 权限，这种创建用户的方法也是 MySQL 官方推荐的方法，而且在 MySQL 8.0 中不再支持使用 GRANT 语句一并创建用户。语句如下：

```
mysql> CREATE USER 'repl'@'%.example.com' IDENTIFIED BY 'password';
mysql> GRANT REPLICATION SLAVE ON *.* TO 'repl'@'%.example.com';
```

14.2.2 GTID 复制

在全新初始化场景下搭建 GTID 复制的复制拓扑，其过程中的关键步骤如图 14-2 所示。

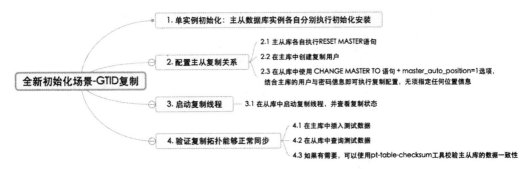

图 14-2

接下来，我们将介绍在全新初始化场景下搭建 GTID 复制的复制拓扑的整个过程。

14.2.2.1　单实例的安装

这里省略初始化安装数据库实例的步骤，其与 14.2.1.1 节"单实例的安装"中的类似。但在初始化数据库实例之前，需要先配置好 GTID 相关的变量，可参考 14.4.2 节"GTID 复制模式的变量模板"。

14.2.2.2　配置主从复制关系

假设即将配置复制关系的两个数据库实例，都已完成初始化安装且已启动，主库 server_id 为 3306102，从库 server_id 为 3306162。现在就可以配置复制关系了。

```
# 主从库各自先清理二进制日志
mysql> reset master;
Query OK, 0 rows affected (0.08 sec)
```

在主库中创建复制用户（如果复制用户已存在，则跳过此步骤）：

```
mysql> grant replication slave on *.* to repl@'%' identified by 'replpass';
Query OK, 0 rows affected, 1 warning (0.00 sec)
```

在从库中使用 CHANGE MASTER TO 语句将复制的配置指向主库：

```
# 使用 CHANGE MASTER TO 语句将复制的配置指向主库 10.10.30.161，这里与基于二进制日志位置的复制的配
置方法不一样，使用选项 master_auto_position = 1 替换掉选项 master_log_file 和 master_log_pos
mysql> change master to master_host='10.10.30.161', master_user='repl', master_
password='replpass', master_port=3306, master_auto_position=1;
Query OK, 0 rows affected, 2 warnings (0.01 sec)
......
```

在从库中启动复制线程：

```
mysql> start slave;
Query OK, 0 rows affected (0.04 sec)
```

在从库中查看复制线程的工作状态：

```
# 这里使用 SHOW SLAVE STATUS 语句查看复制线程的工作状态
mysql> show slave status\G
*************************** 1. row ***************************
```

```
            Slave_IO_State: Waiting for master to send event
               Master_Host: 10.10.30.161
               Master_User: repl
               Master_Port: 3306
             Connect_Retry: 60
           Master_Log_File: mysql-bin.000001
       Read_Master_Log_Pos: 154
            Relay_Log_File: mysql-relay-bin.000002
             Relay_Log_Pos: 367
     Relay_Master_Log_File: mysql-bin.000001
# 当看到两个 Yes 时,就表示复制线程启动正常
          Slave_IO_Running: Yes
         Slave_SQL_Running: Yes
......
```

现在,基于 GTID 复制模式的一个简单的一主一从复制拓扑就搭建完成了,但还需要执行一个重要的验证步骤,其与 14.2.1.2 节"配置主从复制关系"中的类似,这里不再赘述。但是,需要留意在复制状态中是否能看到 GTID 信息,如下:

```
# 通过 SHOW SLAVE STATUS 语句的输出可以发现,Retrieved_Gtid_Set 和 Executed_Gtid_Set 列出现了 GTID 值
mysql> show slave status\G
*************************** 1. row ***************************
            Slave_IO_State: Waiting for master to send event
               Master_Host: 10.10.30.161
               Master_User: repl
......
# 如果 GTID 复制模式正常工作,那么在 Retrieved_Gtid_Set 和 Executed_Gtid_Set 列中就可以看到 GTID 的状态信息
        Retrieved_Gtid_Set: f3372787-0719-11e8-af1f-0025905b06da:1-3
         Executed_Gtid_Set: f3372787-0719-11e8-af1f-0025905b06da:1-3
......
```

如果需要搭建双主复制,只需要将主从角色互换,重复上述搭建步骤即可。但在双主复制拓扑中,需要留意一些变量的设定,详见 14.4.2 节 "GTID 复制模式的变量模板"。

注意:本节的注意事项与 14.2.1.2 节 "配置主从复制关系" 的相同。

14.3 已有数据场景

已有数据场景下的复制拓扑搭建步骤与全新初始化场景下的有所不同,关键的不同点在于,前者需要考虑已有数据的备份和模拟业务的在线问题。

对于备份,我们在本章中选用了 MySQL 安装包中自带的 mysqldump 工具和 Percona 的 XtraBackup 工具:

- 关于 mysqldump 工具的详细介绍,可参考《千金良方:MySQL 性能优化金字塔法则》一书的第 48 章 "MySQL 主流备份工具之 mysqldump 详解"。

- 关于 XtraBackup 工具的详细介绍，可参考《千金良方：MySQL 性能优化金字塔法则》一书的第 49 章"MySQL 主流备份工具之 XtraBackup 详解"。

对于模拟业务在线，我们在本章中选用了 sysbench 工具来持续对主库增加写入压力。关于 sysbench 工具的详细介绍，可参考《千金良方：MySQL 性能优化金字塔法则》一书的第 44 章"sysbench 数据库压测工具"。

14.3.1 传统复制

在已有数据场景下搭建传统复制的复制拓扑，其过程中的关键步骤如图 14-3 所示。

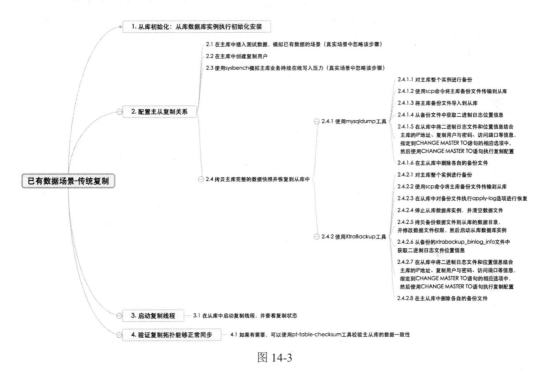

图 14-3

接下来，我们将介绍在已有数据场景下搭建传统复制的复制拓扑的整个过程。

14.3.1.1 单实例的安装

这里省略初始化数据库的步骤，其与 14.2.1.1 节"单实例的安装"中的类似，但在初始化数据库实例之前，需要先配置好传统复制相关的变量，详见 14.4.1 节"传统复制模式的变量模板"。

14.3.1.2 配置主从复制关系

假设即将配置复制关系的从库数据库实例已经完成初始化安装且已启动，主库 server_id 为 3306102，从库 server_id 为 3306162。现在就可以配置复制关系了。注意，这里千万不能像全新初始化场景那样去执行 RESET MASTER 语句，这会导致用于记录数据变更的二进制日志全部被清理。

首先，在主库中插入一些测试数据，用于模拟主库已有数据的场景，由于是模拟场景，所以在真实环境中需要忽略该步骤。

```
# 登录主库，创建 sysbench 工具需要用到的数据库和用户
mysql> create database sysbench;
Query OK, 1 row affected (0.00 sec)

mysql> grant all on *.* to sysbench@'%' identified by 'sysbench';
Query OK, 0 rows affected, 1 warning (0.00 sec)

# 使用 sysbench 造数
[root@localhost ~]# sysbench --db-driver=mysql --time=180 --report-interval=1 --mysql-host=10.10.30.161  --mysql-port=3306  --mysql-user=sysbench  --mysql-password= sysbench --mysql-db=sysbench --tables=8 --table-size=5000 --db-ps-mode=disable oltp_read_write prepare --threads=8
......
Creating table 'sbtest2'...
......
Creating table 'sbtest6'...
Inserting 5000 records into 'sbtest6'
......
Inserting 5000 records into 'sbtest5'
Creating a secondary index on 'sbtest1'...
......
Creating a secondary index on 'sbtest5'...
```

在主库中创建复制用户（如果已存在复制用户，则跳过该步骤）：

```
mysql> grant replication slave on *.* to repl@'%' identified by 'replpass';
Query OK, 0 rows affected, 1 warning (0.00 sec)
```

使用 sysbench 工具模拟主库业务在线持续写入的场景（这是模拟步骤，在真实环境中应忽略）：

```
    [root@localhost  ~]#  sysbench  --db-driver=mysql  --time=99999  --report-interval=1
--mysql-host=10.10.30.161 --mysql-port=3306 --mysql-user=sysbench --mysql-password=sysbench
--mysql-db=sysbench --tables=8 --table-size=5000 --db-ps-mode=disable oltp_read_write run
--threads=1
......
Threads started!
[ 1s ] thds: 1 tps: 216.53 qps: 4344.60 (r/w/o: 3044.41/585.73/714.45) lat (ms,95%): 5.18 err/s: 0.00 reconn/s: 0.00
[ 2s ] thds: 1 tps: 243.11 qps: 4862.21 (r/w/o: 3403.55/737.34/721.33) lat (ms,95%): 4.57 err/s: 0.00 reconn/s: 0.00
[ 3s ] thds: 1 tps: 254.99 qps: 5104.79 (r/w/o: 3570.85/866.96/666.97) lat (ms,95%): 4.25 err/s: 0.00 reconn/s: 0.00
[ 4s ] thds: 1 tps: 268.01 qps: 5360.19 (r/w/o: 3752.13/964.03/644.02) lat (ms,95%): 4.03 err/s: 0.00 reconn/s: 0.00
[ 5s ] thds: 1 tps: 268.00 qps: 5360.00 (r/w/o: 3752.00/952.00/656.00) lat (ms,95%): 3.96 err/s: 0.00 reconn/s: 0.00
```

......

现在，我们需要将主库中的数据拷贝一份到从库中（该过程也称为获取主库的数据快照）。我们需要选择一个合适的备份工具，上文中提到选用的是 MySQL 安装包中自带的 mysqldump 工具和 Percona 的 XtraBackup 工具。接下来，将分别介绍使用这两个工具的完整步骤（在实际的生产环境中，只需要选择其中一个即可。通常我们建议使用 XtraBackup，因为它备份速度快且有更可靠的数据一致性保障）。

1. 使用 mysqldump 工具

使用如下命令对主库的整个实例进行备份（需要使用具备 SUPER 权限的超级管理员账号，避免因为权限不足而导致某些数据库对象无法导出）：

```
[root@localhost ~]# mysqldump -h 10.10.30.161 -usysbench -psysbench -P3306 --single-transaction --master-data=2 --triggers --routines --events --all-databases > /data/backup_`date +%F_%H_%M_%S`.sql
[root@localhost ~]# ls -lh /data/backup_*.sql
-rw-r--r-- 1 root root 171M Jun 16 14:33 /data/backup_2019-06-16_14_32_59.sql
```

在主库中使用 scp 命令将 mysqldump 工具导出的数据文件传输到从库服务器（这里指的是 10.10.30.162）：

```
[root@localhost ~]# scp /data/backup_2019-06-16_14_32_59.sql 10.10.30.162:/data/
......
```

在从库中导入 backup_2019-06-16_14_32_59.sql 文件（需要使用超级管理员账号，避免因为权限不足而导致某些数据库对象无法导入）：

```
[root@localhost ~]# mysql -uroot -ppassword -P3306 < /data/backup_2019-06-16_14_32_59.sql
```

在从库中，从 backup_2019-06-16_14_32_59.sql 文件中获取二进制日志文件位置信息（这个位置信息通常在 30 行以内，不要在整个文件去查找该位置信息）：

```
[root@localhost ~]# head -100 /data/backup_2019-06-16_14_32_59.sql |grep -i 'change master'
-- CHANGE MASTER TO MASTER_LOG_FILE='mysql-bin.000002', MASTER_LOG_POS=465587537;
```

在上述步骤中获取的含有二进制日志文件位置信息的 CHANGE MASTER TO 语句文本的基础上，补充用于指定主库 IP 地址和复制账号信息的选项，然后登录从库数据库中执行，如下：

```
mysql> change master to master_host='10.10.30.161', master_user='repl', master_password='replpass',master_port=3306,MASTER_LOG_FILE='mysql-bin.000002', MASTER_LOG_POS=465587537;
Query OK, 0 rows affected, 2 warnings (0.01 sec)
```

在从库中启动复制，并查看复制状态信息：

```
mysql> start slave;
Query OK, 0 rows affected (0.10 sec)
```

```
mysql> show slave status\G
*************************** 1. row ***************************
               Slave_IO_State: Queueing master event to the relay log
                  Master_Host: 10.10.30.161
                  Master_User: repl
                  Master_Port: 3306
                Connect_Retry: 60
              Master_Log_File: mysql-bin.000003
          Read_Master_Log_Pos: 107911432
               Relay_Log_File: mysql-relay-bin.000002
                Relay_Log_Pos: 5462500
        Relay_Master_Log_File: mysql-bin.000002
# 当看到两个 Yes 时，就表示复制线程正常工作
             Slave_IO_Running: Yes
            Slave_SQL_Running: Yes
......
```

在主库与从库中，各自删除 /data/backup_2019-06-16_14_32_59.sql 文件：

```
[root@localhost ~]# rm -f /data/backup_2019-06-16_14_32_59.sql
```

2. 使用 XtraBackup 工具

使用如下命令对主库的整个实例进行备份（指定的数据库用户要有足够的权限，否则可能因为权限不足而导致备份失败）：

```
# 在 XtraBackup 8.0 之前，主要使用安装包中提供的 innobackupex，XtraBackup 8.0 及之后的版本安装
包中不再提供 innobackupex，统一使用 xtrabackup（innobackupex 是一个为了支持其他非 InnoDB 引擎，对
xtrabackup 做了上层封装的工具。innobackupex 在执行备份时一刀切式采用了 FLUSH TABLES WITH READ LOCK
语句加全局读锁，该语句在极端情况下，容易因获取锁超时而备份失败或者对备份实例造成严重影响。在 MySQL 8.0 之
后，由于 Oracle MySQL 官方已经不再建议使用其他非 InnoDB 引擎，而且对使用非 InnoDB 引擎的字典表统一变更为
使用 InnoDB 引擎，因此，XtraBackup 为了适配该调整，也为了减少使用 FLUSH TABLES WITH READ LOCK 语句
加全局读锁带来的负面影响，不再提供 innobackupex，而是对 xtrabackup 做了优化。当备份目标实例中只有 InnoDB
引擎表时，可以利用 MySQL 8.0 中 InnoDB 引擎的备份锁特性以及事务特性实现不加全局读锁的备份，当备份目标实例
中存在非 InnoDB 引擎表时，才采用 FLUSH TABLES WITH READ LOCK 语句加全局读锁，以便获得一致性备份）
[root@localhost ~]# innobackupex --defaults-file=/etc/my.cnf --user=sysbench --password=
sysbench --no-timestamp --stream=tar ./ | cat - > /data/backup_`date +%Y%m%d`.tar.gz
......
190616 15:37:32 Executing UNLOCK TABLES
190616 15:37:32 All tables unlocked
190616 15:37:32 Backup created in directory '/root'
MySQL binlog position: filename 'mysql-bin.000005', position '295879979'
190616 15:37:32 [00] Streaming backup-my.cnf
190616 15:37:32 [00] ...done
190616 15:37:32 [00] Streaming xtrabackup_info
190616 15:37:32 [00] ...done
xtrabackup: Transaction log of lsn (26745685322) to (26995787873) was copied.
190616 15:37:33 completed OK!
```

在主库中使用 scp 命令将 XtraBackup 工具导出的数据文件传输到从库服务器（这里指的是 10.10.30.162）：

```
[root@localhost ~]# scp /data/backup_20190616.tar.gz 10.10.30.162:/data/
......
```

在从库中解压备份文件，并执行 apply-log 选项，将 redo log 中的数据页变更日志应用到数据库文件中：

```
# 创建一个用于存放恢复数据的目录 test
[root@localhost ~]# mkdir /data/test
[root@localhost ~]# cd /data/test/
[root@localhost test]# mv ../backup_20190616.tar.gz ./
[root@localhost test]# ls -lh
total 2.8G
-rw-r--r-- 1 root root 2.8G Jun 16 15:46 backup_20190616.tar.gz

# 解压
[root@localhost test]# tar xf backup_20190616.tar.gz
[root@localhost test]# ll
total 5357964
-rw-r--r-- 1 root root 2921137152 Jun 16 15:46 backup_20190616.tar.gz
-rw-rw---- 1 root root        466 Jun 16 15:37 backup-my.cnf
drwxr-xr-x 2 root root       4096 Jun 16 15:54 employees
-rw-rw---- 1 root root 2147483648 Jun 16 15:26 ibdata1
drwxr-xr-x 2 root root       4096 Jun 16 15:54 mysql
drwxr-xr-x 2 root root       8192 Jun 16 15:54 performance_schema
drwxr-xr-x 2 root root       8192 Jun 16 15:54 sys
drwxr-xr-x 2 root root       4096 Jun 16 15:54 sysbench
-rw-rw---- 1 root root   10485760 Jun 16 15:26 undo001
......
-rw-rw---- 1 root root   10485760 Jun 16 15:26 undo016
-rw-rw---- 1 root root         27 Jun 16 15:37 xtrabackup_binlog_info
-rw-rw---- 1 root root        121 Jun 16 15:37 xtrabackup_checkpoints
-rw-rw---- 1 root root        527 Jun 16 15:37 xtrabackup_info
-rw-rw---- 1 root root  250105344 Jun 16 15:37 xtrabackup_logfile

# 执行 redo log 的应用
[root@localhost test]# innobackupex --apply-log --use-memory=1G ./
......
InnoDB: Starting shutdown...
InnoDB: Shutdown completed; log sequence number 26995789352
190616 15:56:35 completed OK!
```

由于 XtraBackup 工具拷贝的是完整的数据文件，所以从库中的所有数据文件就都不需要了（它们都必须删除，否则，执行后续的 --copy-back 选项拷贝数据文件到从库的 datadir 时，会报目录非空的错误）。现在，我们将从库的数据库进程停止，并清空数据目录（必须删除的文件有：共享表空间文件、数据表空间文件、undo log、redo log、binlog、relay log 等文件）。

```
# 停止从库数据库进程
[root@localhost test]# service mysqld stop
```

```
Shutting down MySQL.. SUCCESS!

# 清空从库的数据文件（由于本例中是按照生产规范进行配置的，各个数据文件分目录存放，所以需要清空下面这
些目录。如果在你的环境中未分目录存放，只需要清空系统变量 datadir 指定的目录即可）
[root@localhost test]# rm -rf /data/mysqldata1/{binlog,innodb_log,innodb_ts,mydata,
relaylog,undo}/*
```

在从库中用于临时数据恢复的目录下，将数据文件拷贝到 my.cnf 配置文件指定的数据目录：

```
# 先删除临时数据恢复目录下的备份压缩包文件（因为我们已经不再需要它了），否则它在后续步骤中会被一同拷贝
到目标数据库的 datadir 下
[root@localhost test]# rm -f backup_20190616.tar.gz

# 执行拷贝
[root@localhost test]# innobackupex --defaults-file=/etc/my.cnf --copy-back ./
......
190616 16:04:56 [01] ...done
190616 16:04:56 completed OK!
```

在从库中，将系统变量 datadir 指定目录下的所有数据文件的宿主用户改为 mysql 用户（因为 mysqld 进程将使用 mysql 用户启动，如果不修改，将导致启动失败。可以在 my.cnf 配置文件中指定由哪个系统用户启动 mysqld 进程）：

```
[root@localhost test]# chown mysql.mysql /data/mysqldata1/ -R
```

启动从库数据库进程：

```
[root@localhost test]# service mysqld start
Starting MySQL.... SUCCESS!
```

在从库的临时数据恢复目录 test 下，读取 xtrabackup_binlog_info 文件中的内容，该内容就是配置复制时所需的二进制日志文件位置信息：

```
[root@localhost test]# cat xtrabackup_binlog_info
mysql-bin.000005    295879979
```

在上述步骤中获取的含有二进制日志文件位置信息的 CHANGE MASTER TO 语句文本的基础上，补充主库的 IP 地址和用于复制的账号信息，然后登录从库数据库，执行：

```
mysql> CHANGE MASTER TO master_host='10.10.30.161', master_user='repl', master_
password='replpass',master_port=3306,MASTER_LOG_FILE='mysql-bin.000005',
MASTER_LOG_POS=295879979;
Query OK, 0 rows affected, 2 warnings (0.03 sec)
```

在从库中启动复制线程，并查看复制线程的状态信息：

```
mysql> start slave;
Query OK, 0 rows affected (0.12 sec)

mysql>show slave status\G
*************************** 1. row ***************************
```

```
                Slave_IO_State: Queueing master event to the relay log
                   Master_Host: 10.10.30.161
                   Master_User: repl
                   Master_Port: 3306
                 Connect_Retry: 60
               Master_Log_File: mysql-bin.000005
           Read_Master_Log_Pos: 438403370
                Relay_Log_File: mysql-relay-bin.000002
                 Relay_Log_Pos: 3580641
         Relay_Master_Log_File: mysql-bin.000005
# 当看到两个 Yes 时，就表示复制线程正常工作
              Slave_IO_Running: Yes
             Slave_SQL_Running: Yes
......
```

在从库中删除用于恢复临时数据的目录：

```
[root@localhost ~]# rm -rf /data/test/
```

在主库中删除临时备份的文件：

```
[root@localhost ~]# rm -f /data/backup_20190616.tar.gz
```

现在，我们就利用 mysqldump 或者 XtraBackup 备份工具将主库已有的数据备份到了从库，而且通过该备份数据搭建完传统复制的一主一从复制拓扑。此时，可以再观察一段时间，如果从库中的数据能够正常地持续同步，则表明工作稳定。当然，如果有必要，还可以使用 pt-table-checksum 工具对主从库数据的一致性进行校验，有兴趣的读者可自行研究该工具。

最后，停止 sysbench 工具（此为模拟步骤，在真实环境中可忽略）。

如果需要搭建双主复制，则可按照如下步骤操作：

（1）在从库中使用 STOP SLAVE 语句停止复制进程。

（2）使用 FLUSH TABLE WITH READ LOCK 语句加全局读锁。

（3）使用 SHOW MASTER STATUS 语句获取从库的二进制日志位置信息。

（4）在主库中使用 CHANGE MASTER TO 语句反向指定从库的 IP 地址等信息，最后启动复制线程。

提示：需要留意关于双主复制拓扑场景的一些变量设定，详见 14.4.1 节"传统复制模式的变量模板"。

注意：
- 在 14.2.1.2 节"配置主从复制关系"中提到的注意事项这里同样适用。另外，在业务在线的情况下，配置复制关系时需要注意两点，详情可参考微信公众号"沃趣技术"中的文章《mysqldump 与 innobackupex 备份过程你知多少（一）》和《mysqldump 与 innobackupex 备份过程你知多少（二）》。

- 主库中的非事务引擎表在执行备份期间不能发生数据写入，否则可能破坏备份数据的一致性，导致后续在从库中启动复制时报错。针对该问题的解决方法：若使用的是 mysqldump 工具，可以通过去除 --single-transaction 选项，使得备份全程加锁；而 XtraBackup 工具则不存在该问题。

- 在执行数据备份期间，不能对表结构执行变更，否则可能造成备份数据缺失或数据文件拷贝过程报错而中止。针对该问题的解决办法：若使用的是 mysqldump 工具，可以通过规范来规避或者去除 --single-transaction 选项，使得备份全程加锁；若使用的是 XtraBackup 工具，则只能从规范上进行规避，无法全程加锁。

14.3.2　GTID 复制

在已有数据场景下 GTID 复制拓扑，其过程中的关键步骤如图 14-4 所示。

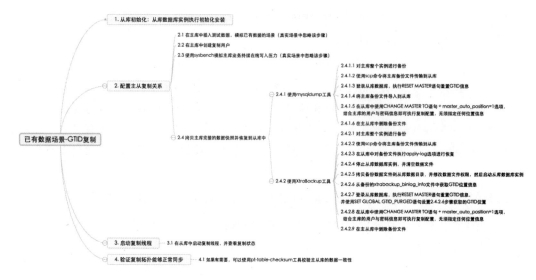

图 14-4

接下来，我们将介绍在已有数据场景下，搭建 GTID 复制的复制拓扑的整个过程。

14.3.2.1　单实例的安装

这里省略初始化数据库的步骤，其与 14.2.1.1 "单实例的安装"中的类似，但在初始化数据库实例之前，需要先配置好 GTID 相关的变量，GTID 的变量配置详见 14.4.2 节 "GTID 复制模式的变量模板"。

14.3.2.2　配置主从复制关系

假设即将配置复制关系的主库和从库数据库实例都已经完成初始化安装且已启动，主库 server_id 为 3306102，从库 server_id 为 3306162。现在，就可以配置复制关系了（注意：这里仍然不能执行 RESET MASTER 语句，因为会导致用于记录数据变更的二进制日志全部被清理）。

首先，在主库中插入一些测试数据，用于模拟主库已有数据的场景（与 14.3.1.2 节 "配置主从复制关系" 中插入测试数据的步骤相同，此处省略）。

在主库中创建复制用户（如果已存在复制用户，则跳过该步骤）：

```
mysql> grant replication slave on *.* to repl@'%' identified by 'replpass';
Query OK, 0 rows affected, 1 warning (0.00 sec)
```

使用 sysbench 工具模拟主库业务在线持续写入的场景（与 14.3.1.2 节 "配置主从复制关系" 中的模拟业务写入命令相同，此处省略）。

现在，需要将主库中的数据拷贝一份到从库中。这里，同样针对 MySQL 安装包中自带的 mysqldump 工具和 Percona 的 XtraBackup 工具分别进行介绍。

1. 使用 mysqldump 工具

使用如下命令对主库的整个实例进行备份（需要使用超级管理员账号，避免因为权限不足而导致某些数据库对象无法导出）：

```
[root@localhost ~]# mysqldump -h 10.10.30.161 -usysbench -psysbench -P3306 --single-transaction --master-data=2 --triggers --routines --events --all-databases > /data/backup_`date +%F_%H_%M_%S`.sql
[root@localhost ~]# ls -lh /data/backup_*.sql
-rw-r--r-- 1 root root 171M Jun 17 09:26 /data/backup_2019-06-17_09_26_29.sql
```

在主库中，使用 scp 命令将 mysqldump 工具导出的数据文件传输到从库服务器（这里指的是 10.10.30.162）：

```
[root@localhost ~]# scp /data/backup_2019-06-17_09_26_29.sql 10.10.30.162:/data/
......
```

在从库中，先找出备份文件 backup_2019-06-17_09_26_29.sql 中的 GTID 信息（这里找出的 GTID 信息是供我们观察和确认用的，没有其他用途）：

```
[root@localhost ~]# head -100 /data/backup_2019-06-17_09_26_29.sql |grep -i gtid
-- GTID state at the beginning of the backup
SET @@GLOBAL.GTID_PURGED='f3372787-0719-11e8-af1f-0025905b06da:1-11322';
```

在从库中导入 backup_2019-06-17_09_26_29.sql 文件（需要使用超级管理员账号，避免因为权限不足而导致某些数据库对象无法导入）：

```
# 从库导入数据之前，需要先登录数据库中执行 RESET MASTER，重置从库自己的 GTID 信息，否则无法执行备份文件中的 SET @@GLOBAL.GTID_PURGED='f3372787-0719-11e8-af1f-0025905b06da:1-11322' 语句（这一点与传统复制配置过程不同）
mysql> reset master;
Query OK, 0 rows affected (0.01 sec)

# 在命令行中使用如下命令导入备份文件数据。注意，由于 mysqldump 工具的备份文件中包含 SET @@GLOBAL.GTID_PURGED='f3372787-0719-11e8-af1f-0025905b06da:1-11322' 语句，所以无须手工执行该语句
[root@localhost ~]# mysql -uroot -ppassword -P3306 < /data/backup_2019-06-17_09_26_29.sql
```

登录从库数据库，执行如下 CHANGE MASTER TO 语句（可以看到，GTID 复制不再

需要二进制日志文件的位置信息，这里也不需要指定 GTID 信息，这就是 GTID 复制模式的魅力所在）：

```
mysql> CHANGE MASTER TO master_host='10.10.30.161',master_user='repl',master_password='replpass',master_port=3306,master_auto_position=1;
Query OK, 0 rows affected, 2 warnings (0.01 sec)
```

在从库中启动复制线程，并查看复制线程的状态信息：

```
mysql> start slave;
Query OK, 0 rows affected (0.06 sec)

mysql> show slave status\G
*************************** 1. row ***************************
               Slave_IO_State: Queueing master event to the relay log
                  Master_Host: 10.10.30.161
                  Master_User: repl
                  Master_Port: 3306
                Connect_Retry: 60
              Master_Log_File: mysql-bin.000002
          Read_Master_Log_Pos: 199188917
               Relay_Log_File: mysql-relay-bin.000002
                Relay_Log_Pos: 5025010
        Relay_Master_Log_File: mysql-bin.000002
# 当看到两个 Yes 时，表示复制线程正常工作
             Slave_IO_Running: Yes
            Slave_SQL_Running: Yes
......
```

在主从库中删除 /data/backup_2019-06-17_09_26_29.sql 文件：

```
[root@localhost ~]# rm -f /data/backup_2019-06-17_09_26_29.sql
```

2. 使用 XtraBackup 工具

使用如下命令对主库的整个实例进行备份（指定的数据库用户要有足够的权限，否则可能因为权限不足而导致备份失败）：

```
[root@localhost ~]# innobackupex --defaults-file=/etc/my.cnf --user=sysbench --password=sysbench --no-timestamp --stream=tar ./ | cat - > /data/backup_`date +%Y%m%d`.tar.gz
......
190617 10:14:08 Executing UNLOCK TABLES
190617 10:14:08 All tables unlocked
190617 10:14:08 Backup created in directory '/root'
MySQL binlog position: filename 'mysql-bin.000003', position '110564334', GTID of the last change 'f3372787-0719-11e8-af1f-0025905b06da:1-285745'
190617 10:14:08 [00] Streaming backup-my.cnf
190617 10:14:08 [00]        ...done
190617 10:14:08 [00] Streaming xtrabackup_info
190617 10:14:08 [00]        ...done
xtrabackup: Transaction log of lsn (27680685468) to (27732800768) was copied.
```

```
190617 10:14:09 completed OK!
```

在主库中,使用 scp 命令将 XtraBackup 工具导出的数据文件传输到从库服务器(这里指的是 10.10.30.162):

```
[root@localhost ~]# scp /data/backup_20190617.tar.gz 10.10.30.162:/data/
......
```

在从库中解压备份文件,并执行 apply-log 选项,将 redo log 中的数据页变更日志应用到数据库文件中:

```
# 创建一个用于存放恢复数据的目录 test
[root@localhost ~]# mkdir /data/test
[root@localhost ~]# cd /data/test/
[root@localhost test]# mv ../backup_20190617.tar.gz ./
[root@localhost test]# ls -lh
total 2.6G
-rw-r--r-- 1 root root 2.6G Jun 17 10:15 backup_20190617.tar.gz

# 解压
[root@localhost test]# tar xf backup_20190617.tar.gz
[root@localhost test]# ll
total 4971272
-rw-r--r-- 1 root root 2723149824 Jun 17 10:15 backup_20190617.tar.gz
-rw-rw---- 1 root root        466 Jun 17 10:14 backup-my.cnf
drwxr-xr-x 2 root root       4096 Jun 17 10:16 employees
-rw-rw---- 1 root root 2147483648 Jun 17 10:12 ibdata1
drwxr-xr-x 2 root root       4096 Jun 17 10:16 mysql
drwxr-xr-x 2 root root       8192 Jun 17 10:16 performance_schema
drwxr-xr-x 2 root root       8192 Jun 17 10:16 sys
drwxr-xr-x 2 root root       4096 Jun 17 10:16 sysbench
-rw-rw---- 1 root root   10485760 Jun 17 10:12 undo001
......
-rw-rw---- 1 root root   10485760 Jun 17 10:12 undo016
-rw-rw---- 1 root root         73 Jun 17 10:14 xtrabackup_binlog_info
-rw-rw---- 1 root root        121 Jun 17 10:14 xtrabackup_checkpoints
-rw-rw---- 1 root root        600 Jun 17 10:14 xtrabackup_info
-rw-rw---- 1 root root   52118016 Jun 17 10:14 xtrabackup_logfile

# 执行 redo log 的应用
[root@localhost test]# innobackupex --apply-log --use-memory=1G ./
......
InnoDB: Starting shutdown...
InnoDB: Shutdown completed; log sequence number 27732802088
190617 10:17:32 completed OK!
```

现在,我们将从库的数据库进程停止,并清空数据目录(必须删除的文件有:共享表空间、数据表空间、undo log、redo log、binlog、relay log):

```
# 停止从库数据库进程
```

```
[root@localhost test]# service mysqld stop
Shutting down MySQL.. SUCCESS!

# 清空从库的数据文件
[root@localhost test]# rm -rf /data/mysqldata1/ {binlog,innodb_log,innodb_ts,mydata,
relaylog,undo}/*
```

将从库中临时数据恢复目录下的数据文件拷贝到 my.cnf 配置文件指定的目录中：

```
# 先删除临时数据恢复目录下的备份压缩包文件（因为已经不再需要它了），否则会在后续步骤中它会被一同拷贝到
目标数据库的 datadir 选项指定的目录下
[root@localhost test]# rm -f backup_20190617.tar.gz

# 执行拷贝
[root@localhost test]# innobackupex --defaults-file=/etc/my.cnf --copy-back ./
......
190617 10:19:08 [01] ...done
190617 10:19:08 completed OK!
```

在从库中，将数据库系统变量 datadir 指定目录下所有数据文件的宿主用户改为 mysql 用户（如果不修改，将导致 mysqld 进程启动失败，因为该进程会使用 mysql 用户启动。可以在 my.cnf 配置文件中指定由哪个系统用户启动 mysqld 进程）：

```
[root@localhost test]# chown mysql.mysql /data/mysqldata1/ -R
```

启动从库数据库进程：

```
[root@localhost test]# service mysqld start
Starting MySQL.... SUCCESS!
```

在从库的临时数据恢复目录 test 下，读取 xtrabackup_binlog_info 文件中的内容：

```
[root@localhost test]# cat xtrabackup_binlog_info
mysql-bin.000003  110564334  f3372787-0719-11e8-af1f-0025905b06da:1-285745
# 在主库启用 GTID 复制模式的情况下，该文件中会同时包含二进制日志文件位置信息和 GTID 信息，这里只需要截
取 GTID 信息串 "f3372787-0719-11e8-af1f-0025905b06da:1-285745" 即可
```

用上述步骤中截取的 GTID 信息串生成如下所示的 SET GLOBAL GTID_PURGED 语句，并在从库中执行：

```
# 执行 RESET MASTER 语句,否则可能会因为从库带有自己的 GTID 信息而导致后续的 SET GLOBAL GTID_PURGED
= 'xx'语句无法执行
mysql> reset master;
Query OK, 0 rows affected (0.00 sec)

# 执行 SET GLOBAL GTID_PURGED = 'xx'语句
mysql> SET GLOBAL GTID_PURGED='f3372787-0719-11e8-af1f-0025905b06da:1-285745';
Query OK, 0 rows affected (0.00 sec)
```

登录从库，执行如下 CHANGE MASTER TO 语句：

```
mysql> CHANGE MASTER TO master_host='10.10.30.161',master_user='repl',master_password=
'replpass',master_port=3306,master_auto_position=1;
```

```
Query OK, 0 rows affected, 2 warnings (0.02 sec)
```

在从库中启动复制线程，并查看复制线程的状态信息：

```
mysql> start slave;
Query OK, 0 rows affected (0.06 sec)

mysql> show slave status\G
*************************** 1. row ***************************
               Slave_IO_State: Queueing master event to the relay log
                  Master_Host: 10.10.30.161
                  Master_User: repl
                  Master_Port: 3306
                Connect_Retry: 60
              Master_Log_File: mysql-bin.000003
          Read_Master_Log_Pos: 275295205
               Relay_Log_File: mysql-relay-bin.000002
                Relay_Log_Pos: 3864228
        Relay_Master_Log_File: mysql-bin.000003
# 当看到两个 Yes 时，表示复制线程正常工作
             Slave_IO_Running: Yes
            Slave_SQL_Running: Yes
......
```

在从库中删除用于恢复临时数据的目录：

```
[root@localhost ~]# rm -rf /data/test/
```

在主库中删除临时备份的文件：

```
[root@localhost ~]# rm -f /data/backup_20190617.tar.gz
```

现在，我们已经通过 mysqldump 或者 XtraBackup 备份工具成功将主库已有的数据备份到从库，而且通过该备份数据成功搭建 GTID 复制的一主一从复制拓扑。此时，可以再观察一段时间，如果从库中的数据能够正常地持续同步，则表明其工作稳定。当然，如有必要，还可以使用 pt-table-checksum 工具校验主从库数据的一致性，有兴趣的读者可自行研究该工具。

最后，停止用 sysbench 工具加压（此为模拟步骤，在真实环境中可忽略）。

如果需要搭建双主复制，则直接在主库中使用 CHANGE MASTER TO 语句反向指定从库的 IP 地址等信息，启动复制线程即可。

提示：需要留意关于双主复制拓扑场景的一些变量设定，详见 14.4.2 节 "GTID 复制模式的变量模板"。

注意：
- 14.3.1 节中的注意事项在本节也适用。
- 如果复制拓扑已经处于运行状态，而且是基于二进制日志文件位置的复制模式，想要将其转换为 GTID 复制模式，可参考第 17 章 "复制模式切换"。

14.4 变量模板

关于变量模板中所有服务端系统变量的含义详解，可参考《千金良方：MySQL 性能优化金字塔法则》一书的附录 C "MySQL 常用配置变量和状态变量详解"。

14.4.1 传统复制模式的变量模板

1. 主库[1]

```
[root@localhost ~]# cat /etc/my.cnf
[client]
loose_default-character-set = utf8
port=3306
socket=/home/mysql/data/mysqldata1/sock/mysql.sock
user=admin

[mysqldump]
quick
max_allowed_packet = 2G
default-character-set = utf8

[mysql]
no-auto-rehash
show-warnings
prompt="\\u@\\h : \\d \\r:\\m:\\s> "
default-character-set = utf8

[myisamchk]
key_buffer = 512M
sort_buffer_size = 512M
read_buffer = 8M
write_buffer = 8M

[mysqlhotcopy]
interactive-timeout

[mysqld_safe]
user=mysql
open-files-limit = 65535

[mysqld]
read_only=on
default-storage-engine = INNODB
character-set-server=utf8
```

---

[1] 这是生产环境模板变量，对服务器的硬件配置要求较高，如果使用该模板在低端配置的服务器中做练习，则需要调整与路径、磁盘文件大小及内存 buffer 相关的变量。

```
collation_server = utf8_bin

user=mysql
port=3306
socket=/home/mysql/data/mysqldata1/sock/mysql.sock
pid-file=/home/mysql/data/mysqldata1/sock/mysql.pid
datadir=/home/mysql/data/mysqldata1/mydata
tmpdir=/home/mysql/data/mysqldata1/tmpdir

skip-name-resolve
skip_external_locking

lower_case_table_names=1
event_scheduler=0
back_log=512
default-time-zone='+8:00'

max_connections = 3000
max_connect_errors=99999
max_allowed_packet = 64M
max_heap_table_size = 8M
max_length_for_sort_data = 16k

wait_timeout=172800
interactive_timeout=172800

net_buffer_length = 8K
read_buffer_size = 2M
read_rnd_buffer_size = 2M
sort_buffer_size = 2M
join_buffer_size = 4M
binlog_cache_size = 2M

table_open_cache = 4096
table_definition_cache = 4096
thread_cache_size = 512
tmp_table_size = 8M

query_cache_size=0
query_cache_type=OFF
explicit_defaults_for_timestamp=ON
metadata_locks_hash_instances=32

### Logs related settings ###
log-error=/home/mysql/data/mysqldata1/log/error.log
long_query_time = 1
slow_query_log
slow_query_log_file=/home/mysql/data/mysqldata1/slowlog/slow-query.log
```

```
log_slow_slave_statements

### Replication related settings ###
#### For Master
# 用于在同一个复制拓扑中唯一标识每个数据库实例，有效值为 1 ~ 4294967295（$2^{32}-1$）
server-id=3306102
# 主从复制是基于写入二进制日志中的数据逻辑变更记录来实现数据同步的，所以必须启用二进制日志记录功能，
否则无法实现主从复制。要注意，该系统变量为只读变量，启用/禁用该系统变量时需要重启数据库实例方可生效
log-bin=/home/mysql/data/mysqldata1/binlog/mysql-bin
binlog-format=ROW
binlog-checksum=CRC32
max_binlog_size = 512M
expire_logs_days=15
sync_binlog=1
master-verify-checksum=1
# 如果需要搭建双主复制拓扑，则需要设置下面两个自增变量
auto_increment_increment=2
auto_increment_offset=1

#### For Slave
relay-log=/home/mysql/data/mysqldata1/relaylog/mysql-relay-bin
relay-log-recovery=1
slave-sql-verify-checksum=1
log_bin_trust_function_creators=1
log_slave_updates=1
slave-net-timeout=10

### MyISAM Specific options ###
key_buffer_size = 8M
bulk_insert_buffer_size = 8M
myisam_sort_buffer_size = 64M
myisam_max_sort_file_size = 10G
myisam_repair_threads = 1
myisam_recover_options=force

### InnoDB Specific options ###
#### Data options
innodb_data_home_dir = /home/mysql/data/mysqldata1/innodb_ts
innodb_data_file_path = ibdata1:2048M:autoextend
innodb_file_per_table
innodb_file_format = barracuda
innodb_file_format_max = barracuda
innodb_file_format_check = ON
innodb_strict_mode = 1
innodb_flush_method = O_DIRECT
innodb_checksum_algorithm=crc32
innodb_autoinc_lock_mode=2
```

```
#### Buffer Pool options
innodb_buffer_pool_size = 9300M
innodb_buffer_pool_instances=8
innodb_max_dirty_pages_pct = 75
innodb_adaptive_flushing = ON
innodb_flush_neighbors = 0
innodb_lru_scan_depth = 4096
#innodb_change_buffering = inserts
innodb_old_blocks_time = 1000

#### Redo options
innodb_log_group_home_dir = /home/mysql/data/mysqldata1/innodb_log
innodb_log_buffer_size = 64M
innodb_log_file_size = 2G
innodb_log_files_in_group = 2
innodb_flush_log_at_trx_commit = 1
innodb_fast_shutdown = 2
innodb_support_xa = ON

#### Transaction options
innodb_thread_concurrency = 64
innodb_lock_wait_timeout = 120
innodb_rollback_on_timeout = 1
transaction_isolation = READ-COMMITTED

#### IO options
innodb_read_io_threads = 8
innodb_write_io_threads = 16
innodb_io_capacity = 20000
innodb_use_native_aio = 1

#### Undo options
innodb_undo_directory = /home/mysql/data/mysqldata1/undo/
innodb_undo_tablespaces=16
innodb_purge_threads = 4
innodb_purge_batch_size = 512
innodb_max_purge_lag = 65536

#### >=5.6 mariadb warning
table_open_cache_instances = 2
master-info-repository=TABLE
relay-log-info-repository=TABLE
slave_pending_jobs_size_max=128M
binlog-rows-query-log-events=1

#### >=5.7 mariadb error
#super_read_only=on
slave_parallel_workers=16
```

```
slave_parallel_type=LOGICAL_CLOCK
log_timestamps=SYSTEM
slave_preserve_commit_order=ON


#### MySQL >=5.7 || MariaDB >=10.2
secure_file_priv=null
innodb_undo_log_truncate=ON
innodb_adaptive_hash_index_parts=32
```

2. 从库

```
# 系统变量server-id指定的值需要保证在整个复制拓扑中是唯一的,即不能出现两个不同数据库实例使用系统变
量server-id指定相同的值
  server-id=3306162
# 一主一从的复制拓扑中,从库可不启用系统变量log-bin,这样从库可以降低磁盘吞吐量,但如果有备份需求或
者该实例用于双主复制中的备主,则需要启用log-bin
  log-bin=/home/mysql/data/mysqldata1/binlog/mysql-bin

# 如果需要搭建双主复制拓扑,则需要设置下面两个自增变量
  auto_increment_increment=2
  auto_increment_offset=2
```

提示：如果要配置多线程复制，则不同的多线程复制类型，相关的系统变量的设定也不同，详情可参考第 6 章 "多线程复制"。

14.4.2　GTID 复制模式的变量模板

1. 主库

在 14.4.1 节中主库的变量模板基础上，增加如下变量：

```
gtid-mode=on
enforce-gtid-consistency=true
# 如果需要搭建双主架构,主库根据需要可以启用或禁用如下变量
log_slave_updates=on
```

2. 从库

在 14.4.1 节中从库的变量模板基础上，增加如下变量：

```
gtid-mode=on
enforce-gtid-consistency=true
# 根据需要可以启用或禁用如下变量
log_slave_updates=on
```

提示：

- 对于系统变量 log_slave_updates，从库可以启用，也可以不启用。如启用，则会增加二进制日志文件的写入量；如禁用，则会增加 mysql.gtid_executed 表的数据写入量。关于二者的详细区别，可参考第 4 章 "传统复制与 GTID 复制"。

- 在 MySQL 8.0 中，账号的默认密码插件为 caching_sha2_password。同时，搭建复制

时，从库的 I/O 线程需要使用 SSL 方式与主库进行连接，如果不想配置 SSL，可以在创建复制账号时，指定使用 MySQL 8.0 之前的默认密码插件 mysql_native_password，代码如下：

```
mysql> create user 'repl'@'%' IDENTIFIED WITH mysql_native_password BY 'replpass';
Query OK, 0 rows affected (0.01 sec)

mysql> grant replication slave on *.* to 'repl'@'%';
Query OK, 0 rows affected (0.00 sec)
```

第 15 章 搭建半同步复制

第 14 章详细介绍了如何搭建异步复制。虽然异步复制性能较高，但由于主库在提交事务时，不关心从库是否成功接收对应事务的二进制日志，埋下了安全隐患。当主库发生宕机故障时，无从得知主库中已完成提交的事务对应的二进制日志是否已被从库接收，因此无法保证主库宕机时不丢失数据。如何解决数据丢失的问题呢？目前主流的开源解决方案大致有如下几种。

- 使用共享存储。使用开源或者商业的共享存储作为数据盘，将整个复制拓扑中数据库实例的数据都放到共享存储中。这样，任何一个数据库实例发生宕机故障，数据都不会丢失。只需要另外寻找一台主机，重新挂载宕机的数据库实例的数据目录，启动数据库即可恢复数据。

- 使用集群架构，指的是 PXC（Percona Xtradb Cluster）、MGR（MySQL Group Replication）、MGC（MariaDB Galera Cluster）等开源的集群架构解决方案，它们都需要至少部署 3 个节点来组成一个数据库实例集群。任何一个数据库实例发生宕机故障时，剩余的 2 个节点会进行仲裁，剔除故障节点，集群继续对外提供服务，数据不会丢失。当故障节点恢复后，重新加入集群即可。

- 使用半同步复制（MySQL 5.7 及之后的版本中的半同步复制）。使用增强半同步复制插件实现，能够确保主库中提交事务的二进制日志被从库成功接收。当主库（写节点）发生宕机故障时，将从库提升为写节点（拥有最新数据的从库），继续对外提供服务，数据不会丢失。当发生故障的主库恢复后，重新加入集群即可。

除了上述开源解决方案之外，目前市面上还出现一些采用新概念及新架构的商业解决方案，例如基于 Aurora 架构的 Amazon Aurora、PolarDB、X-DB 等，有兴趣的读者可自行研究。

本章将介绍性价比最高、实现方式最简单、最容易安装及部署的半同步复制解决方案。

15.1 半同步复制插件的安装和配置环境要求

半同步复制是使用额外的插件实现的，并非 MySQL 中内置的，因此必须单独安装相应插件之后才能启用。安装插件后，可以通过与之关联的系统变量来控制半同步复制的启用或禁用。在安装关联的插件之前，与半同步复制相关的系统变量不可用（因为这些系统变量是半同步复制插件携带的）。

要使用半同步复制，必须遵循以下要求：

- MySQL Server 要支持动态加载安装插件。可以通过检查 have_dynamic_loading 系统变量的值来判断（语句为：show variables like 'have_dynamic_loading'）。若为 YES，表示支持动态加载，否则为不支持。。
- 已经配置好异步复制且复制处于可用状态。关于搭建异步复制的详细过程，可参考第 14 章"搭建异步复制"。
- 不能配置多个复制通道（多源复制）。半同步复制只与默认复制通道兼容（准确地说，是半同步复制对多源复制的支持不够好。经测试，发现可以配置并正常拉起多个半同步复制通道，但是启动或停止其中一个通道时，另一个通道会受到影响）。

要设置半同步复制，可以使用以下几个指令：

- INSTALL PLUGIN：安装半同步复制插件。
- SET GLOBAL：修改半同步复制系统变量。例如，主库修改半同步复制的超时等待时间。
- STOP SLAVE、START SLAVE：重启已有的复制，如果是在已运行的异步复制基础上安装半同步复制插件，则安装及配置完之后需要重启复制（重启 I/O 线程即可），否则插件不生效。

提示：执行以上指令时用户需要有 SUPER 权限。

15.2 半同步复制插件的安装和配置

15.2.1 关键步骤

半同步复制插件的安装和配置过程中的关键步骤如图 15-1 所示。

图 15-1

15.2.2 详细过程

要查看已经安装了哪些插件，可以使用 SHOW PLUGINS 语句，或查询 information_schema.plugins 表，如下：

```
# 使用 SHOW PLUGINS 语句查看
mysql> show plugins;
......

# 查看 information_schema.plugins 表
mysql> select plugin_name, plugin_status from information_schema.plugins where plugin_name like '%semi%';
......
```

MySQL 发行的安装包中提供了用于安装半同步复制的插件库文件，插件库文件分为主库端和从库端。在配置半同步复制时，要在主从库中分别安装对应主从角色的半同步复制插件，插件位于 MySQL 系统变量 plugin_dir 指定的目录下（如果不在默认的插件目录下，则需要在启动 MySQL Server 之前通过系统变量 plugin_dir 来指定），插件库文件名的前缀分别为 semisync_master 和 semisync_slave，但库文件名的后缀因平台而异（例如，对于 UINX 和类 UNIX 系统为.so，对于 Windows 系统为.dll）。

要加载半同步复制插件，请在主从库中分别执行如下语句（根据不同的平台调整库文件名后缀）：

```
# 主库
mysql> install plugin rpl_semi_sync_master soname 'semisync_master.so';
Query OK, 0 rows affected (0.04 sec)

# 每个从库
mysql> install plugin rpl_semi_sync_slave soname 'semisync_slave.so';
Query OK, 0 rows affected (0.02 sec)

# 如果安装插件时出现如下报错信息，则必须安装 libimf
mysql> install plugin rpl_semi_sync_master soname 'semisync_master.so';
ERROR 1126 (HY000): Can't open shared library
'/usr/local/mysql/lib/plugin/semisync_master.so'
(errno: 22 libimf.so: cannot open shared object file:
No such file or directory)

## 可以在 Oracle MySQL 官方手册中搜索关键字 libimf，根据相关的提示找到获取 libimf 安装包的下载链接
## 提示：如果半同步复制插件无法完成初始化，则可以检查数据库错误日志以获取诊断消息
```

检查半同步复制插件的安装信息：

```
# 主库
## 从 SHOW PLUGINS 语句输出的 Name 为 rpl_semi_sync_master 的信息行中可以看到,Status 为 ACTIVE,
表示半同步复制插件在主库中安装成功，且当前处于可用状态
mysql> show plugins;
```

```
+--------------+----------+-------------+--------------------+---------+
| Name | Status | Type | Library | License |
+--------------+----------+-------------+--------------------+---------+
......
| rpl_semi_sync_master |ACTIVE|REPLICATION|semisync_master.so | GPL |
+--------------+----------+-------------+--------------------+---------+
45 rows in set (0.00 sec)
```

查看 information_schema.plugins 表。从输出信息中可以看到，PLUGIN_STATUS 为 ACTIVE，表示半同步复制插件在主库中安装成功，且当前处于可用状态

```
mysql> select plugin_name, plugin_status from information_schema.plugins where plugin_name like '%semi%';
+----------------------+---------------+
| PLUGIN_NAME          | PLUGIN_STATUS |
+----------------------+---------------+
| rpl_semi_sync_master | ACTIVE        |
+----------------------+---------------+
1 row in set (0.00 sec)
```

从库
SHOW PLUGINS 语句
```
mysql> show plugins;
+-----------+----------+--------------+--------------------+---------+
| Name      | Status   | Type         | Library            | License |
+-----------+----------+--------------+--------------------+---------+
......
| rpl_semi_sync_slave|ACTIVE|REPLICATION | semisync_slave.so | GPL |
+-----------+----------+--------------+--------------------+---------+
45 rows in set (0.01 sec)
```

information_schema.plugins 表
```
mysql> select plugin_name, plugin_status from information_schema.plugins where plugin_name like '%semi%';
+----------------------+---------------+
| PLUGIN_NAME          | PLUGIN_STATUS |
+----------------------+---------------+
| rpl_semi_sync_slave  | ACTIVE        |
+----------------------+---------------+
1 row in set (0.00 sec)
```

半同步复制插件安装后，默认是禁用的，必须在主从库分别启用半同步复制插件且都生效之后，才能在复制拓扑中启用半同步复制。如果只在一端启用，则半同步复制最终会被降级运行为异步复制。要启用或禁用半同步复制插件，可以通过修改半同步复制的系统变量来实现，如下：

```
# 在数据库实例运行时，要启用主库端的半同步复制插件，可以通过修改如下系统变量来实现
mysql> set global rpl_semi_sync_master_enabled=1;
Query OK, 0 rows affected (0.00 sec)
```

```
# 在数据库实例运行时,要启用从库端的半同步复制插件,可以通过修改如下系统变量来实现
mysql> set global rpl_semi_sync_slave_enabled =1;
Query OK, 0 rows affected (0.01 sec)

# 系统变量 rpl_semi_sync_master_enabled 和 rpl_semi_sync_slave_enabled 的值为 1 时,表示启用
半同步复制;为 0 时表示禁用,默认情况下其值为 0
```

检查半同步复制插件是否被启用:

```
# 对于主库,必须看到状态变量 Rpl_semi_sync_master_status 的值为 ON,否则就表示半同步复制插件在主
库中未生效
mysql> show status like '%semi%status%';
+-----------------------------+-------+
| Variable_name               | Value |
+-----------------------------+-------+
| Rpl_semi_sync_master_status | ON    |
+-----------------------------+-------+
1 row in set (0.01 sec)

# 对于从库,必须看到状态变量 Rpl_semi_sync_slave_status 的值为 ON,否则表示半同步复制插件在从库中
未生效
mysql> show status like '%semi%status%';
+----------------------------+-------+
| Variable_name              | Value |
+----------------------------+-------+
| Rpl_semi_sync_slave_status | OFF   |
+----------------------------+-------+
1 row in set (0.01 sec)
```

确保在复制拓扑中异步复制已经处于运行状态,否则,请先按照步骤设置好异步复制,详情可参考第 14 章。

在本章的示例中,我们是基于第 14 章的异步复制环境来配置半同步复制的,配置到这一步时,由于异步复制已经处于运行状态,从库的复制线程并不能自动感知到半同步复制插件的配置变化。所以,如上文中对状态变量 Rpl_semi_sync_slave_status 的介绍那样,该状态变量的值此时为 OFF,表示半同步复制插件目前在从库中未生效。

现在,必须重启从库的 I/O 线程,否则半同步复制不生效。重启 I/O 线程之后,从库会重新连接到主库,并注册为半同步的从库(如果是全新搭建的复制拓扑,可以忽略该步骤。但是,半同步复制的配置如果发生在异步复制的配置之后,则仍然需要此步骤)。

```
# 在从库中执行重启 I/O 线程的语句
mysql> stop slave io_thread;
Query OK, 0 rows affected (0.01 sec)

mysql> start slave io_thread;
Query OK, 0 rows affected (0.00 sec)
```

```
mysql> show status like '%semi%status%';
+----------------------------+-------+
| Variable_name              | Value |
+----------------------------+-------+
| Rpl_semi_sync_slave_status | ON    |
+----------------------------+-------+
1 row in set (0.00 sec)

# 如果从库 I/O 线程已在运行中，并且在配置半同步复制之后没有重新启动，则主库会继续使用异步复制的方式向
从库发送二进制日志
```

至此，我们就构建了一个半同步复制拓扑，但还需要在主库中写一些测试数据，做最后的验证，确认这些数据是否是以半同步复制的方式被同步到从库的。详情可参考 15.3 节 "半同步复制工作状态的验证"。

在 MySQL Server 启动之前，可以将控制半同步复制的系统变量设置为命令行选项或 my.cnf 配置文件。配置文件中列出的设置将在每次 MySQL Server 启动时生效。例如，可以在主库端和从库端的 my.cnf 文件中设置，如下所示：

```
# 在主库端:
[mysqld]
rpl_semi_sync_master_enabled=1
rpl_semi_sync_master_timeout=1000 # 1 second

# 在每个从库端:
[mysqld]
rpl_semi_sync_slave_enabled=1
```

提示：

- 文中提到的所有配置变量的含义，可参考《千金良方：MySQL 性能优化金字塔法则》一书附录 C "MySQL 常用配置变量和状态变量详解"。

- 如果你的复制拓扑中存在双主（或者需要更高效、便捷地切换主从角色），则可以在两个从库中都安装主从库的半同步复制插件。这样，在主从复制的角色互换时，就不再需要额外的半同步复制配置步骤，在半同步复制拓扑下快速、便捷地实现主从角色的互换（例如，从库切换为主库时，需要安装主库的半同步复制插件，主库切换为从库时，需要安装从库的半同步复制插件，否则无法继续使用半同步复制在主从库之间同步数据）。

- 我们还可以直接在 my.cnf 配置变量文件中使用系统变量 plugin_load，使数据库实例启动时直接加载半同步复制插件，而无须手工配置，类似如下语句：

```
# 主库
[root@localhost ~]# cat my.cnf
......
[mysqld]
plugin_load = "rpl_semi_sync_master=semisync_master.so" # 然后，将半同步复制相关
的配置变量直接写在该项配置变量之后即可生效
```

```
# 从库
[root@localhost ~]# cat my.cnf
......
[mysqld]
plugin_load = "rpl_semi_sync_slave=semisync_slave.so"   # 同上
```

15.3 半同步复制工作状态的验证

在主库中插入测试数据：

```
mysql> create database test;
Query OK, 1 row affected (0.00 sec)

mysql> use test
Database changed

mysql> create table test(id int);
Query OK, 0 rows affected (0.02 sec)

mysql> insert into test values(1),(2),(3);
Query OK, 3 rows affected (0.00 sec)
Records: 3  Duplicates: 0  Warnings: 0

mysql> select * from test;
+------+
| id   |
+------+
|  1   |
|  2   |
|  3   |
+------+
3 rows in set (0.00 sec)
```

在主库中查看半同步复制状态（这里指的是数据的半同步状态，而不是半同步复制插件的工作状态）：

```
# 如果发现状态变量 Rpl_semi_sync_master_no_tx 不为 0,则表示有数据使用异步复制的方式被同步到从库。
由于是在线配置，而且生效需要一定时间，所以在半同步复制生效之前，可能存在一些事务通过异步复制的方式同步到从
库，这是正常现象（也就是说，状态变量 Rpl_semi_sync_master_no_tx 不为 0 是正常的）。只要该值没有持续增加，
就无须担心。但如果值持续增加，则需要检查复制的配置是否得当，或者主从库之间的复制网络是否存在异常中断，导致
半同步复制被降级为异步复制
mysql> show status like '%semi%tx%';
+-----------------------------------------+-------+
| Variable_name                           | Value |
+-----------------------------------------+-------+
| Rpl_semi_sync_master_no_tx              | 0     |
| Rpl_semi_sync_master_tx_avg_wait_time   | 645   |
| Rpl_semi_sync_master_tx_wait_time       | 1937  |
```

```
| Rpl_semi_sync_master_tx_waits | 3 |
| Rpl_semi_sync_master_yes_tx | 3 |
+----------------------------------------+-------+
5 rows in set (0.00 sec)
```

在从库中检查数据同步情况：

```
mysql> use test
Database changed
mysql> show tables;
+----------------+
| Tables_in_test |
+----------------+
| test |
+----------------+
1 row in set (0.00 sec)

mysql> select * from test;
+------+
| id |
+------+
| 1 |
| 2 |
| 3 |
+------+
3 rows in set (0.00 sec)
```

注意：

- 系统变量 rpl_semi_sync_master_wait_no_slave 用于控制在系统变量 rpl_semi_sync_master_timeout 设置的超时周期内，当连接主库的从库数量降为 0 时（即在主库中，所有从库与主库建立的复制线程连接都被断开。状态变量 Rpl_semi_sync_master_clients 可以实时显示正常连接主库的从库数量），主库是否要继续等待从库的 ACK 消息（如果超过系统变量 rpl_semi_sync_master_timeout 设置的时间，主库未收到从库的 ACK 消息，就会将半同步复制降级为异步复制）。其默认值为 ON，表示在系统变量 rpl_semi_sync_master_timeout 设置的超时周期内，如果连接主库的从库数量降为 0，主库仍然继续等待从库的 ACK 消息。根据 MySQL 5.7 增强半同步复制的原理，我们知道，当主库未收到设定从库数量的中继日志的 ACK 消息时（默认情况下，主库只需要收到一个从库的 ACK 消息即可提交事务。系统变量 rpl_semi_sync_master_wait_for_slave_count 可以设置必须接收多少数量的从库 ACK 消息，主库的事务才进行存储引擎层的提交），主库的事务是不会被提交的（默认情况下，超过 10 s 还未收到从库 ACK 消息时，半同步复制会转换为异步复制）。

- 在一些对主从库数据一致性要求较高，且对主库可用性要求不高的场景下，可以采用增加系统变量 rpl_semi_sync_master_timeout 的值，保持系统变量 rpl_semi_sync_master_wait_no_slave 为 ON 的方案，这样就能够确保，即使从库发生异常或者主从库之间的网络发生异常，主库也不会将半同步复制降级为异步复制，从而保证主从库数据的一致性。但是当主从库之间的复制网络发生故障时，可能导致主库写事务长时间处于阻塞状态。

- 在一些对数据一致性要求不高，且对主库可用性要求高的场景下，可以采用增加系统变量 rpl_semi_sync_master_timeout 的值（这里增加该变量的值是为了在从库能正常连接主库的情况下，尽可能地使用半同步复制），同时修改系统变量 rpl_semi_sync_master_wait_no_slave 为 OFF 的方案。当连接主库的从库数量降为 0 时，主库立即将半同步复制降级为异步复制，这样就能够确保主库的写业务能够迅速恢复。当有任意从库与主库恢复复制线程的连接时，主库会将异步复制提升为半同步复制。但该方案存在丢失数据的风险，因为所有从库丢失与主库的连接时，无法判定是什么类型的故障，在该未知故障下，主库将半同步复制降级为异步复制，一旦主库再**发生宕机故障，将可能丢失数据**。

第 16 章　通过扩展从库以提高复制性能

在大多数 OLTP 或 OLAP 业务系统中，对数据库的操作通常读多写少（特殊场景除外。例如，数据采集、数据恢复等是写多读少的场景），所以，随着业务的增长，原有复制拓扑中的数据库实例渐渐难以支撑读访问流量（假定复制拓扑中的主库提供写服务，所有从库提供读服务），迫切需要提高数据库的读访问能力。那么，我们就需要一个方便、快捷、轻量的读访问能力扩展方案。主流的方案有如下几种：

- 横向扩展（scale out）：在原复制拓扑的基础上对提供读服务的从库进行扩展（即在此基础上增加从库的数量），对原有复制拓扑几乎无影响，这无疑是首选方案。
- 纵向扩展（scale up）：升级原复制拓扑中的读库服务器的硬件配置（例如，增加更多的 CPU 和内存、提高 I/O 能力，使用更好的磁盘设备及更高带宽的网络设备）。但这通常需要停机，对原复制拓扑有一定影响，而且复制拓扑中所有数据库实例的软硬件配置通常要求保持一致；否则，如果只升级从库而不升级主库，当发生主从切换（主从库角色互换）时，仍然会遭遇读能力瓶颈（因为主从切换之后，读库使用的是未升级过的主库硬件配置）。
- 调整架构：你也许需要基于更长远的考虑来设计解决方案，比如可能需要重新设计数据库架构，而这通常要进行多方考量与评估，对原复制拓扑的影响较大，所以这不是一个很好的解决临时问题的方案（但如果是为应对不断增长的业务需求而做一些未来规划，调整架构却是一个好方案）。例如，实施架构调整的时间窗口可能比较长（因为可能需要将数据从旧架构迁移到新架构），费用和成本可能会大幅增加（因为可能会增加一大批服务器）。

本章以横向扩展为例，介绍整个方案的实施过程。

16.1　操作环境

这里只列出与复制拓扑规划相关的信息，其余信息可参考第 14 章 "搭建异步复制"。

1. 服务器列表

- 原主库（Master）：10.10.30.161
- 原从库 1（Slave1）：10.10.30.162

- 新从库 2（Slave2）：10.10.30.163
- 新从库 3（Slave3）：10.10.30.164

2. 复制拓扑规划

在原一主一从的复制拓扑基础上（原主库和原从库 1 构成的原复制拓扑），扩展为一主三从的复制拓扑。

提示：本章只演示在已经运行的一主一从复制拓扑中，分别使用 mysqldump 和 XtraBackup 备份工具对从库执行备份，并用备份数据来扩展从库的步骤。对于其他情形的详细操作步骤，可参考第 14 章 "搭建异步复制"。

16.2　横向扩展

在横向扩展方案中，我们通过扩展多个从库来分摊数据库的查询（读）负载。这里讨论的横向扩展方案是将数据从一个主库复制到一个或多个从库，因此该横向扩展方案只适合读多写少的场景，大多数网站的业务都属于此类别（浏览网站、阅读文章、发帖等）。

本节将以一主一从扩展为一主三从的复制拓扑为例进行介绍（在第 14 章所讲解的复制拓扑基础上进行扩展），最终实现的复制拓扑（1 个主库承载写业务，3 个从库承载读业务，并实现读能力的负载均衡）如图 16-1 所示（参考 Oracle MySQL 官方资料）。

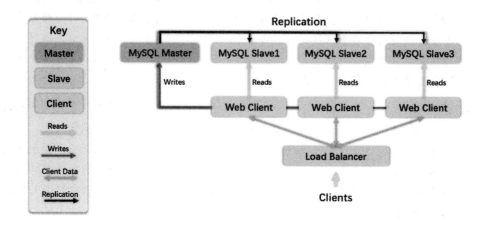

图 16-1

在图 16-1 中，来自 Client（客户端）的请求（Writes 和 Reads 请求）发送到 Load Balancer（负载均衡器），负载均衡器在多个 Web Client（Web 客户端）之间分发客户端数据。Web 客户端发出的 Writes（写入）请求被转发到 MySQL Master（MySQL 主库），Web 客户端发出的 Reads（读取）请求被负载均衡到 3 个 MySQL 从库（MySQL Slave1、MySQL Slave2、MySQL Slave3）。数据的变更将从 MySQL 主库同步到 3 个 MySQL 从库。

16.2.1 扩展从库的简要步骤

第 14 章详细介绍了如何搭建复制拓扑，但都是通过在主库上执行备份实现的，本节将介绍如何通过在从库上执行备份来完成对从库的扩展（该过程也可以称为"从库克隆"）。

利用 mysqldump 备份数据扩展从库，其简要步骤如图 16-2 所示。

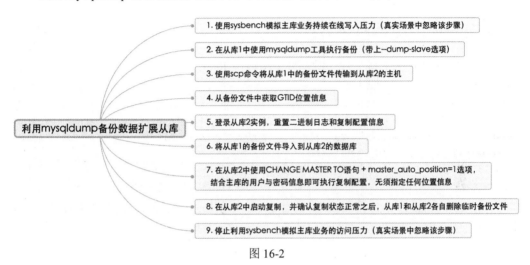

图 16-2

利用 XtraBackup 备份数据扩展从库，其简要步骤如图 16-3 所示。

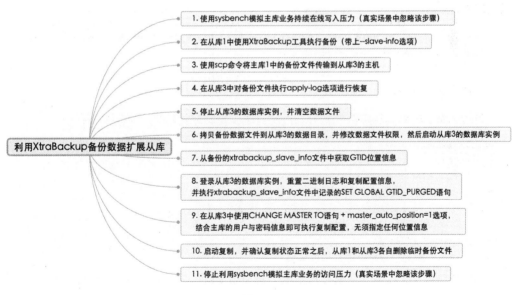

图 16-3

16.2.2 扩展从库的详细过程

由于需要扩展 2 个从库，而我们打算使用 mysqldump 和 XtraBackup 两个工具，所以下面将分别演示如何使用这两个工具在从库 1 中执行备份，然后分别使用其备份来扩展从库 2 和从库 3。

16.2.2.1 使用 mysqldump 工具执行备份

首先，使用 sysbench 对主库（Master）加压（此为模拟步骤，真实环境中应忽略）。

在从库 1 的服务器中使用 mysqldump 工具执行备份，命令如下：

```
# 如果是采用传统复制，则--dump-slave 选项至关重要。使用该选项之后，在从库中获取的二进制日志文件位置
就是从库中的 SQL 线程重放的位置（而不是从库自身的二进制日志文件位置），该位置是相对于主库而言的
[root@localhost ~]# mysqldump -h 10.10.30.162 -usysbench -psysbench -P3306 --single-
transaction --master-data=2 --dump-slave --triggers --routines --events --all-databases >
/data/backup_`date +%F_%H_%M_%S`.sql
[root@localhost ~]# ls -lh /data/backup_*
-rw-r--r-- 1 root root 171M Jul 7 16:09 /data/backup_2019-07-07_16_09_18.sql
```

在从库 1（Slave1）的服务器中，将备份文件传送到从库 2（Slave2）的服务器，命令如下：

```
[root@localhost ~]# scp /data/backup_2019-07-07_16_09_18.sql 10.10.30.163:/data/
......
```

在从库 2 的服务器中，获取备份文件中的位置信息（由于这里采用了 GTID 复制，所以只获取 GTID 的位置信息），命令如下：

```
[root@localhost ~]# head -100 /data/backup_2019-07-07_16_09_18.sql |grep -i gtid
-- GTID state at the beginning of the backup
SET @@GLOBAL.GTID_PURGED='f3372787-0719-11e8-af1f-0025905b06da:1-532638';
```

在从库 2 的服务器中，初始化数据库实例并启动（具体步骤略）。

登录从库 2 的数据库实例，重置二进制日志和复制配置信息，语句如下：

```
mysql> reset master;
Query OK, 0 rows affected (0.00 sec)

mysql> stop slave;
Query OK, 0 rows affected, 1 warning (0.00 sec)

mysql> reset slave all;
Query OK, 0 rows affected (0.00 sec)
```

将备份数据导入从库 2 的数据库实例，命令如下：

```
[root@localhost ~]# mysql -uadmin -ppassword -h10.10.30.163 -P3306 < /data/backup_
2019-07-07_16_09_18.sql
......
```

登录从库 2 的数据库实例，使用 CHANGE MASTER TO 语句配置复制，语句如下：

```
mysql> change master to master_host='10.10.30.161', master_user='repl', master_password='replpass',master_port=3306,master_auto_position=1;
Query OK, 0 rows affected, 2 warnings (0.01 sec)
```

登录从库 2 的数据库实例，启动复制线程，并查看其状态，语句如下：

```
mysql> start slave;
Query OK, 0 rows affected (0.02 sec)

mysql> show slave status\G
*************************** 1. row ***************************
               Slave_IO_State: Queueing master event to the relay log
                  Master_Host: 10.10.30.161
                  Master_User: repl
                  Master_Port: 3306
                Connect_Retry: 60
              Master_Log_File: mysql-bin.000006
          Read_Master_Log_Pos: 232928763
               Relay_Log_File: mysql-relay-bin.000002
                Relay_Log_Pos: 16211485
        Relay_Master_Log_File: mysql-bin.000006
             Slave_IO_Running: Yes
            Slave_SQL_Running: Yes
......
        Seconds_Behind_Master: 587
......
           Retrieved_Gtid_Set: f3372787-0719-11e8-af1f-0025905b06da:532639-580232
            Executed_Gtid_Set: f3372787-0719-11e8-af1f-0025905b06da:1-539856
                Auto_Position: 1
         Replicate_Rewrite_DB:
                 Channel_Name:
           Master_TLS_Version:
1 row in set (0.00 sec)
```

停止 sysbench 对主库的加压（此为模拟步骤，真实环境中应忽略）。

在从库 1 和从库 2 的服务器中，分别删除临时备份文件，命令如下：

```
# 从库 1
[root@localhost ~]# rm -f /data/backup_2019-07-07_16_09_18.sql

# 从库 2
[root@localhost ~]# rm -f /data/backup_2019-07-07_16_09_18.sql
```

16.2.2.2 使用 XtraBackup 工具执行备份

首先，使用 sysbench 对主库（Master）加压（此为模拟步骤，真实环境中应忽略）。

在从库 1（Slave1）的服务器中使用 XtraBackup 工具执行备份（当然，也可以在从库 2 中执行备份，这么做在实际生产环境中可能更友好），命令如下：

```
# 如果是采用传统复制，则--slave-info选项至关重要。使用该选项之后，在从库中获取的二进制日志文件位置
就是从库中SQL线程重放的位置（而不是从库自身的二进制日志文件位置），该位置是相对于主库而言的
[root@localhost ~]# innobackupex --defaults-file=/etc/my.cnf --user=sysbench --password=
sysbench --no-timestamp --slave-info --stream=tar ./ | cat - > /data/backup_`date +%Y%m%d`.
tar.gz
    190707 16:26:33 innobackupex: Starting the backup operation
    ......
    MySQL binlog position: filename 'mysql-bin.000003', position '60876242', GTID of the last
change 'f3372787-0719-11e8-af1f-0025905b06da:1-742743'
    MySQL slave binlog position: master host '10.10.30.161', purge list 'f3372787-0719-11e8-
af1f-0025905b06da:1-742743'
    ......
    190707 16:26:46 completed OK!


[root@localhost ~]# ls -lh /data/backup_*
-rw-r--r-- 1 root root 2.6G Jul 7 16:26 /data/backup_20190707.tar.gz
```

在从库1的服务器中，将备份文件传送到从库3（Slave3）的服务器，命令如下：

```
[root@localhost ~]# scp /data/backup_20190707.tar.gz 10.10.30.164:/data/
......
```

在从库3的服务器中，解压并对备份文件执行apply-log操作（将备份数据恢复到一致的状态），命令如下：

```
# 创建一个用于恢复临时数据的目录test
[root@localhost ~]# mkdir /data/test
[root@localhost ~]# cd /data/test/
[root@localhost test]# mv ../backup_20190707.tar.gz ./
[root@localhost test]# ls -lh
total 2.6G
-rw-r--r-- 1 root root 2.6G Jul 7 16:28 backup_20190707.tar.gz

# 解压
[root@localhost test]# tar xf backup_20190707.tar.gz
[root@localhost test]# ll
total 4967780
-rw-r--r-- 1 root root 2721479680 Jul 7 16:28 backup_20190707.tar.gz
-rw-rw---- 1 root root 466 Jul 7 16:26 backup-my.cnf
drwxr-xr-x 2 root root 272 Jul 7 16:32 employees
-rw-rw---- 1 root root 2147483648 Jul 7 16:25 ibdata1
drwxr-xr-x 2 root root 4096 Jul 7 16:32 mysql
drwxr-xr-x 2 root root 8192 Jul 7 16:32 performance_schema
drwxr-xr-x 2 root root 8192 Jul 7 16:32 sys
drwxr-xr-x 2 root root 4096 Jul 7 16:32 sysbench
drwxr-xr-x 2 root root 52 Jul 7 16:32 test
-rw-rw---- 1 root root 10485760 Jul 7 16:25 undo001
......
-rw-rw---- 1 root root 10485760 Jul 7 16:25 undo016
-rw-rw---- 1 root root 72 Jul 7 16:26 xtrabackup_binlog_info
```

```
-rw-rw---- 1 root root       121 Jul 7 16:26 xtrabackup_checkpoints
-rw-rw---- 1 root root       612 Jul 7 16:26 xtrabackup_info
-rw-rw---- 1 root root  50213376 Jul 7 16:26 xtrabackup_logfile
-rw-rw---- 1 root root       112 Jul 7 16:26 xtrabackup_slave_info

# 执行 redo log 的应用
[root@localhost test]# innobackupex --apply-log --use-memory=1G ./
......
InnoDB: Starting shutdown...
InnoDB: Shutdown completed; log sequence number 28557834792
190707 16:33:23 completed OK!
```

在从库 3 的服务器中，初始化安装数据库实例（具体步骤略。只要求初始化安装，不要求数据库实例处于运行状态）。

停止从库 3 的服务器中的数据库实例，并清空数据目录，命令如下：

```
# 停止从库数据库进程（如果正在运行的话）
[root@localhost test]# service mysqld stop
Shutting down MySQL.. SUCCESS!

# 清空从库的数据文件（由于本例是按照生产规范进行配置的，对各个数据文件做了分目录存放，所以需要清空下面这些目录，如果在你的环境中未配置分目录存放，只需要清空系统变量 datadir 指定的目录即可）
[root@localhost test]# rm -rf /data/mysqldata1/{binlog,innodb_log, innodb_ts, mydata, relaylog,undo}/*
```

在从库 3 的服务器中，将临时数据恢复的目录下的数据文件拷贝到 my.cnf 配置文件指定的数据目录中（系统变量 datadir 指定的目录），命令如下：

```
# 先删除用于恢复临时数据的目录下的备份压缩包文件（因为这里已经不再需要它了），否则在后续步骤中它会被一同拷贝到目标数据库的系统变量 datadir 指定的目录下
[root@localhost test]# rm -f backup_20190707.tar.gz

# 执行拷贝
[root@localhost test]# innobackupex --defaults-file=/etc/my.cnf --copy-back ./
......
190707 16:46:01 [01] ...done
190707 16:46:01 completed OK!
```

在从库 3 的服务器中，将数据库系统变量 datadir 指定目录下的所有数据文件的宿主用户修改为 mysql 用户（因为 mysqld 进程将使用 mysql 用户启动，如果不修改，将导致启动失败。可以在 my.cnf 配置文件中指定由哪个系统用户启动 mysqld 进程），命令如下：

```
[root@localhost test]# chown mysql.mysql /data/mysqldata1/ -R
```

在从库 3 的服务器中，启动数据库实例，命令如下：

```
[root@localhost test]# service mysqld start
```

```
# 由于是在从库 1 的服务器上执行的备份，该数据库实例是一个已经处于运行中的从库，所以，备份中将包含这些
复制配置信息，在数据库实例启动时，复制进程会随着数据库实例一起启动。你可能不希望发生这样的事情，那么可以在
```

my.cnf 中添加 skip_slave_start 选项，以避免复制线程随着数据库实例一并启动。但是，要记得配置完复制之后，在 my.cnf 配置文件中删除该选项，以防止下次重启数据库时，复制线程不能随着 mysqld 进程一并启动（在执行正常的数据库重启时，你可能希望复制线程能够随着数据库实例一并启动）。

在从库 3 的服务器中，获取备份文件中的位置信息，命令如下：

```
[root@localhost test]# cat xtrabackup_slave_info
SET GLOBAL gtid_purged='f3372787-0719-11e8-af1f-0025905b06da:1-742743';
CHANGE MASTER TO MASTER_AUTO_POSITION=1
```

事实上，从 xtrabackup_binlog_info 文件中获取 GTID 位置也是可以的，因为获取的结果相同（但如果是传统复制，则只能从 xtrabackup_slave_info 文件中获取，因为获取的结果完全不同。xtrabackup_slave_info 文件中记录的是从库的 SQL 线程应用的位置，是对应于主库二进制日志的位置，而 xtrabackup_binlog_info 文件中记录的是从库自身的二进制日志位置）。

```
[root@localhost test]# cat xtrabackup_binlog_info
mysql-bin.000003  60876242  f3372787-0719-11e8-af1f-0025905b06da:1-742743
```

登录从库 3 的数据库实例，重置二进制日志和复制配置信息，并执行上述步骤中获取的语句 "SET GLOBAL gtid_purged='f3372787-0719-11e8-af1f-0025905b06da:1-742743';"。代码如下：

```
mysql> reset master;
Query OK, 0 rows affected (0.01 sec)

mysql> stop slave;
Query OK, 0 rows affected, 1 warning (0.00 sec)

mysql> reset slave all;
Query OK, 0 rows affected (0.03 sec)

mysql> set global gtid_purged='f3372787-0719-11e8-af1f-0025905b06da:1-742743';
Query OK, 0 rows affected (0.00 sec)
```

登录从库 3 的数据库实例，使用 CHANGE MASTER TO 语句配置复制，并启动复制线程（通过从库的备份执行扩展从库的操作时，此步骤可忽略，只有在主库中执行备份才需要执行此步骤），语句如下：

```
mysql> change master to master_host='10.10.30.161',master_user='repl',master_password='replpass',master_port=3306,master_auto_position=1;
Query OK, 0 rows affected, 2 warnings (0.03 sec)

mysql> start slave;
Query OK, 0 rows affected (0.09 sec)
```

登录从库 3 的数据库实例，查看复制线程的状态，语句如下：

```
mysql> show slave status\G
*************************** 1. row ***************************
               Slave_IO_State: Queueing master event to the relay log
                  Master_Host: 10.10.30.161
                  Master_User: repl
```

```
            Master_Port: 3306
          Connect_Retry: 60
        Master_Log_File: mysql-bin.000008
    Read_Master_Log_Pos: 114928565
         Relay_Log_File: mysql-relay-bin.000002
          Relay_Log_Pos: 110864836
  Relay_Master_Log_File: mysql-bin.000007
       Slave_IO_Running: Yes
      Slave_SQL_Running: Yes
......
  Seconds_Behind_Master: 7917
......
      Retrieved_Gtid_Set: f3372787-0719-11e8-af1f-0025905b06da:742744-1001974
       Executed_Gtid_Set: f3372787-0719-11e8-af1f-0025905b06da:1-791772
           Auto_Position: 1
    Replicate_Rewrite_DB:
            Channel_Name:
      Master_TLS_Version:
1 row in set (0.00 sec)
```

停止 sysbench 对主库的加压（此为模拟步骤，真实环境中应忽略）。

在从库 1 和从库 3 的服务器中，分别删除临时备份文件，命令如下：

```
# 从库 1
[root@localhost ~]# rm -f /data/backup_20190707.tar.gz
```

```
# 从库 3
[root@localhost ~]# rm -rf /data/test/
```

提示：

- 在启用多线程复制的从库中执行备份时，如果从库处于高并发压力下（并行应用的事务数量较多），建议先停止 SQL 线程，然后执行备份，等备份完成之后再重新启动 SQL 线程，否则容易出现锁死现象（注意，不是死锁，是锁死，即正常的业务访问的事务和备份执行加锁的语句之间形成相互的锁等待现象，需要人工介入处理）。但如果执行备份的从库承载着读访问任务，停止 SQL 线程通常会影响读业务，所以在具有多个从库的复制拓扑中，通常建议将其中一个从库作为"备份专用"库，不对外提供读访问。如果由于一些特殊原因无法这样做（例如，复制拓扑中只有一主一从，从库的读访问不能中断的情况），可以在业务低峰期，在主库或者从库中执行备份。
- 复制拓扑中的从库数量需要控制在合理的范围内，通常不建议超过 10 个。超过这个数量之后，主从库之间用于复制的网络带宽将可能成为瓶颈（在绝大多数高并发 OLTP 场景中，I/O 线程的平均吞吐流量参考值大约为 10 MB/s，但特殊情况除外，需要根据具体问题具体分析。例如，如果存在并行大事务，用于复制的网络带宽可能被迅速占满。另外，如果复制与业务共用网络，则此时可能因为复制占用太多网络带宽，导致业务访问受到影响，尤其是主库的写业务。对于这个问题，我们将在 16.3 节"提高复制性能"中介绍一个简单的优化方案）。如果由于一些特殊原因，复制拓扑中的从库不得不超过 10 个，则可以考虑使用系统变量 slave_compressed_protocol 启用对主从库之间传输的二进制日志的压缩，但启用压缩之后会增加 CPU 的开销（系统变量 slave_compressed_protocol 的有效值为 ON 或 OFF，默认为 OFF。但要注意，从 MySQL 8.0 开始不建议使用该系统变量，在将来的版本中它将被移除，有兴趣的读者可自行研究）。

- 虽然 GTID 的机制能够让从库自动查找所需的二进制日志，也能够实现自动跳过 GTID 重复的事务，但是在一些特殊场景下需要一些关键步骤的配合才能确保不发生意外。例如，在有压力的复制拓扑中，使用 XtraBackup 工具执行备份时，由于备份拷贝的是物理文件，而且不拷贝二进制日志，所以，必须执行 SET GLOBAL gtid_purged 语句指定与备份数据一致的 GTID 位置，否则在高压力的复制拓扑中通过从库备份数据来实现从库扩展时，启动复制时可能出现如下问题：

  - 启动复制时报告主库的二进制日志被 purge（清理）: Last_IO_Error: Got fatal error 1236 from master when reading data from binary log: 'The slave is connecting using CHANGE MASTER TO MASTER_AUTO_POSITION = 1, but the master has purged binary logs containing GTIDs that the slave requires.'。
  - 启动复制时报告初始化 relay log info 的结构错误: Slave failed to initialize relay log info structure from the repository。

16.2.3 配置从库的读负载均衡

负载均衡的配置通常有如下几种主流解决方案：

- 使用智能 DNS（Domain Name Server，域名服务器）实现（可能需要付费）。正常工作时，程序使用 DNS 域名地址访问数据库，DNS 解析会随机分配一条解析记录执行读访问，当某个从库出现故障时，DNS 解析记录会被剔除，当故障恢复时，DNS 解析记录会被重新加进来。
- 使用开源的 LVS + Keepalived 解决方案实现（免费）。正常工作时，程序使用 VIP 地址访问数据库，LVS（Linux Virtual Server，Linux 虚拟服务器）会随机分配一个 real server（这里指的就是数据库服务器的 IP 地址）执行访问，当某个从库出现故障时，对应的 real server 会被剔除，当故障恢复时，real server 会被重新加进来。
- 使用读写分离中间件实现（可使用开源的中间件，也可以使用付费的商业中间件），但读写分离的中间件不仅实现了读负载均衡，还可以实现读写分离，也就是说，使用该方案，可能需要对应用进行改造，读访问和写访问使用统一的访问入口（使用同一个 IP 地址访问）。

综上所述，到底选择哪个方案更合适，需要结合实际应用场景，具体问题具体分析。

关于负载均衡组件的安装及部署过程，本节不做介绍，有兴趣的读者可自行研究。

16.3 提高复制性能

随着连接到主库的从库数量的增加，不仅其占用的主库带宽会增加，而且对主库的负载影响（尽管很小）也会增大，因为每个从库都会与主库创建一个连接。如果一个主库上连接着大量从库，而且该主库又忙于处理各种客户端的请求，那么此时的复制性能就可能会受影响。为了提高复制性能，可以创建一个更深层次的复制结构，使主库只被复制到一个从库，其他的从库与这个从库建立连接进行复制，如图 16-4 所示（参考 Oracle MySQL 官方资料）。

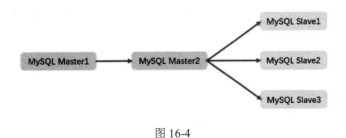

图 16-4

在图 16-4 中，MySQL Master1 用于接受写请求，其中的数据变更被复制到 MySQL Master2 中，MySQL Master2 将来自 MySQL Master1 的数据变更复制到 MySQL Slave1、MySQL Slave2、MySQL Slave3。为了实现这个复制拓扑，需要按照如下步骤进行配置：

- Master1 是主要的主库，所有数据更改和更新都写入 Master1 数据库，因此必须启用二进制日志记录功能。
- Master2 是 Master1 的从库，它为复制拓扑中的其余从库提供复制功能。Master2 是唯一允许连接到 Master1 的 Server，Master2 也必须启用二进制日志记录功能，并且还需要启用--log-slave-updates 选项（或在 my.cnf 中启用系统变量 log_slave_updates），使得 Master1 的二进制日志在 Master2 上重放之后也写入 Master2 自己的二进制日志，以便将这些数据变更复制到真正的从库（Slave1、Slave2 和 Slave3）。
- Slave1、Slave3 和 Slave3 在该复制拓扑中是真正承担读业务访问的从库，数据从 Master2 中复制，但实际上数据的真正来源是 Master1。

上述复制拓扑减少了真正承担业务写访问的主库上的客户端连接负载和网络接口负载，能在一定程度上提高主库的性能。

如果复制拓扑中的从库复制延迟太长，可以使用如下一些方法来优化从库的复制性能：

- 可以将从库的中继日志文件和数据文件放在不同的磁盘设备上，使用--relay-log 选项（或在 my.cnf 中使用系统变量 relay_log）指定中继日志存放到其他路径下。
- 可以在从库中使用复制过滤选项，将不同的数据库复制到不同的从库，详情可参考

第 23 章"将不同的数据库复制到不同的实例"。

- 如果不关心从库是否支持事务，则可以在从库中使用非事务存储引擎（例如 MyISAM 引擎）。

- 不承载高可用切换的从库，可以关闭--log-slave-updates 选项（或在 my.cnf 中关闭系统变量 log_slave_updates），以避免将重放的来自主库的二进制日志写入从库自身的二进制日志，浪费从库的 I/O 设备性能。

如果有多个从库，在每个从库上重复以上步骤进行配置。

提示：

- 如图 16-4 所示，为提高复制性能，实际上是增加了一个用于承接所有从库连接的中间从库。通常，我们还可以在这个中间从库与主库之间搭建双主复制拓扑，用于承载主库的高可用切换，即作为该主库的备用主库。要实现这个复制拓扑，需要新增一个从库，而且要调整复制拓扑。关于增加一个从库的操作步骤，可参考 16.2.2 节"扩展从库的详细过程"，对于调整复制拓扑的操作步骤，可参考第 18 章"复制拓扑的在线调整"。

- 如果新建的从库数据来源于一个已经运行的从库，则备份数据中已包含复制配置信息，在启动新的从库 mysqld 进程时，需要加上--skip-slave-start 选项，使得新从库在启动实例时不启动复制线程。当新的从库启动之后，在使用 CHANGE MASTER TO 语句配置复制之前，执行 RESET SLAVE ALL 语句清理之前的复制配置信息。

- 如果未设置--server-id 选项（或在 my.cnf 中未设置系统变量 server_id），或者设置的值与复制拓扑中的其他实例冲突，则从库无法连接到主库，会报告类似如下错误：

```
Warning: You should set server-id to a non-0 value if master_host is set; we will force server id to 2, but this MySQL server will not act as a slave.
```

第 17 章 复制模式的切换

MySQL 5.5 及其之前的版本，由于不支持 GTID 机制，所以它们使用的都是传统复制（即基于二进制日志文件和位置的复制）。MySQL 5.6 中引入了 GTID 机制，该机制有众多优点（详情可参考第 4 章、第 12 章等章节，以及下文中将要演示的复制模式切换过程，这里先不展开介绍），为 MySQL 管理者的维护工作带来了极大便利，所以大多数用户都会选择切换到 GTID 复制模式（即基于 GTID 的复制，下文统一称为"GTID 复制"）。而除了一些特殊的应用场景之外，通常很少需要从 GTID 复制切换到传统复制。

在第 1 章中，我们将传统复制和 GTID 复制称为"数据同步方法"，但通常我们更喜欢将其称为"复制模式"。本章将对这两种复制模式中的一些概念以及它们的相互切换过程进行详细介绍。

提示：

本章中的大多数操作步骤都需要由具有 root 权限或 SUPER 权限的 MySQL 用户完成。在离线切换复制模式的过程中，如果关闭数据库实例时使用了 mysqladmin shutdown 命令，则用户必须有 SUPER 权限或 SHUTDOWN 权限。

17.1 操作环境信息

假设我们已经按照第 14 章中介绍的方法，搭建了传统复制的一主一从的复制拓扑（这里只列出复制拓扑规划的相关信息，其余信息可参考第 14 章"搭建异步复制"）：

- 主库（master）：10.10.30.161
- 从库（slave）：10.10.30.162

17.2 复制模式的相关概念

为了能安全执行复制模式的在线变更，首先需要了解与复制模式相关的一些关键概念。本节将简单介绍这些概念，并对在线变更复制模式的相关变量的兼容性进行说明。

MySQL 中可用的复制模式，依赖不同的技术来识别被记录的事务。复制模式使用的事务类型如下：

- GTID 事务：由 GTID 标识，GTID 的格式为 SOURCE_ID:TRANSACTION_ID（详情可参考 4.2.1.1 节）。在二进制日志中，每个 GTID 事务使用 Gtid_log_event 来记录 GTID 信息（记录为类似 SET @@SESSION.GTID_NEXT= '5de97c46-a496-11e9-824b-0025905b06da:3755'的语句），它总是位于记录 BEGIN 语句的 Query_log_event 之前。GTID 事务可以使用 GTID 信息来自动查找事务，也可以使用二进制日志文件和二进制日志位置寻址。当整个复制拓扑启用 GTID 且使用 GTID 自动寻址时，就表示使用了 GTID 复制模式。

- 匿名事务：未启用 GTID，写库中的新事务不会分配 GTID。MySQL 5.7.6 及更高版本在二进制日志的每个匿名事务中使用 Anonymous_gtid_log_event 来记录匿名事务的信息（记录为类似 SET @@SESSION.GTID_NEXT= 'ANONYMOUS'的语句），它总是位于记录 BEGIN 语句的 Query_log_event 之前。由于不具备 GTID 复制模式的自动定位功能，所以在从库中配置复制时，必须人工指定向主库请求的起始二进制日志文件和位置。如果整个复制拓扑均未启用 GTID，就表示使用的是传统复制模式（如果启用了 GTID 但未启用 GTID 自动寻址，而是使用二进制日志文件和位置来定位，则表示使用的仍然是传统复制模式。但它与未启用 GTID 的传统复制模式不同，启用 GTID 之后，利用 GTID 机制能自动跳过重复的事务。通常，除非有特殊需求，否则不应该这么做）。

系统变量 gtid_mode 和 enforce_gtid_consistency 在 MySQL 5.7 中都是可以动态修改的（在 MySQL 5.6 中是只读的，要使修改的值生效，需要重启 Server），以此来实现复制模式的在线变更，但是只有具有 SUPER 权限的用户才能修改它们的值。

- 在所有 MySQL 版本中，gtid_mode 可以设置为 ON 或 OFF，表示启用或禁用 GTID 事务。当 gtid_mode = ON 时，只能复制 GTID 事务；而当 gtid_mode = OFF 时，只能复制匿名事务。

- 从 MySQL 5.7.6 起，系统变量 gtid_mode 有两个附加状态值：OFF_PERMISSIVE 和 ON_PERMISSIVE。当 gtid_mode = OFF_PERMISSIVE 时，主库新产生的事务是匿名的事务，允许复制拓扑中的从库重放 GTID 事务和匿名事务。当 gtid_mode = ON_PERMISSIVE 时，主库新产生的事务使用 GTID，允许复制拓扑中的从库重放 GTID 事务和匿名事务。这意味着系统变量 gtid_mode 被设置为 OFF_PERMISSIVE 和 ON_PERMISSIVE 时，匿名事务和 GTID 事务在同一个复制拓扑可能同时存在。例如，gtid_mode = ON 的主库产生的 GTID 事务可能会被复制到 gtid_mode = ON_PERMISSIVE 的从库中。

- 系统变量 gtid_mode 的有效值列表按照顺序（这里姑且将该顺序称为正向顺序，与之相反的顺序，称为反向顺序）依次为：OFF、OFF_PERMISSIVE、ON_PERMISSIVE、ON。

  注意：系统变量 gtid_mode 在 MySQL 5.7 中新增了 OFF_PERMISSIVE、

ON_PERMISSIVE 值之后，其值只能按照上述顺序逐一变更（启用 GTID 时，按照正向顺序依次变更；关闭 GTID 时，按照反向顺序依次变更）。例如，假设系统变量 gtid_mode 当前被设置为 OFF_PERMISSIVE，则可以更改为 OFF 或 ON_PERMISSIVE，但不能设置为 ON。这是为了确保 Server 能够正确处理网络上从匿名事务更改为 GTID 事务的过程。当 gtid_mode 的值在 ON 和 OFF 之间切换时，GTID 状态（位置）是持久的（GTID 位置记录在 InnoDB 存储引擎的 mysql.gtid_executed 表中）。这样可以确保 GTID 位置在系统变量 gtid_mode 的值发生变化期间被持久化，使得无论 gtid_mode 的值如何变化，都不会丢失 GTID 位置信息。

从 MySQL 5.7.6 开始，为了兼容 GTID 复制和传统复制中的事务状态记录，对原有的查看相关事务状态的表或系统变量做了一些调整，使得无论系统变量 gtid_mode 被设置为何值，它们都能够正确显示相关的状态信息。这些调整包括：在 performance_schema.replication_connection_status 表中新增 RECEIVED_TRANSACTION_SET 字段，新增 mysql.gtid_executed 表和系统变量 gtid_purged，在 SHOW SLAVE STATUS 语句输出的结果中新增 Retrieved_Gtid_Set 和 Executed_Gtid_Set 选项（Retrieved_Gtid_Set 和 Executed_Gtid_Set 选项在启用 GTID 时输出 GTID 的相关信息；未启用 GTID 时，就为空），在 performance_schema.replication_applier_status_by_worker 表中新增 LAST_SEEN_TRANSACTION 字段（启用 GTID 时，LAST_SEEN_TRANSACTION 字段显示一个 GTID 字符串；不启用 GTID 时，该字段为 ANONYMOUS）。

```
# performance_schema.replication_applier_status_by_worker 表的 LAST_SEEN_TRANSACTION 字
段记录着每个 Worker 线程的 GTID 状态信息
mysql> select * from performance_schema.replication_applier_status_by_worker\G
*************************** 1. row ***************************
         CHANNEL_NAME: 
            WORKER_ID: 1
            THREAD_ID: 44
        SERVICE_STATE: ON
 LAST_SEEN_TRANSACTION: f3372787-0719-11e8-af1f-0025905b06da:6933582
    LAST_ERROR_NUMBER: 0
   LAST_ERROR_MESSAGE: 
 LAST_ERROR_TIMESTAMP: 0000-00-00 00:00:00
*************************** 2. row ***************************
......
*************************** 16. row ***************************
         CHANNEL_NAME: 
            WORKER_ID: 16
            THREAD_ID: 61
        SERVICE_STATE: ON
 LAST_SEEN_TRANSACTION: 
    LAST_ERROR_NUMBER: 0
   LAST_ERROR_MESSAGE: 
```

```
        LAST_ERROR_TIMESTAMP: 0000-00-00 00:00:00
16 rows in set (0.00 sec)

# 系统变量 gtid_purged 记录在当前数据库实例中已被清理的二进制日志所包含的 GTID SET 范围
mysql> show variables like '%gtid%';
+---------------------+-----------------------------------------------+
| Variable_name       | Value                                         |
+---------------------+-----------------------------------------------+
......
| gtid_purged         | f3372787-0719-11e8-af1f-0025905b06da:1-285745 |
| session_track_gtids | OFF                                           |
+---------------------+-----------------------------------------------+
8 rows in set (0.01 sec)

# mysql.gtid_executed 表中的 source_uuid 字段记录当前数据库实例的数据对应的 GTID SET 范围
mysql> select * from mysql.gtid_executed;
+--------------------------------------+----------------+--------------+
| source_uuid                          | interval_start | interval_end |
+--------------------------------------+----------------+--------------+
| f3372787-0719-11e8-af1f-0025905b06da | 1              | 6890019      |
+--------------------------------------+----------------+--------------+
1 row in set (0.00 sec)

# SHOW SLAVE STATUS 语句输出信息中的 Retrieved_Gtid_Set 和 Executed_Gtid_Set 选项，表示当前的
I/O 线程接收的 GTID SET 范围和 SQL 线程应用的 GTID SET 范围
mysql> show slave status\G;
*************************** 1. row ***************************
               Slave_IO_State: Waiting for master to send event
                  Master_Host: 10.10.30.161
                  Master_User: repl
                  Master_Port: 3306
                Connect_Retry: 60
              Master_Log_File: mysql-bin.000033
          Read_Master_Log_Pos: 138292017
               Relay_Log_File: mysql-relay-bin.000092
                Relay_Log_Pos: 138292230
        Relay_Master_Log_File: mysql-bin.000033
             Slave_IO_Running: Yes
            Slave_SQL_Running: Yes
......
        Seconds_Behind_Master: 0
......
             Master_Server_Id: 3306102
                  Master_UUID: f3372787-0719-11e8-af1f-0025905b06da
......
            Retrieved_Gtid_Set: f3372787-0719-11e8-af1f-0025905b06da:1172802-6933582
```

```
        Executed_Gtid_Set: f3372787-0719-11e8-af1f-0025905b06da:1-6933582
           Auto_Position: 1
......
1 row in set (0.00 sec)
```

启用 GTID 时（gtid_mode = ON 时），复制拓扑中的主库提供了使用 GTID 来自动定位二进制日志文件和二进制日志位置的功能，在从库中使用 CHANGE MASTER TO MASTER_AUTO_POSITION = 1，可使主库自动定位从库所需的二进制日志文件和二进制日志位置（主库根据从库 I/O 线程建立连接时注册的 GTID SET 范围等信息自动定位）。

注意：自动定位功能依赖于 GTID，并且与匿名事务不兼容。简单来说，自动定位指的是从库使用自身的 GTID SET 自动在主库中寻找所需的二进制日志，而不需要手工指定二进制日志文件和位置信息。如果启用自动定位时碰到匿名事务，则会发生错误。所以需要确保，启用自动定位之后，复制拓扑的任何数据库实例的任何地方都不存在匿名事务。

主库和从库上的系统变量 gtid_mode 值的组合兼容性如表 17-1 所示。有些组合会导致自动定位功能失效，而对于有些组合，此功能生效。表中的"Y"代表兼容[1]，"N"代表不兼容，"*"代表能够使用自动定位功能。从表 17-1 中可以看到，要使自动定位功能生效，从库系统变量 gtid_mode 的值必须为 ON，而主库系统变量 gtid_mode 的值可以是 OFF_PERMISSIVE、ON_PERMISSIVE 与 ON 中的任意一个。

表 17-1

从库 \ 主库	OFF	OFF_PERMISSIVE	ON_PERMISSIVE	ON
OFF	Y	Y	N	N
OFF_PERMISSIVE	Y	Y	Y	Y*
ON_PERMISSIVE	Y	Y	Y	Y*
ON	N	N	Y	Y*

系统变量 gtid_mode 的值还会影响系统变量 gtid_next 的值，二者值的组合兼容性如表 17-2 所示。其中，ANONYMOUS 代表自动生成匿名事务；Error 代表发生错误，不兼容且不能通过 SET GTID_NEXT 语句设置；SOURCE_ID:TRANSACTION_ID 表示使用语句可以指定 GTID_NEXT 的值；New GTID 表示自动产生一个新的事务 GTID，当二进制日志记录功能关闭且 gtid_next 被设置为 AUTOMATIC 时，不会生成 GTID。

表 17-2

gtid_next \ gtid_mode	AUTOMATIC,且开启二进制日志	AUTOMATIC,且关闭二进制日志	匿名事务 ANONYMOUS	指定一个 SOURCE_ID: TRANSACTION_ID
OFF	ANONYMOUS	ANONYMOUS	ANONYMOUS	Error

---

[1] 兼容的意思是，主从复制能够正常进行，不会出现报错信息。

续表

gtid_mode \ gtid_next	AUTOMATIC，且开启二进制日志	AUTOMATIC，且关闭二进制日志	匿名事务 ANONYMOUS	指定一个 SOURCE_ID: TRANSACTION_ID
OFF_PERMISSIVE	ANONYMOUS	ANONYMOUS	ANONYMOUS	SOURCE_ID: TRANSACTION_ID
ON_PERMISSIVE	New GTID	ANONYMOUS	ANONYMOUS	SOURCE_ID: TRANSACTION_ID
ON	New GTID	ANONYMOUS	Error	SOURCE_ID: TRANSACTION_ID

关于系统变量 gtid_mode、gtid_next、enforce_gtid_consistency 的详细含义，可参考《千金良方：MySQL 性能优化金字塔法则》的附录 C "MySQL 常用配置变量和状态变量详解"。

关于传统复制与 GTID 复制的更多内容，可参考第 4 章。

提示：GTID 事务和匿名事务在二进制日志中的事件记录内容样本如下，请留意打印内容中的 GTID 关键字和 Anonymous_GTID 关键字，这个位置对应的事件就是上文中提到的 Gtid_log_event 和 Anonymous_gtid_log_event。

```
# GTID 事务
# at 5639710
#170509  16:49:46  server  id  3306241  end_log_pos  5639775  CRC32
0xff3e826d    GTID    last_committed=6523   sequence_number=6525
SET @@SESSION.GTID_NEXT= '2016f827-2d98-11e7-bb1e-00163e407cfb:2910568'/*!*/;
# at 5639775
#170509  16:49:46  server  id  3306241  end_log_pos  5639854  CRC32
0xa8a606e1    Query    thread_id=18721 exec_time=0    error_code=0
SET TIMESTAMP=1494319786/*!*/;
BEGIN
/*!*/;
……
#170509 16:49:46 server id 3306241  end_log_pos 5640429 CRC32 0x55298459    Xid =
10347561
COMMIT/*!*/;

# 匿名事务
# at 1297
#170509 16:49:46 server id 3306241  end_log_pos 1362 CRC32 0xf7a870df    Anonymous_GTID
last_committed=0       sequence_number=2
SET @@SESSION.GTID_NEXT= 'ANONYMOUS'/*!*/;
# at 1362
#170509    16:49:46    server    id    3306241    end_log_pos    1441    CRC32
0xd686b479   Query    thread_id=18721 exec_time=0    error_code=0
SET TIMESTAMP=1494319786/*!*/;
BEGIN
/*!*/;
```

```
......
#170509 16:49:46 server id 3306241  end_log_pos 2016 CRC32 0xc5712634   Xid = 10347603
COMMIT/*!*/;
```

17.3 传统复制在线变更为 GTID 复制

17.3.1 简要步骤

将复制模式从传统复制在线变更为 GTID 复制的前提条件：

- 复制拓扑中的所有数据库 Server 都必须使用 MySQL 5.7.6 或更高版本。
- 变更前，所有数据库 Server 的系统变量 gtid_mode 都设置为 OFF。

传统复制在线变更为 GTID 复制过程中的关键步骤如图 17-1 所示。

图 17-1

17.3.2 详细过程

首先，使用 sysbench 对主库持续加压，不需要加太大压力，只需要有持续不断的数据变更即可（此为模拟步骤，真实环境中应忽略）。

（1）在复制拓扑中每个数据库 Server 上执行如下语句（我们的样例拓扑中只有一主一从，所以这里指的就是分别在主库和从库中执行，下文对此不再赘述）。保证所有数据库 Server 在执行该语句之后，再正常复制 60 s，然后查看所有数据库 Server 的错误日志，确保所有数据库 Server 的错误日志中不会产生警告信息，如果出现了警告，请按照警告中的提示调整你的应用程序（这是第一个重要步骤，必须确保在进入下一个步骤之前在错误日志中不会生成警告）。

```
# 在所有数据库 Server 中执行如下语句，设置 ENFORCE_GTID_CONSISTENCY = WARN
mysql> set @@global.enforce_gtid_consistency = WARN;
Query OK, 0 rows affected (0.00 sec)

mysql> show variables like 'enforce_gtid_consistency';
+--------------------------+-------+
| Variable_name            | Value |
+--------------------------+-------+
| enforce_gtid_consistency | WARN  |
+--------------------------+-------+
1 row in set (0.00 sec)

mysql> select sleep(60);
+-----------+
| sleep(60) |
+-----------+
|         0 |
+-----------+
1 row in set (59.99 sec)

# 查看所有数据库 Server 中的错误日志，看是否有警告信息（从下面的结果中可以看到，主从库中都没有发现警告信息）
## 主库
[root@localhost binlog]# grep -iE 'warn|note' /data/mysqldata1/log/error.log |tail -100
2019-07-13T15:27:19.679526+08:00 18 [Note] Changed ENFORCE_GTID_CONSISTENCY from ON to WARN.

## 从库
[root@localhost backup]# grep -iE 'warn|note' /data/mysqldata1/log/error.log |tail -100
2019-07-13T15:27:32.525957+08:00 24 [Note] Changed ENFORCE_GTID_CONSISTENCY from ON to WARN.
2019-07-13T15:28:50.038907+08:00 26 [Note] Multi-threaded slave statistics for channel '': seconds elapsed = 120; events assigned = 1886209; worker queues filled over overrun level = 0; waited due a Worker queue full = 0; waited due the total size = 0; waited at clock conflicts = 1179227400 waited (count) when Workers occupied = 0 waited when Workers occupied = 0
```

（2）在每个数据库 Server 上执行如下语句（主从库各自执行）：

```
mysql> set @@global.enforce_gtid_consistency = ON;
Query OK, 0 rows affected (0.00 sec)

mysql> show variables like 'enforce_gtid_consistency';
+--------------------------+-------+
| Variable_name            | Value |
+--------------------------+-------+
| enforce_gtid_consistency | ON    |
+--------------------------+-------+
1 row in set (0.00 sec)
```

（3）在每个数据库 Server 上执行如下语句（主从库各自执行）：

```
mysql> set @@global.gtid_mode = OFF_PERMISSIVE;
Query OK, 0 rows affected (0.01 sec)

mysql> show variables like 'gtid_mode';
+---------------+----------------+
| Variable_name | Value          |
+---------------+----------------+
| gtid_mode     | OFF_PERMISSIVE |
+---------------+----------------+
1 row in set (0.01 sec)
```

根据系统变量 gtid_mode 的解释，当其被设置为 OFF_PERMISSIVE 时，写库新的事务是匿名事务，查看一下主从库当前的 GTID 值是否增加
主库（从下面的结果来看，状态变量 gtid_executed 为空，表示未出现 GTID 事务，结果正常）
```
mysql> show global variables like 'gtid_executed';
+---------------+-------+
| Variable_name | Value |
+---------------+-------+
| gtid_executed |       |
+---------------+-------+
1 row in set (0.00 sec)
```

从库（从下面的结果来看，状态变量 gtid_executed 为空，表示未出现 GTID 事务，结果正常）
```
mysql> show global variables like 'gtid_executed';
+---------------+-------+
| Variable_name | Value |
+---------------+-------+
| gtid_executed |       |
+---------------+-------+
1 row in set (0.00 sec)
```

（4）在每个数据库 Server 上执行如下语句（主从库各自执行）：

```
mysql> set @@global.gtid_mode = ON_PERMISSIVE;
Query OK, 0 rows affected (0.01 sec)

mysql> show variables like 'GTID_MODE';
+---------------+---------------+
| Variable_name | Value         |
+---------------+---------------+
| gtid_mode     | ON_PERMISSIVE |
+---------------+---------------+
1 row in set (0.00 sec)
```

根据系统变量 gtid_mode 的解释，当其被设置为 ON_PERMISSIVE 时，写库新的事务是 GTID 事务，查看一下主从库当前的 GTID 值是否增加
主库（从下面的结果来看，状态变量 gtid_executed 已有 GTID 信息，表示出现 GTID 事务，结果正常）
```
mysql> show global variables like 'gtid_executed';
```

```
+----------------+------------------------------------------+
| Variable_name  | Value                                    |
+----------------+------------------------------------------+
| gtid_executed  | f3372787-0719-11e8-af1f-0025905b06da:1-6590|
+----------------+------------------------------------------+
1 row in set (0.01 sec)

## 从库（我们对主库使用 sysbench 持续加压，由于人工查看存在时差，从主库上抓取的 GTID SET 可能比从库的小）
mysql> show global variables like 'gtid_executed';
+----------------+------------------------------------------+
| Variable_name  | Value                                    |
+----------------+------------------------------------------+
| gtid_executed  | f3372787-0719-11e8-af1f-0025905b06da:1-6570 |
+----------------+------------------------------------------+
1 row in set (0.00 sec)
```

（5）在每个数据库 Server 上，等待状态变量 ONGOING_ANONYMOUS_TRANSACTION_COUNT 的值变更为 0（主从库须分别查看），可以使用如下语句查看（注意，查看这个值的时候，如果多次查询，值有可能有时候为 0，有时候不为 0，这是正常的，只要查看到有一次为 0 即可）。必须等待所有 Server 的 ONGOING_ANONYMOUS_TRANSACTION_COUNT 都被查询到有为 0 的情况才可以继续进行下一步。

```
mysql> show status like 'ongoing_anonymous_transaction_count';
+-------------------------------------+-------+
| Variable_name                       | Value |
+-------------------------------------+-------+
| Ongoing_anonymous_transaction_count | 1     |
+-------------------------------------+-------+
1 row in set (0.00 sec)
......
mysql> show status like 'ongoing_anonymous_transaction_count';
+-------------------------------------+-------+
| Variable_name                       | Value |
+-------------------------------------+-------+
| Ongoing_anonymous_transaction_count | 0     |   #发现值为 0
+-------------------------------------+-------+
1 row in set (0.00 sec)
```

（6）如果你的复制拓扑中有用于备份的数据库 Server，那么在所有 Server 上完成第 5 步之后（也就是执行第 6 步之前），再在备份 Server 上执行 FLUSH LOGS 语句，手动触发一次备份操作，以便做备份恢复操作时，能够基于时间点进行恢复。然后，清理除最新的二进制日志文件之外的二进制日志文件（这是第 2 个重要步骤，你必须确保整个复制拓扑中的任何地方都不存在匿名事务，包括没有同步到其他从库的二进制日志或者备份中。因为执行完下一个步骤之后，复制拓扑中不再允许存在匿名事务，否则复制会报错并中止）。

```
# 查询主库的二进制日志文件和二进制日志文件位置
mysql> show master status;
```

```
+-------------------+--------------+-----------------+------------------+---------------+
|File|Position | Binlog_Do_DB | Binlog_Ignore_DB |Executed_Gtid_Set |
+-------------------+--------------+-----------------+------------------+---------------+
| mysql-bin.000006 | 403442852 |  |  | f3372787-0719-11e8-af1f-0025905b06da:1-177975 |
+-------------------+--------------+-----------------+------------------+---------------+
1 row in set (0.00 sec)
```

所有从库执行语句 SELECT MASTER_POS_WAIT('mysql-bin.000006','403442852');必须返回0，不为0时重复执行这个语句

```
mysql> select master_pos_wait('mysql-bin.000006','403442852');
+-------------------------------------------------+
| master_pos_wait('mysql-bin.000006','403442852') |
+-------------------------------------------------+
| 0 |
+-------------------------------------------------+
1 row in set (0.00 sec)
```

在备份 Server 中执行 FLUSH LOGS 语句（建议所有数据库 Server 都执行该语句，这样就可以立即触发将 GTID 记录到 mysql.gtid_executed 表中）

主库

```
mysql> flush logs;
Query OK, 0 rows affected (0.01 sec)

mysql> show master status;
+-------------------+-----------+--------------+------------------+-----------------+
|File|Position| Binlog_Do_DB | Binlog_Ignore_DB | Executed_Gtid_Set |
+-------------------+-----------+--------------+------------------+-----------------+
| mysql-bin.000007 | 10001658 |  |  | f3372787-0719-11e8-af1f-0025905b06da:1-226338 |
+-------------------+-----------+--------------+------------------+-----------------+
1 row in set (0.00 sec)
```

从库

```
mysql> flush logs;
Query OK, 0 rows affected (0.01 sec)

mysql> show master status;
+-------------------+-----------+--------------+------------------+-----------------+
|File|Position| Binlog_Do_DB | Binlog_Ignore_DB | Executed_Gtid_Set |
+-------------------+-----------+--------------+------------------+-----------------+
| mysql-bin.000007 | 19050462 |  |  | f3372787-0719-11e8-af1f-0025905b06da:1-229720 |
+-------------------+-----------+--------------+------------------+-----------------+
1 row in set (0.00 sec)
```

在备份 Server 中执行备份数据（操作步骤略，如果没有备份专用 Server，至少需要选择一个从库进行备份）
……

在备份 Server 中执行备份二进制日志（操作步骤略，如果没有备份专用 Server，至少需要选择一个从库进行备份）

```
......
# 在主从库中都清理除最新的二进制日志之外的其他二进制日志文件。可用 PURGE BINARY LOGS TO
'mysql-bin.xx';语句来清理
## 主库（需要等到所有从库的 SQL 线程应用到 mysql-bin.000007 二进制日志文件的位置，即从库中执行 SHOW
SLAVE STATUS 语句时，其输出结果中 Relay_Master_Log_File 选项值包含的二进制日志文件名中的数字编号要大
于或等于 000007，否则主库不能清理二进制日志。若强行清理，会导致存在复制延迟的从库发现其未读取的二进制日志
被清理而造成从库的复制中断）
mysql> purge binary logs to 'mysql-bin.000007';
Query OK, 0 rows affected (0.25 sec)

## 从库
mysql> purge binary logs to 'mysql-bin.000007';
Query OK, 0 rows affected (0.27 sec)
......
```

（7）在每个数据库 Server 上执行如下语句（主从库各自执行）：

```
mysql> set @@global.gtid_mode = ON;
Query OK, 0 rows affected (0.01 sec)

mysql> show variables like 'gtid_mode';
+---------------+-------+
| Variable_name | Value |
+---------------+-------+
| gtid_mode     | ON    |
+---------------+-------+
1 row in set (0.00 sec)
```

（8）在每个数据库 Server 上，将 gtid_mode = ON 和 enforce_gtid_consistency = ON 添加到 my.cnf 文件中：

```
[root@localhost ~]# cat /etc/my.cnf
[mysqld]
gtid_mode = ON
enforce_gtid_consistency = ON
......
```

（9）现在可以保证复制拓扑中所有事务都具有 GTID。此时，需要在复制拓扑的每个从库上执行以下操作，以便启用 GTID 自动定位（如果使用了多源复制，请对每个通道执行如下操作）：

```
# 单通道复制需要执行的步骤如下
mysql> stop slave;
Query OK, 0 rows affected (0.01 sec)

# 不加复制通道名称时，对应的是默认的复制通道
mysql> change master to master_auto_position = 1;
Query OK, 0 rows affected (0.01 sec)
```

```
mysql> start slave;
Query OK, 0 rows affected (0.02 sec)

mysql> show slave status\G
*************************** 1. row ***************************
               Slave_IO_State: Waiting for master to send event
                  Master_Host: 10.10.30.161
                  Master_User: repl
                  Master_Port: 3306
                Connect_Retry: 60
              Master_Log_File: mysql-bin.000009
          Read_Master_Log_Pos: 15392781
               Relay_Log_File: mysql-relay-bin.000002
                Relay_Log_Pos: 6834419
        Relay_Master_Log_File: mysql-bin.000008
             Slave_IO_Running: Yes
            Slave_SQL_Running: Yes
......
        Seconds_Behind_Master: 41
......
           Retrieved_Gtid_Set: f3372787-0719-11e8-af1f-0025905b06da:522715-535545
            Executed_Gtid_Set: f3372787-0719-11e8-af1f-0025905b06da:1-525773
                Auto_Position: 1
......
1 row in set (0.00 sec)

# 如果使用了多源复制，则在对每个复制通道各自执行如下操作时需要带上 FOR CHANNEL 子句
STOP SLAVE [FOR CHANNEL 'channel'];
CHANGE MASTER TO MASTER_AUTO_POSITION = 1 [FOR CHANNEL 'channel'];
START SLAVE [FOR CHANNEL 'channel'];
```

至此，我们就完成了从传统复制到 GTID 复制的切换，可停止 sysbench 的加压（此为模拟步骤，真实环境中可忽略）。

如果你还需要校验主从库数据的一致性，那么主从复制的延迟不能过大，否则无法进行（在真实场景中，这个步骤需要停止业务）。使用 pt-table-checksum 校验主从库数据的一致性。

注意：以上变更步骤的每一步都至关重要，且顺序不能颠倒。

17.4 GTID 复制在线变更为传统复制

17.4.1 简要步骤

将 GTID 复制在线变更为传统复制的前提条件：

- 复制拓扑中的所有数据库 Server 都必须使用 MySQL 5.7.6 或更高版本。

- 变更前，所有数据库 Server 的系统变量 gtid_mode 都设置为 ON。

将 GTID 复制在线变更为传统复制过程中的关键步骤如图 17-2 所示。

图 17-2

提示：通常不需要将 GTID 复制模式变更为传统复制，除非有特殊需求。

17.4.2 详细过程

首先，使用 sysbench 对主库持续加压，不需要加太大压力，只需要有持续不断的数据变更即可（此为模拟步骤，真实环境中可忽略）。

（1）在每个从库上执行以下操作（所有从库各自执行），如果使用了多源复制，需要在每个通道上执行以下操作：

```
# 单通道中执行如下步骤
mysql> show global variables like 'gtid_executed';
+---------------+----------------------------------------------+
| Variable_name | Value                                        |
+---------------+----------------------------------------------+
| gtid_executed | f3372787-0719-11e8-af1f-0025905b06da:1-625291 |
+---------------+----------------------------------------------+
1 row in set (0.00 sec)

mysql> stop slave;
Query OK, 0 rows affected (0.00 sec)

# 查看 Master_Log_File、Read_Master_Log_Pos、Relay_Log_File、Relay_Log_Pos 的值
mysql> show slave status\G
```

```
*************************** 1. row ***************************
             Slave_IO_State:
                Master_Host: 10.10.30.161
                Master_User: repl
                Master_Port: 3306
              Connect_Retry: 60
            Master_Log_File: mysql-bin.000009
        Read_Master_Log_Pos: 226032830
             Relay_Log_File: mysql-relay-bin.000004
              Relay_Log_Pos: 226032978
......
```

将SHOW SLAVE STATUS语句获取的Relay_Log_File、Relay_Log_Pos值拼接成如下语句，并在从库中执行（CHANGE MASTER TO 语句中 Master_Log_File、Master_Log_Pos 选项的值可以不指定，但如果不指定 Relay_Log_File、Relay_Log_Pos 选项的值，一旦从库存在复制延迟，则未应用的中继日志会被清理，可能导致复制过程出现异常）

```
mysql> change master to relay_log_file='mysql-relay-bin.000004', relay_log_pos= 226032978, master_auto_position = 0;
Query OK, 0 rows affected (0.02 sec)

mysql> start slave;
Query OK, 0 rows affected (0.02 sec)

mysql> show slave status\G
*************************** 1. row ***************************
             Slave_IO_State: Waiting for master to send event
                Master_Host: 10.10.30.161
                Master_User: repl
                Master_Port: 3306
              Connect_Retry: 60
            Master_Log_File: mysql-bin.000009
        Read_Master_Log_Pos: 415901892
             Relay_Log_File: mysql-relay-bin.000005
              Relay_Log_Pos: 17287778
      Relay_Master_Log_File: mysql-bin.000009
           Slave_IO_Running: Yes
          Slave_SQL_Running: Yes
......
         Retrieved_Gtid_Set: f3372787-0719-11e8-af1f-0025905b06da:522715-712225
          Executed_Gtid_Set: f3372787-0719-11e8-af1f-0025905b06da:1-636299
              Auto_Position: 0
......
1 row in set (0.00 sec)
```

如果使用了多源复制，则在每个复制通道中按照如下步骤操作
```
STOP SLAVE [FOR CHANNEL 'channel'];
CHANGE MASTER TO RELAY_LOG_FILE='relay_log_file', RELAY_LOG_POS=pos, MASTER_AUTO_POSITION = 0 [FOR CHANNEL 'channel'];
```

```
START SLAVE [FOR CHANNEL 'channel'];
```

（2）在每个数据库 Server 上执行如下语句（主从库各自执行）：

```
mysql> set @@global.gtid_mode = ON_PERMISSIVE;
Query OK, 0 rows affected (0.01 sec)

mysql> show variables like '%gtid%';
+--------------------------------+------------------------------------------+
| Variable_name                  | Value                                    |
+--------------------------------+------------------------------------------+
| binlog_gtid_simple_recovery    | ON                                       |
| enforce_gtid_consistency       | ON                                       |
| gtid_executed_compression_period | 1000                                   |
| gtid_mode                      | ON_PERMISSIVE                            |
| gtid_next                      | AUTOMATIC                                |
| gtid_owned                     |                                          |
| gtid_purged                    | f3372787-0719-11e8-af1f-0025905b06da:1-221926 |
| session_track_gtids            | OFF                                      |
+--------------------------------+------------------------------------------+
8 rows in set (0.00 sec)
```

（3）在每个数据库 Server 上执行如下语句（主从库各自执行）：

```
mysql> set @@global.gtid_mode = OFF_PERMISSIVE;
Query OK, 0 rows affected (0.00 sec)

mysql> show variables like '%gtid%';
+--------------------------------+------------------------------------------+
| Variable_name                  | Value                                    |
+--------------------------------+------------------------------------------+
| binlog_gtid_simple_recovery    | ON                                       |
| enforce_gtid_consistency       | ON                                       |
| gtid_executed_compression_period | 1000                                   |
| gtid_mode                      | OFF_PERMISSIVE                           |
| gtid_next                      | AUTOMATIC                                |
| gtid_owned                     |                                          |
| gtid_purged                    | f3372787-0719-11e8-af1f-0025905b06da:1-221926 |
| session_track_gtids            | OFF                                      |
+--------------------------------+------------------------------------------+
8 rows in set (0.00 sec)

# 在所有数据库 Server 中查看当前 GTID SET，确保主库的最新 GTID 都已同步到所有从库
## 主库
mysql> show global variables like 'gtid_executed';
+---------------+------------------------------------------------+
| Variable_name | Value                                          |
+---------------+------------------------------------------------+
| gtid_executed | f3372787-0719-11e8-af1f-0025905b06da:1-777234  |
+---------------+------------------------------------------------+
```

```
1 row in set (0.01 sec)

## 从库
mysql> show global variables like 'gtid_executed';
+----------------+------------------------------------------------+
| Variable_name  | Value                                          |
+----------------+------------------------------------------------+
| gtid_executed  | f3372787-0719-11e8-af1f-0025905b06da:1-777234 |
+----------------+------------------------------------------------+
1 row in set (0.00 sec)
```

（4）在每个数据库 Server 上，等待变量@@GLOBAL.GTID_OWNED 的值变为空字符串。可以使用如下语句查看（主从库各自查看。注意，当多次查询时，可能有时是空值，有时为非空值，但只要有一次查询到为空值即可）：

```
mysql> select @@global.gtid_owned;
+---------------------+
| @@global.gtid_owned |
+---------------------+
|                     |
+---------------------+
1 row in set (0.00 sec)
```

（5）等待主库的所有二进制日志同步到所有从库，且完成应用：

```
# 在主库上使用 SHOW MASTER STATUS 语句查看二进制日志位置
mysql> show master status;
+-----------------+-----------+--------------+------------------+-----------------------------------------------+
|File|Position| Binlog_Do_DB | Binlog_Ignore_DB | Executed_Gtid_Set |
+-----------------+-----------+--------------+------------------+-----------------------------------------------+
|mysql-bin.000012 | 127428335 |  |  | f3372787-0719-11e8-af1f-0025905b06da:1-777234 |
+-----------------+-----------+--------------+------------------+-----------------------------------------------+
1 row in set (0.00 sec)

# 在所有从库上使用 select master_pos_wait('mysql-bin.000012','127428335');语句，必须返回 0
值，表示主库中所有的 GTID 事务已经复制到从库，不为 0 时重复执行这个语句
mysql> select master_pos_wait('mysql-bin.000012','127428335');
+-------------------------------------------------+
| master_pos_wait('mysql-bin.000012','127428335') |
+-------------------------------------------------+
| 0 |
+-------------------------------------------------+
1 row in set (0.00 sec)
```

（6）如果将二进制日志用于除复制之外的其他任何操作（例如进行基于时间点的备份或还原），则需要等待具有 GTID 的事务完全同步到其他所有从库（在从库执行上述验证步骤，即可确认主库中的 GTID 事务是否已完全同步到从库）。然后，在备份 Server 中执行 FLUSH LOGS，再手动触发一次备份操作。这里的备份操作指的是备份二进制日志。因为在完全切换为 GTID 复制模式时，需要清理匿名事务的二进制日志，否则在 GTID 复制模式

中读取到匿名事务时会报错。在执行清理之前，手工备份匿名事务的二进制日志到一个指定位置以备后用。

最后，清理最新二进制日志之前的所有二进制日志。这是第一个重要步骤，下一步就不能使用包含 GTID 事务的二进制日志了，所以在继续进行下一步之前，必须确保复制拓扑中的任何地方都不存在 GTID 事务，也没有同步到其他从库的二进制日志和备份。

```
# 在备份 Server 中执行 FLUSH LOGS 语句（建议在所有数据库 Server 中都执行，这样就可以立即触发将 GTID
记录到 mysql.gtid_executed 表中）
## 主库
mysql> select * from mysql.gtid_executed;
+--------------------------------------+----------------+--------------+
| source_uuid                          | interval_start | interval_end |
+--------------------------------------+----------------+--------------+
| f3372787-0719-11e8-af1f-0025905b06da | 1              | 777234       |
+--------------------------------------+----------------+--------------+
1 row in set (0.00 sec)

mysql> flush logs;
Query OK, 0 rows affected (0.01 sec)

mysql> show master status;
+----------------+----------+--------------+------------------+-------------------+
|File|Position| Binlog_Do_DB | Binlog_Ignore_DB | Executed_Gtid_Set |
+----------------+----------+--------------+------------------+-------------------+
|mysql-bin.000013 | 2675254 | | | f3372787-0719-11e8-af1f-0025905b06da:1-777234 |
+----------------+----------+--------------+------------------+-------------------+
1 row in set (0.00 sec)

## 从库
mysql> flush logs;
Query OK, 0 rows affected (0.01 sec)

mysql> show master status;
+----------------+----------+--------------+------------------+-------------------+
|File|Position| Binlog_Do_DB | Binlog_Ignore_DB | Executed_Gtid_Set |
+----------------+----------+--------------+------------------+-------------------+
|mysql-bin.000012 | 2109938 | | | f3372787-0719-11e8-af1f-0025905b06da:1-777234 |
+----------------+----------+--------------+------------------+-------------------+
1 row in set (0.00 sec)

# 在备份 Server 中执行备份数据（具体步骤略。如果没有备份专用的 Server，至少需要选择一个从库进行备份）
……

# 在备份 Server 中执行备份二进制日志（具体步骤略。如果没有备份专用的 Server，至少需要选择一个从库进行备份）
……
```

```
# 在主从库中都清理除最新二进制日志之外的其他二进制日志文件，可用 PURGE BINARY LOGS TO 'mysql-
bin.xx';语句清理
## 主库（需要等所有从库的 SQL 线程应用到 mysql-bin.000013 二进制日志文件，即从库执行 SHOW SLAVE
STATUS 语句输出中 Relay_Master_Log_File 选项值的二进制日志文件的文件名编号需要大于或等于 000013，否则
主库不能清理二进制日志，会造成复制有延迟的从库的复制中断）
mysql> purge binary logs to 'mysql-bin.000013';
Query OK, 0 rows affected (0.32 sec)

## 从库
mysql> purge binary logs to 'mysql-bin.000012';
Query OK, 0 rows affected (0.35 sec)
......
```

（7）在每个数据库 Server 上执行如下语句（主从库各自执行）：

```
mysql> set @@global.gtid_mode = OFF;
Query OK, 0 rows affected (0.01 sec)

mysql> show variables like '%gtid%';
+--------------------------------+-------------------------------------------+
| Variable_name                  | Value                                     |
+--------------------------------+-------------------------------------------+
| binlog_gtid_simple_recovery    | ON                                        |
| enforce_gtid_consistency       | ON                                        |
| gtid_executed_compression_period | 1000                                    |
| gtid_mode                      | OFF                                       |
| gtid_next                      | AUTOMATIC                                 |
| gtid_owned                     |                                           |
| gtid_purged                    | f3372787-0719-11e8-af1f-0025905b06da:1-777234 |
| session_track_gtids            | OFF                                       |
+--------------------------------+-------------------------------------------+
8 rows in set (0.00 sec)
```

（8）在每个数据库 Server 上的 my.cnf 文件中设置 gtid_mode = OFF 和 enforce_gtid_consistency = OFF，如果要降级到早期版本的 MySQL，现在可以使用正常的降级过程：

```
[root@localhost ~]# cat /etc/my.cnf
[mysqld]
gtid_mode = OFF
enforce_gtid_consistency = OFF
......
```

查看所有从库的复制状态（多次查询同一个从库，此时你会发现 GTID SET 值不再增加）：

```
mysql> show slave status\G
*************************** 1. row ***************************
               Slave_IO_State: Waiting for master to send event
                  Master_Host: 10.10.30.161
                  Master_User: repl
                  Master_Port: 3306
```

```
            Connect_Retry: 60
          Master_Log_File: mysql-bin.000016
      Read_Master_Log_Pos: 106222973
           Relay_Log_File: mysql-relay-bin.000021
            Relay_Log_Pos: 106182380
    Relay_Master_Log_File: mysql-bin.000016
         Slave_IO_Running: Yes
        Slave_SQL_Running: Yes
......
       Retrieved_Gtid_Set: f3372787-0719-11e8-af1f-0025905b06da:522715-777234
        Executed_Gtid_Set: f3372787-0719-11e8-af1f-0025905b06da:1-777234
            Auto_Position: 0
......
```

至此，我们就完成了从 GTID 复制到传统复制的在线切换，可停止用 sysbench 加压（此为模拟步骤，在真实环境中可忽略）。

如果你还需要校验主从库数据的一致性，那么主从复制的延迟不能过大，否则无法进行（在真实场景中这里需要停止业务）。使用 pt-table-checksum 来校验主从库数据的一致性。

注意：以上变更步骤的每一步都至关重要，且顺序不能颠倒。

17.5　GTID 复制离线变更为传统复制

17.5.1　简要步骤

将 GTID 复制离线变更为传统复制的前提条件：

- 复制拓扑中的所有数据库 Server 都必须使用 MySQL 5.7.6 或更高版本。
- 变更前，所有数据库 Server 的系统变量 gtid_mode 都设置为 ON。

将 GTID 复制离线变更为传统复制过程中的关键步骤如图 17-3 所示。

图 17-3

提示：通常不需要将 GTID 复制变更为传统复制，除非有特殊需求。

17.5.2 详细过程

停止业务的所有读/写操作，并等待所有从库追赶上主库的数据（比较 Master_Log_File 和 Relay_Master_Log_File 的值，Read_Master_Log_Pos 和 Exec_Master_Log_Pos 的值，若二者相等，则说明从库已经同步主库数据）。

```
mysql> show slave status\G
*************************** 1. row ***************************
               Slave_IO_State: Waiting for master to send event
                  Master_Host: 10.10.30.161
                  Master_User: repl
                  Master_Port: 3306
                Connect_Retry: 60
              Master_Log_File: mysql-bin.000016
          Read_Master_Log_Pos: 172545168
               Relay_Log_File: mysql-relay-bin.000021
                Relay_Log_Pos: 172545381
        Relay_Master_Log_File: mysql-bin.000016  # Relay_Master_Log_File 与 Master_Log_
File 值相等
             Slave_IO_Running: Yes
            Slave_SQL_Running: Yes
......
          Exec_Master_Log_Pos: 172545168  # Exec_Master_Log_Pos 与 Read_Master_Log_Pos
的值相等，说明从库已经完全追赶上主库的数据
......
```

把系统变量 gtid_mode 和 enforce_gtid_consistency 都设置为 OFF 并加入配置文件：

```
[root@localhost ~]# cat /etc/my.cnf
[mysqld]
gtid_mode=OFF
enforce_gtid_consistency=OFF
......
```

在所有的主从数据库中执行二进制日志的滚动，并备份滚动之前的二进制日志，因为后面需要清理这些二进制日志，建议将其备份，以防万一后续还可能要使用它们：

```
# 登录所有数据库实例，执行日志滚动
mysql> flush binary logs;
Query OK, 0 rows affected (0.02 sec)

# 获取最新的二进制日志文件名（如果不做二进制日志的备份，则无须执行此操作。通常，建议在配置过定时数据
备份策略的数据库实例上将二进制日志一同备份）
mysql> show master status;
+----------+----------+--------------+------------------+-------------------+
|File|Position| Binlog_Do_DB | Binlog_Ignore_DB | Executed_Gtid_Set |
+----------+----------+--------------+------------------+-------------------+
```

```
|mysql-bin.000018 |  194 | |  | f3372787-0719-11e8-af1f-0025905b06da:1-777234 |
+-----------------+------+--+--+------------------------------------------------+
1 row in set (0.00 sec)
```

获取的最新二进制日志文件名为 mysql-bin.000018,则需要执行备份与清理的二进制日志文件为 mysql-bin.000017 及其之前的所有二进制日志

```
[root@localhost binlog]# ll
total 1364076
-rw-r----- 1 mysql mysql 536871118 Jul 13 17:12 mysql-bin.000013
-rw-r----- 1 mysql mysql 536872355 Jul 13 17:27 mysql-bin.000014
-rw-r----- 1 mysql mysql 150501691 Jul 13 17:31 mysql-bin.000015
-rw-r----- 1 mysql mysql 172545191 Jul 13 19:20 mysql-bin.000016
-rw-r----- 1 mysql mysql       241 Jul 13 19:26 mysql-bin.000017
-rw-r----- 1 mysql mysql       194 Jul 13 19:26 mysql-bin.000018
-rw-r----- 1 mysql mysql       312 Jul 13 19:26 mysql-bin.index

[root@localhost binlog]# mkdir /data/backup
[root@localhost binlog]# \cp -ar mysql-bin.0000{13,14,15,16,17} /data/backup/
[root@localhost binlog]# ll /data/backup
total 1364068
-rw-r----- 1 mysql mysql 536871118 Jul 13 17:12 mysql-bin.000013
-rw-r----- 1 mysql mysql 536872355 Jul 13 17:27 mysql-bin.000014
-rw-r----- 1 mysql mysql 150501691 Jul 13 17:31 mysql-bin.000015
-rw-r----- 1 mysql mysql 172545191 Jul 13 19:20 mysql-bin.000016
-rw-r----- 1 mysql mysql       241 Jul 13 19:26 mysql-bin.000017

# 清理mysql-bin.000018之前的所有二进制日志
mysql> purge binary logs to 'mysql-bin.000018';
Query OK, 0 rows affected (0.00 sec)

mysql> show binary logs;
+------------------+-----------+
| Log_name         | File_size |
+------------------+-----------+
| mysql-bin.000018 |       194 |
+------------------+-----------+
1 row in set (0.00 sec)
```

重新启动所有数据库 Server(对主从库分别操作):

```
# 先重启主库,然后重启从库
[root@localhost ~]# service mysqld restart
Shutting down MySQL........... SUCCESS!
Starting MySQL..... SUCCESS!

# 登录所有数据库 Server 查看系统变量 gtid_mode 和 enforce_gtid_consistency 是否已关闭,从下面的输出信息中可以看到 gtid_mode 和 enforce_gtid_consistency 都为 OFF,表示关闭成功
mysql> show variables like '%gtid%';
+------------------------+---------------------------------------------+
```

```
| Variable_name                         | Value                                    |
+---------------------------------------+------------------------------------------+
| binlog_gtid_simple_recovery           | ON                                       |
| enforce_gtid_consistency              | OFF                                      |
| gtid_executed_compression_period      | 1000                                     |
| gtid_mode                             | OFF                                      |
| gtid_next                             | AUTOMATIC                                |
| gtid_owned                            |                                          |
| gtid_purged                           | f3372787-0719-11e8-af1f-0025905b06da:1-777234 |
| session_track_gtids                   | OFF                                      |
+---------------------------------------+------------------------------------------+
8 rows in set (0.01 sec)
```

登录所有从库,设置 master_auto_position = 0,关闭自动定位:

```
mysql> stop slave;
Query OK, 0 rows affected (0.00 sec)

# 查看 Master_Log_File、Read_Master_Log_Pos、Relay_Log_File、Relay_Log_Pos 的值
mysql> show slave status\G
*************************** 1. row ***************************
               Slave_IO_State:
                  Master_Host: 10.10.30.161
                  Master_User: repl
                  Master_Port: 3306
                Connect_Retry: 60
              Master_Log_File: mysql-bin.000019
          Read_Master_Log_Pos: 194
               Relay_Log_File: mysql-relay-bin.000029
                Relay_Log_Pos: 367
        Relay_Master_Log_File: mysql-bin.000019
......
```

将 SHOW SLAVE STATUS 语句获取的 Relay_Log_File、Relay_Log_Pos 的值拼接成如下语句,并在从库中执行(CHANGE MASTER TO 语句的 Master_Log_File、Master_Log_Pos 选项的值可以不指定,但如果不指定 Relay_Log_File、Relay_Log_Pos 选项的值,而从库存在复制延迟,则未应用的中继日志会被清理,可能导致复制出现异常)

```
mysql> change master to relay_log_file='mysql-relay-bin.000029', relay_log_pos=367,
master_auto_position = 0;
Query OK, 0 rows affected (0.02 sec)

mysql> start slave;
Query OK, 0 rows affected (0.02 sec)

mysql> show slave status\G
*************************** 1. row ***************************
               Slave_IO_State: Waiting for master to send event
                  Master_Host: 10.10.30.161
                  Master_User: repl
```

```
                Master_Port: 3306
              Connect_Retry: 60
            Master_Log_File: mysql-bin.000019
        Read_Master_Log_Pos: 194
             Relay_Log_File: mysql-relay-bin.000030
              Relay_Log_Pos: 320
      Relay_Master_Log_File: mysql-bin.000019
           Slave_IO_Running: Yes
          Slave_SQL_Running: Yes
......
          Retrieved_Gtid_Set:
          Executed_Gtid_Set: f3372787-0719-11e8-af1f-0025905b06da:1-777234
              Auto_Position: 0
......
1 row in set (0.00 sec)
```

```
# 如果使用了多源复制，则在每个复制通道中按照如下步骤操作
STOP SLAVE [FOR CHANNEL 'channel'];
CHANGE MASTER TO RELAY_LOG_FILE='relay_log_file',RELAY_LOG_POS=pos,MASTER_AUTO_POSITION
= 0 [FOR CHANNEL 'channel'];
START SLAVE [FOR CHANNEL 'channel'];
```

在主库中使用 sysbench 加压 1 分钟，然后登录所有从库，查看是否能正常同步数据：

```
# 对主库用 sysbench 加压（命令略）

# 查看从库的复制状态
mysql> show slave status\G
*************************** 1. row ***************************
             Slave_IO_State: Waiting for master to send event
                Master_Host: 10.10.30.161
                Master_User: repl
                Master_Port: 3306
              Connect_Retry: 60
            Master_Log_File: mysql-bin.000019
        Read_Master_Log_Pos: 12044225
             Relay_Log_File: mysql-relay-bin.000030
              Relay_Log_Pos: 11894729
      Relay_Master_Log_File: mysql-bin.000019
           Slave_IO_Running: Yes      # I/O 线程状态正常
          Slave_SQL_Running: Yes      # SQL 线程状态正常
......
         Exec_Master_Log_Pos: 11894603
......
       Seconds_Behind_Master: 0
......
          Retrieved_Gtid_Set:
          Executed_Gtid_Set: f3372787-0719-11e8-af1f-0025905b06da:1-777234
              Auto_Position: 0
```

```
......
1 row in set (0.00 sec)
```

至此，我们就完成了从 GTID 复制到传统复制的离线切换。如果有需要，可以在主库使用 pt-table-checksum 校验主从库的数据一致性（从以下结果中可以看到，DIFFS 列全部为 0，表示主从库的数据此时是一致的）。

```
[root@localhost ~]# pt-table-checksum --nocheck-replication-filters --no-check- binlog-
format --replicate=xiaoboluo.checksums h=localhost,u=admin,p=password,P=3306
        TS        ERRORS DIFFS  ROWS    CHUNKS SKIPPED TIME   TABLE
......
0508T23:32:51  0   0    1       1      0       0.008  qfsys.qfsys_heartbeat
05-08T23:33:05 0   0    6154192 24     0       14.461 sbtest.sbtest1
......
```

17.6 传统复制离线变更为 GTID 复制

17.6.1 简要步骤

将传统复制离线变更为 GTID 复制的前提条件：

- 复制拓扑中的所有数据库 Server 都必须使用 MySQL 5.7.6 或更高版本。
- 变更前，所有数据库 Server 的系统变量 gtid_mode 都设置为 OFF。

将传统复制离线变更为 GTID 复制过程中的关键步骤如图 17-4 所示。

图 17-4

17.6.2 详细过程

停止业务的所有读/写操作，并等待所有从库追赶上主库的数据（比较 Master_Log_File 和 Relay_Master_Log_File 的值，Read_Master_Log_Pos 和 Exec_Master_Log_Pos 的值，若二者相等，则说明从库已经同步主库数据）：

```
mysql> show slave status\G
```

```
*************************** 1. row ***************************
               Slave_IO_State: Waiting for master to send event
                  Master_Host: 10.10.30.161
                  Master_User: repl
                  Master_Port: 3306
                Connect_Retry: 60
              Master_Log_File: mysql-bin.000019
          Read_Master_Log_Pos: 80353936
               Relay_Log_File: mysql-relay-bin.000030
                Relay_Log_Pos: 80354062
        Relay_Master_Log_File: mysql-bin.000019  # Relay_Master_Log_File 与 Master_Log_
File 值相等
             Slave_IO_Running: Yes
            Slave_SQL_Running: Yes
......
           Exec_Master_Log_Pos: 80353936 # Exec_Master_Log_Pos 与 Read_Master_Log_Pos 值
相等，说明从库已经完全追赶上主库的数据
......
```

把系统变量 gtid_mode 和 enforce_gtid_consistency 都设置为 ON，并加入配置文件：

```
[root@localhost ~]# cat /etc/my.cnf
[mysqld]
gtid_mode=ON
enforce_gtid_consistency=ON
......
```

在所有的主从库中都对二进制日志执行日志滚动，并备份滚动之前的二进制日志，因为后面需要清理这些二进制日志，建议将其备份，以防万一后续可能还要使用它们：

```
# 登录所有数据库实例，执行日志滚动
mysql> flush binary logs;
Query OK, 0 rows affected (0.02 sec)

# 获取最新的二进制日志文件名（如果实例不对二进制日志做备份，则无须执行此操作。通常建议在配置过定时数
据备份的数据库上执行）
mysql> show master status;
+------------------+----------+--------------+------------------+--------------+
|File|Position| Binlog_Do_DB | Binlog_Ignore_DB | Executed_Gtid_Set |
+------------------+----------+--------------+------------------+--------------+
|mysql-bin.000020|194|||f3372787-0719-11e8-af1f-0025905b06da:1-777234 |
+------------------+----------+--------------+------------------+--------------+
1 row in set (0.00 sec)

# 获取的最新文件名为 mysql-bin.000020，则需要执行备份与清理的二进制日志文件为 mysql-bin.000019
及其之前的所有二进制日志
[root@localhost binlog]# ll
total 78484
-rw-r----- 1 mysql mysql      217 Jul 13 19:34 mysql-bin.000018
-rw-r----- 1 mysql mysql 80353983 Jul 13 19:52 mysql-bin.000019
```

```
-rw-r----- 1 mysql mysql 194 Jul 13 19:52 mysql-bin.000020
-rw-r----- 1 mysql mysql 156 Jul 13 19:52 mysql-bin.index

[root@localhost binlog]# mkdir /data/backup
[root@localhost binlog]# \cp -ar mysql-bin.0000{18,19} /data/backup/
[root@localhost binlog]# ll /data/backup/
total 78476
-rw-r----- 1 mysql mysql 217 Jul 13 19:34 mysql-bin.000018
-rw-r----- 1 mysql mysql 80353983 Jul 13 19:52 mysql-bin.000019

# 清理mysql-bin.000020之前的所有二进制日志
mysql> purge binary logs to 'mysql-bin.000020';
Query OK, 0 rows affected (0.00 sec)


mysql> show binary logs;
+------------------+-----------+
| Log_name         | File_size |
+------------------+-----------+
| mysql-bin.000020 | 194       |
+------------------+-----------+
1 row in set (0.00 sec)
```

重新启动所有数据库 Server（主从库各自操作）：

```
# 先重启主库，然后重启从库
[root@localhost ~]# service mysqld restart
Shutting down MySQL............ SUCCESS!
Starting MySQL..... SUCCESS!

# 登录所有数据库Server，查看系统变量gtid_mode和enforce_gtid_consistency是否已开启。可以看到
gtid_mode和enforce_gtid_consistency都为ON
mysql> show variables like '%gtid%';
+--------------------------------+----------------------------------------------+
| Variable_name                  | Value                                        |
+--------------------------------+----------------------------------------------+
| binlog_gtid_simple_recovery    | ON                                           |
| enforce_gtid_consistency       | ON                                           |
| gtid_executed_compression_period | 1000                                       |
| gtid_mode                      | ON                                           |
| gtid_next                      | AUTOMATIC                                    |
| gtid_owned                     |                                              |
| gtid_purged                    | f3372787-0719-11e8-af1f-0025905b06da:1-777234 |
| session_track_gtids            | OFF                                          |
+--------------------------------+----------------------------------------------+
8 rows in set (0.00 sec)
```

登录所有从库，设置 master_auto_position = 1，启动自动定位：

```
mysql> stop slave;
Query OK, 0 rows affected (0.00 sec)
```

```
mysql> change master to master_auto_position=1;
Query OK, 0 rows affected (0.01 sec)

mysql> start slave;
Query OK, 0 rows affected (0.03 sec)

mysql> show slave status\G
*************************** 1. row ***************************
               Slave_IO_State: Waiting for master to send event
                  Master_Host: 10.10.30.161
                  Master_User: repl
                  Master_Port: 3306
                Connect_Retry: 60
              Master_Log_File: mysql-bin.000023
          Read_Master_Log_Pos: 194
               Relay_Log_File: mysql-relay-bin.000002
                Relay_Log_Pos: 367
        Relay_Master_Log_File: mysql-bin.000023
             Slave_IO_Running: Yes
            Slave_SQL_Running: Yes
......
            Retrieved_Gtid_Set:
            Executed_Gtid_Set: f3372787-0719-11e8-af1f-0025905b06da:1-777234
                Auto_Position: 1
......
1 row in set (0.00 sec)
```

使用 sysbench 加压 1 分钟，登录所有从库查看是否能正常同步数据：

```
# 对主库用 sysbench 加压（命令略）

# 查看从库的复制状态
mysql> show slave status\G
*************************** 1. row ***************************
               Slave_IO_State: Waiting for master to send event
                  Master_Host: 10.10.30.161
                  Master_User: repl
                  Master_Port: 3306
                Connect_Retry: 60
              Master_Log_File: mysql-bin.000023
          Read_Master_Log_Pos: 3307747
               Relay_Log_File: mysql-relay-bin.000002
                Relay_Log_Pos: 3196837
        Relay_Master_Log_File: mysql-bin.000023
             Slave_IO_Running: Yes
            Slave_SQL_Running: Yes
......
            Retrieved_Gtid_Set: f3372787-0719-11e8-af1f-0025905b06da:777235-778693
```

```
            Executed_Gtid_Set: f3372787-0719-11e8-af1f-0025905b06da:1-778693
                Auto_Position: 1
......
1 row in set (0.00 sec)
```

至此，我们就完成了从传统复制到 GTID 复制的离线切换。如果有需要，可以在主库上使用 pt-table-checksum 校验主从库的数据一致性（从以下结果中可以看到，DIFFS 列全部为 0，表示主从库的数据此时是一致的）。

```
[root@localhost ~]# pt-table-checksum --nocheck-replication-filters --no-check- binlog-
format --replicate=xiaoboluo.checksums   h=localhost,u=admin,p=password,P=3306
    TS       ERRORS  DIFFS  ROWS    CHUNKS  SKIPPED  TIME   TABLE
......
 0508T23:45:16  0     0      1       1       0       0.008  qfsys.qfsys_heartbeat
 0508T23:45:29  0     0      6181462 23      0       13.341 sbtest.sbtest1
......
```

第 18 章 复制拓扑的在线调整

如果 MySQL 数据库的访问量并不大,那么可以使用一主一从的复制拓扑,主库提供写服务,从库提供读服务,采用 MySQL 5.7 的半同步复制配合高可用软件,即可实现数据的高可用、零丢失。在读访问量较大的场景中,可能需要将复制拓扑扩展为一主多从,甚至双主多从,缓解单个从库的读访问压力(关于从库的扩展,可参考第 16 章"通过扩展从库以提高复制性能")。但扩展复制拓扑也就意味着增加管理难度。例如,在主库发生故障或者在线切换到新主库之后,连接原主库的从库需要进行调整以连接到新主库;或者为了提高复制性能,可能会增加一个中间库,并将复制拓扑调整为级联复制,这时就涉及复制拓扑的在线调整。本章将详细介绍复制拓扑的在线调整步骤。

18.1 操作环境信息

服务器的 IP 地址信息如下:

- Master1(主库):10.10.30.161
- Slave1(从库 1):10.10.30.162
- Slave2(从库 2):10.10.30.163

提示:
复制拓扑中一些复制结构的定义如下:

- 并行复制(水平复制):指的是在一主多从的复制拓扑中,多个从库都连接到一个主库。例如,Master1→Slave1,Master1→Slave2,Slave1 和 Slave2 的主库都是 Master1。

- 串行复制(级联复制、垂直复制):指的是在多主多从的复制拓扑中,存在多个主库,不同的从库可能连接到不同的主库。例如,Master1→Slave1→Slave2,Slave1 的主库是 Master1,Slave2 的主库是 Slave1。

本章只介绍具有多个从库且所有数据库实例都正常运行时的拓扑调整,且仅详细介绍从库自身的调整步骤。对于主库的例行切换(在线切换),可参考第 19 章"主从实例的例行切换";对于主库故障转移,可参考第 20 章"数据库故障转移"。

关于传统复制与 GTID 复制的搭建步骤,详情可参考第 14 章"搭建异步复制"。

- 传统复制：即基于二进制日志文件和位置的复制，需要在配置复制时指定二进制日志文件和二进制日志位置。
- GTID 复制：即基于 GTID 的复制，具备自动定位的功能，在配置复制时不需要指定任何位置。

18.2 传统复制模式下的复制拓扑在线调整

在传统复制模式下，复制拓扑的调整对读访问的影响较大，建议在业务低峰期进行操作。

- 在传统复制模式下，并行复制变更为串行复制的关键步骤如图 18-1 所示。

图 18-1

- 在传统复制模式下，串行复制变更为并行复制的关键步骤如图 18-2 所示。

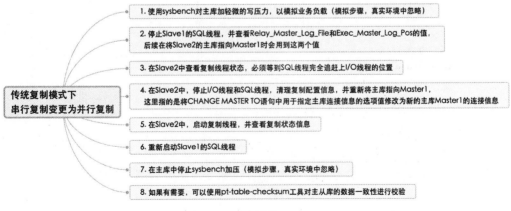

图 18-2

18.2.1　并行复制变更为串行复制

假定已经搭建好一个一主两从的使用传统复制模式的复制拓扑：Master1→Slave1，Master1→Slave2，现在需要将复制拓扑调整为 Master1→Slave1→Slave2。

首先，使用 sysbench 对主库加轻微的写压力，以模拟业务负载（此为模拟步骤，真实环境中可忽略）。

查看 Slave1 和 Slave2 的复制状态信息，确保复制线程正在运行中：

```
# Slave1 的复制状态信息
mysql> show slave status\G
*************************** 1. row ***************************
            Slave_IO_State: Waiting for master to send event
               Master_Host: 10.10.30.161
               Master_User: repl
               Master_Port: 3306
             Connect_Retry: 60
           Master_Log_File: mysql-bin.000001
       Read_Master_Log_Pos: 93315507
            Relay_Log_File: mysql-relay-bin.000002
             Relay_Log_Pos: 93155610
     Relay_Master_Log_File: mysql-bin.000001
          Slave_IO_Running: Yes    # I/O 线程正在运行中
         Slave_SQL_Running: Yes    # SQL 线程正在运行中
......
       Exec_Master_Log_Pos: 93155397
           Relay_Log_Space: 93315927
......
     Seconds_Behind_Master: 1
......
        Retrieved_Gtid_Set:
         Executed_Gtid_Set:
             Auto_Position: 0
......

# Slave2 的复制状态信息
mysql> show slave status\G
*************************** 1. row ***************************
            Slave_IO_State: Waiting for master to send event
               Master_Host: 10.10.30.161
               Master_User: repl
               Master_Port: 3306
             Connect_Retry: 60
           Master_Log_File: mysql-bin.000001
       Read_Master_Log_Pos: 93765486
            Relay_Log_File: mysql-relay-bin.000002
             Relay_Log_Pos: 93704097
```

```
            Relay_Master_Log_File: mysql-bin.000001
                Slave_IO_Running: Yes    # I/O 线程正在运行中
               Slave_SQL_Running: Yes    # SQL 线程正在运行中
......
              Exec_Master_Log_Pos: 93703892
                  Relay_Log_Space: 93765890
......
            Seconds_Behind_Master: 1
......
               Retrieved_Gtid_Set:
                Executed_Gtid_Set:
                    Auto_Position: 0
......
```

在 Slave1 和 Slave2 中分别创建一个测试库（由具有 SUPER 权限的用户进行操作，且从库需要关闭系统变量 super_read_only）：

```
# 在 Slave1 中查看系统变量 super_read_only 的设置，确保其值为 OFF
mysql> show variables like '%super_read%';
+------------------+-------+
| Variable_name    | Value |
+------------------+-------+
| super_read_only  | OFF   |
+------------------+-------+
1 row in set (0.05 sec)

# 临时关闭二进制日志记录功能，并创建一个名为 test_adjustment 的数据库
mysql> set sql_log_bin=0;create database test_adjustment;
Query OK, 0 rows affected (0.00 sec)
Query OK, 1 row affected (0.00 sec)

# 在 Slave2 中查看系统变量 super_read_only 的设置，确保其值为 OFF
mysql> show variables like '%super_read%';
+------------------+-------+
| Variable_name    | Value |
+------------------+-------+
| super_read_only  | OFF   |
+------------------+-------+
1 row in set (0.01 sec)

# 临时关闭二进制日志记录功能，并创建一个名为 test_adjustment 的数据库
mysql> set sql_log_bin=0;create database test_adjustment;
Query OK, 0 rows affected (0.00 sec)
Query OK, 1 row affected (0.00 sec)
```

在主库（Master1）中创建名为 test_adjustment 的数据库：

```
mysql> create database test_adjustment;
Query OK, 1 row affected (0.00 sec)
```

分别查看 Slave1 和 Slave2 的复制状态信息，直到它们的 SQL 线程都出现报错为止。因为在两个从库中都各自预先创建了一个名为 test_adjustment 的数据库，所以如果在主库中也创建名为 test_adjustment 的数据库，则创建 test_adjustment 数据库的操作语句通过复制机制同步到两个从库时，必然会导致 SQL 线程重放建库语句而报错中止。此时两个从库 SQL 线程相对于主库的二进制日志来说，处于同一个位置。因此，实际上这时两个从库的数据是一致的（相同），因为它们报错中止的位置都为应用创建 test_adjustment 数据库的操作语句的位置。

```
# 在 Slave1 中查看复制状态，发现 SQL 线程报告编号为 1007 的错误
mysql> show slave status\G
*************************** 1. row ***************************
               Slave_IO_State: Queueing master event to the relay log
                  Master_Host: 10.10.30.161
                  Master_User: repl
                  Master_Port: 3306
                Connect_Retry: 60
              Master_Log_File: mysql-bin.000001
          Read_Master_Log_Pos: 374214304
               Relay_Log_File: mysql-relay-bin.000002
                Relay_Log_Pos: 332667753
        Relay_Master_Log_File: mysql-bin.000001
             Slave_IO_Running: Yes
            Slave_SQL_Running: No    # SQL 线程因为发生报错而中止了
......
                   Last_Errno: 1007
                   Last_Error: Coordinator stopped because there were error(s) in the worker(s).
The most recent failure being: Worker 1 failed executing transaction 'ANONYMOUS' at master
log mysql-bin.000001, end_log_pos 332816016. See error log and/or performance_schema.
replication_applier_status_by_worker table for more details about this failure or others,
if any.
                 Skip_Counter: 0
          Exec_Master_Log_Pos: 332667540
              Relay_Log_Space: 374214724
......
        Seconds_Behind_Master: NULL
......
               Last_SQL_Errno: 1007
               Last_SQL_Error: Coordinator stopped because there were error(s) in the worker(s).
The most recent failure being: Worker 1 failed executing transaction 'ANONYMOUS' at master
log mysql-bin.000001, end_log_pos 332816016. See error log and/or performance_schema.
replication_applier_status_by_worker table for more details about this failure or others,
if any.
......
           Retrieved_Gtid_Set:
            Executed_Gtid_Set:
                Auto_Position: 0
......
```

因为这里启用了多线程复制，所以从上述 Last_Error 和 Last_SQL_Error 字段的错误信息中无法判断具体是什么错误，可以通过 performance_schema.replication_applier_status_by_worker 表来查看具体的报错信息
```
mysql> select * from performance_schema.replication_applier_status_by_worker where LAST_ERROR_MESSAGE!=''\G
*************************** 1. row ***************************
         CHANNEL_NAME:
            WORKER_ID: 1
            THREAD_ID: NULL
        SERVICE_STATE: OFF
LAST_SEEN_TRANSACTION: ANONYMOUS
    LAST_ERROR_NUMBER: 1007
```
从下面的报错信息可以得知，在 Slave1 中已经存在名为 test_adjustment 的数据库，所以重放来自主库的建库语句时就会报错
```
   LAST_ERROR_MESSAGE: Worker 1 failed executing transaction 'ANONYMOUS' at master log mysql-bin.000001, end_log_pos 332816016; Error 'Can't create database 'test_adjustment'; database exists' on query. Default database: 'test_adjustment'. Query: 'create database test_adjustment'
  LAST_ERROR_TIMESTAMP: 2019-07-21 15:22:50
1 row in set (0.00 sec)
```

在 Slave2 中查看复制状态，发现 SQL 线程报告编号为 1007 的错误
```
mysql> show slave status\G
*************************** 1. row ***************************
               Slave_IO_State: Waiting for master to send event
                  Master_Host: 10.10.30.161
                  Master_User: repl
                  Master_Port: 3306
                Connect_Retry: 60
              Master_Log_File: mysql-bin.000001
          Read_Master_Log_Pos: 455430309
               Relay_Log_File: mysql-relay-bin.000002
                Relay_Log_Pos: 332734755
        Relay_Master_Log_File: mysql-bin.000001
             Slave_IO_Running: Yes
            Slave_SQL_Running: No   # SQL 线程因为发生报错而中止了
......
                   Last_Errno: 1007
                   Last_Error: Coordinator stopped because there were error(s) in the worker(s). The most recent failure being: Worker 1 failed executing transaction 'ANONYMOUS' at master log mysql-bin.000001, end_log_pos 332816016. See error log and/or performance_schema.replication_applier_status_by_worker table for more details about this failure or others, if any.
                 Skip_Counter: 0
          Exec_Master_Log_Pos: 332734550
              Relay_Log_Space: 455430713
......
        Seconds_Behind_Master: NULL
```

```
    ......
        Last_SQL_Errno: 1007
        Last_SQL_Error: Coordinator stopped because there were error(s) in the worker(s).
The most recent failure being: Worker 1 failed executing transaction 'ANONYMOUS' at master
log mysql-bin.000001, end_log_pos 332816016. See error log and/or performance_schema.
replication_applier_status_by_worker table for more details about this failure or others,
if any.
    ......
        Retrieved_Gtid_Set:
        Executed_Gtid_Set:
        Auto_Position: 0
    ......

# 因为这里启用了多线程复制，所以从上述 Last_Error 和 Last_SQL_Error 字段的错误信息中无法判断具体是什
么错误，可以通过 performance_schema.replication_applier_status_by_worker 表来查看具体的报错信息
mysql> select * from performance_schema.replication_applier_status_by_worker where
LAST_ERROR_MESSAGE!=''\G
*************************** 1. row ***************************
         CHANNEL_NAME:
            WORKER_ID: 1
            THREAD_ID: NULL
        SERVICE_STATE: OFF
LAST_SEEN_TRANSACTION: ANONYMOUS
    LAST_ERROR_NUMBER: 1007
# 从下面的报错信息可以得知，在 Slave2 中已经存在名为 test_adjustment 的数据库，所以重放来自主库的建
库语句时就会报错
   LAST_ERROR_MESSAGE: Worker 1 failed executing transaction 'ANONYMOUS' at master log
mysql-bin.000001, end_log_pos 332816016; Error 'Can't create database 'test_adjustment';
database exists' on query. Default database: 'test_adjustment'. Query: 'create database
test_adjustment'
 LAST_ERROR_TIMESTAMP: 2019-07-21 15:22:50
1 row in set (0.00 sec)
```

在 Slave1 中，查看二进制日志文件名和二进制日志位置信息：

```
mysql> show master status;
+------------------+-----------+--------------+------------------+-------------------+
| File             | Position  | Binlog_Do_DB | Binlog_Ignore_DB | Executed_Gtid_Set |
+------------------+-----------+--------------+------------------+-------------------+
| mysql-bin.000001 | 330938671 |              |                  |                   |
+------------------+-----------+--------------+------------------+-------------------+
1 row in set (0.00 sec)
```

现在，在 Slave2 中，停止 I/O 线程和 SQL 线程，并清理复制配置信息，重新将复制指向新的主库（Slave1）：

```
# 停止 I/O 线程和 SQL 线程
mysql> stop slave;
Query OK, 0 rows affected (0.01 sec)
```

```
# 清理复制配置信息
mysql> reset slave all;
Query OK, 0 rows affected (0.31 sec)
```

使用在 Slave1 中查到的二进制日志位置信息，生成并执行如下语句，重新配置复制。注意：在 CHANGE MASTER 语句的配置项中，引号内的值不要有多余的空格，尤其是 master_log_file。如果有多余的空格，启动复制时，系统会报告在主库二进制日志文件的索引文件（Binlog File Index）中找不到该文件，而在报错信息中又很难看出是因为多余的空格造成的，下文对此不再赘述

```
mysql> change master to master_host='10.10.30.162', master_user='repl', master_password='password',master_port=3306,master_log_file='mysql-bin.000001',master_log_pos=330938671;
Query OK, 0 rows affected, 2 warnings (0.02 sec)
```

在 Slave2 中，启动复制线程，并观察其状态：

```
# 启动复制线程
mysql> start slave;
Query OK, 0 rows affected (0.02 sec)

# 查看复制状态
mysql> show slave status\G
*************************** 1. row ***************************
              Slave_IO_State: Waiting for master to send event
                 Master_Host: 10.10.30.162
                 Master_User: repl
                 Master_Port: 3306
               Connect_Retry: 60
             Master_Log_File: mysql-bin.000001
         Read_Master_Log_Pos: 330938671
              Relay_Log_File: mysql-relay-bin.000002
               Relay_Log_Pos: 320
       Relay_Master_Log_File: mysql-bin.000001
            Slave_IO_Running: Yes   # I/O 线程处于运行中
           Slave_SQL_Running: Yes   # SQL 线程处于运行中
......
         Exec_Master_Log_Pos: 330938671
             Relay_Log_Space: 527
......
       Seconds_Behind_Master: 0
......
          Retrieved_Gtid_Set:
           Executed_Gtid_Set:
               Auto_Position: 0
......
```

在 Slave1 和 Slave2 中删除名为 test_adjustment 的数据库（注意，需要在会话级别临时关闭二进制日志记录功能）：

```
# 在 Slave1 中删除名为 test_adjustment 的数据库
mysql> set sql_log_bin=0;drop database test_adjustment;
```

```
Query OK, 0 rows affected (0.00 sec)
Query OK, 0 rows affected (0.00 sec)

# 在 Slave2 中删除名为 test_adjustment 的数据库
mysql> set sql_log_bin=0;drop database test_adjustment;
Query OK, 0 rows affected (0.00 sec)
Query OK, 0 rows affected (0.00 sec)
```

在 Slave1 中重新启动 SQL 线程。此时，在从库中导致冲突报错的 adjustment 数据库已经被删除，因此，来自主库（Master1）二进制日志的删除 adjustment 数据库的语句就能正常执行了：

```
# 启动 SQL 线程
mysql> start slave sql_thread;
Query OK, 0 rows affected (0.03 sec)

# 查看复制状态
mysql> show slave status\G
*************************** 1. row ***************************
               Slave_IO_State: Waiting for master to send event
                  Master_Host: 10.10.30.161
                  Master_User: repl
                  Master_Port: 3306
                Connect_Retry: 60
              Master_Log_File: mysql-bin.000003
          Read_Master_Log_Pos: 496384268
               Relay_Log_File: mysql-relay-bin.000002
                Relay_Log_Pos: 342619926
        Relay_Master_Log_File: mysql-bin.000001
             Slave_IO_Running: Yes
            Slave_SQL_Running: Yes    # 可以看到 SQL 线程已经能正常启动
......
          Exec_Master_Log_Pos: 342619713
              Relay_Log_Space: 1570131071
......
        Seconds_Behind_Master: 2203
......
            Retrieved_Gtid_Set:
             Executed_Gtid_Set:
                 Auto_Position: 0
......


mysql> show databases;
+--------------------+
| Database           |
+--------------------+
| information_schema |
......
| test_adjustment    |   # 此时可以看到，名为 test_adjustment 的数据库通过应用主库的二进制日志又被
```

重新创建了
```
+--------------------+
8 rows in set (0.00 sec)
```

在 Slave2 中，查看复制线程状态信息：

```
mysql> show slave status\G
*************************** 1. row ***************************
              Slave_IO_State: Waiting for master to send event
                 Master_Host: 10.10.30.162
                 Master_User: repl
                 Master_Port: 3306
               Connect_Retry: 60
             Master_Log_File: mysql-bin.000002
         Read_Master_Log_Pos: 26209116
              Relay_Log_File: mysql-relay-bin.000004
               Relay_Log_Pos: 25794613
       Relay_Master_Log_File: mysql-bin.000002
            Slave_IO_Running: Yes   # I/O 线程正常运行中
           Slave_SQL_Running: Yes   # SQL 线程正常运行中
......
         Exec_Master_Log_Pos: 25794400
             Relay_Log_Space: 26209583
......
       Seconds_Behind_Master: 1957
......
           Retrieved_Gtid_Set:
            Executed_Gtid_Set:
                Auto_Position: 0
......

mysql> show databases;
+--------------------+
| Database           |
+--------------------+
| information_schema |
......
| test_adjustment    |   # 可以看到，名为 test_adjustment 的数据库通过应用 Slave1 的二进制日志又被
重新创建了
+--------------------+
8 rows in set (0.00 sec)
```

现在，Slave1 和 Slave2 的复制状态都已经恢复正常。登录主库（Master1），删除名为 test_adjustment 的数据库，然后 Slave1 和 Slave2 会通过级联复制正常删除名为 test_adjustment 的数据库（无须人工干预）：

```
# 在 Master1 中删除名为 test_adjustment 的数据库
mysql> drop database test_adjustment;
Query OK, 0 rows affected (0.00 sec)
```

```
# 然后，查看名为test_adjustment数据库，发现其已经不存在
mysql> show create database test_adjustment;
ERROR 1049 (42000): Unknown database 'test_adjustment'

# Slave1（注：由于在处理复制报错的过程中，模拟业务压力的sysbench持续对主库进行写入操作，此时可能
存在较大延迟，需要等到SQL线程追赶上I/O线程之后，才能查看到Slave1也正常删除了名为test_adjustment
的数据库）
mysql> show create database test_adjustment;
ERROR 1049 (42000): Unknown database 'test_adjustment'

# Slave2（注释同上）
mysql> show create database test_adjustment;
ERROR 1049 (42000): Unknown database 'test_adjustment'
```

主库停止 sysbench 的加压（此为模拟步骤，真实环境中可忽略）。

如果有需要，可以使用 pt-table-checksum 工具对主从库的数据一致性进行校验（应在业务低峰期执行，下文不再赘述）：

```
[root@localhost ~]# pt-table-checksum h='localhost',u='admin',p='password',P=3306
--databases=sysbench --nocheck-replication-filters --replicate=test.checksums --no-check-
binlog-format
Checking if all tables can be checksummed ...
Starting checksum ...
Cannot connect to P=3306,h=10.10.30.162,p=...,u=admin  # 注意，给定的账号应该能使用从库的
IP地址远程连接从库（从库的IP地址默认使用SHOW PROCESSLIST和SHOW SLAVE HOSTS语句来获取，否则此处
会报告无法连接从库的错误信息。以上错误信息就表明使用admin账号无法远程连接到IP地址为10.10.30.162、端
口为3306的数据库Server（从库）
Diffs cannot be detected because no slaves were found. Please read the --recursion-method
documentation for information.
            TS ERRORS DIFFS ROWS CHUNKS SKIPPED TIME TABLE
07-21T16:27:49 0      0     6135 4      0       0.089 sysbench.sbtest1
......
07-21T16:27:49 0      0     6133 1      0       0.029 sysbench.sbtest8
```

至此，我们就完成了传统复制模式下并行复制变更为串行复制的操作。

提示：在本示例中，串行复制（级联复制）需要启用从库 1（Slave1）的系统变量 log_slave_updates，以便从库 1 在重放主库（Master1）的二进制日志之后，将这些内容写入自身的二进制日志，这样才能将这些日志传递给从库 1 自己的从库（Slave2）进行应用（下文中涉及串行复制的要求与此相同，不再赘述）。

18.2.2 串行复制变更为并行复制

假定已经搭建好了一个一主两从的使用传统复制模式的复制拓扑：Master1→Slave1→Slave2，现在需要将复制拓扑调整为 Master1→Slave1，Master1→Slave2（这里我们基于 18.2.1 节中的操作环境来演示本节的调整步骤）。

首先，使用 sysbench 对主库加轻微的写压力，以模拟业务负载（此为模拟步骤，真实环境中可忽略）。

查看 Slave1 和 Slave2 的复制状态信息，确保复制线程正在运行：

```
# 在 Slave1 中查看复制线程状态
mysql> show slave status\G
*************************** 1. row ***************************
             Slave_IO_State: Waiting for master to send event
                Master_Host: 10.10.30.161
                Master_User: repl
                Master_Port: 3306
              Connect_Retry: 60
            Master_Log_File: mysql-bin.000005
        Read_Master_Log_Pos: 39007546
             Relay_Log_File: mysql-relay-bin.000014
              Relay_Log_Pos: 38946550
      Relay_Master_Log_File: mysql-bin.000005
           Slave_IO_Running: Yes   # I/O 线程处于运行中
          Slave_SQL_Running: Yes   # SQL 线程处于运行中
......
        Exec_Master_Log_Pos: 38946337
            Relay_Log_Space: 39008013
......
      Seconds_Behind_Master: 0
......
         Retrieved_Gtid_Set:
          Executed_Gtid_Set:
              Auto_Position: 0
......

# 在 Slave2 中查看复制线程状态
mysql> show slave status\G
*************************** 1. row ***************************
             Slave_IO_State: Waiting for master to send event
                Master_Host: 10.10.30.162
                Master_User: repl
                Master_Port: 3306
              Connect_Retry: 60
            Master_Log_File: mysql-bin.000005
        Read_Master_Log_Pos: 29213262
             Relay_Log_File: mysql-relay-bin.000013
              Relay_Log_Pos: 29145855
      Relay_Master_Log_File: mysql-bin.000005
           Slave_IO_Running: Yes # I/O 线程处于运行中
          Slave_SQL_Running: Yes # SQL 线程处于运行中
......
        Exec_Master_Log_Pos: 29145642
            Relay_Log_Space: 29213729
```

```
......
        Seconds_Behind_Master: 0
......
            Retrieved_Gtid_Set:
            Executed_Gtid_Set:
                Auto_Position: 0
......
```

停止 Slave1 的 SQL 线程，并查看 Relay_Master_Log_File 和 Exec_Master_Log_Pos 的值（记下这两个选项的值，后面在 Slave2 中执行 CHANGE MASTER TO 语句时会用到）：

```
# 停止 SQL 线程
mysql> stop slave sql_thread;
Query OK, 0 rows affected (0.00 sec)

# 查看复制线程状态
mysql> show slave status\G
*************************** 1. row ***************************
               Slave_IO_State: Waiting for master to send event
                  Master_Host: 10.10.30.161
                  Master_User: repl
                  Master_Port: 3306
                Connect_Retry: 60
              Master_Log_File: mysql-bin.000005
          Read_Master_Log_Pos: 98334553
               Relay_Log_File: mysql-relay-bin.000014
                Relay_Log_Pos: 96811999
        Relay_Master_Log_File: mysql-bin.000005  # Relay_Master_Log_File 的值在这里，它表
示相对于 Master1 来说，Slave1 应用了其哪个二进制日志文件
             Slave_IO_Running: Yes
            Slave_SQL_Running: No
......
          Exec_Master_Log_Pos: 96811786   # Exec_Master_Log_Pos 的值在这里，它表示相对于
Master1 来说，Slave1 应用到了 Master1 的哪个位置（这里指的是 Relay_Master_Log_File 显示的二进制日志
文件中的位置）
              Relay_Log_Space: 98335020
......
        Seconds_Behind_Master: NULL
......
           Retrieved_Gtid_Set:
           Executed_Gtid_Set:
                Auto_Position: 0
......
```

在 Slave2 中查看复制线程状态（必须等 SQL 线程完全追赶上 I/O 线程的位置后再查看）：

```
mysql> show slave status\G
*************************** 1. row ***************************
               Slave_IO_State: Waiting for master to send event
                  Master_Host: 10.10.30.162
```

```
                Master_User: repl
                Master_Port: 3306
              Connect_Retry: 60
            Master_Log_File: mysql-bin.000005
        Read_Master_Log_Pos: 83971737
             Relay_Log_File: mysql-relay-bin.000013
              Relay_Log_Pos: 83971950
      Relay_Master_Log_File: mysql-bin.000005    # 这里的值与 Master_Log_File 的值相等，说
明 SQL 线程重放的二进制日志文件与 I/O 线程读取的文件相同，否则需要继续等待
           Slave_IO_Running: Yes
          Slave_SQL_Running: Yes
......
        Exec_Master_Log_Pos: 83971737    # 这里的值与 Read_Master_Log_Pos 的值相等，说明 SQL
线程重放的二进制日志位置与 I/O 线程读取的位置相同，否则需要继续等待
            Relay_Log_Space: 83972204
......
      Seconds_Behind_Master: 0
......
         Retrieved_Gtid_Set:
          Executed_Gtid_Set:
              Auto_Position: 0
......
```

在 Slave2 中，停止 I/O 线程和 SQL 线程，清理复制配置信息，并将复制配置指向 Master1：

```
# 停止复制线程
mysql> stop slave;
Query OK, 0 rows affected (0.00 sec)

# 清理复制配置信息
mysql> reset slave all;
Query OK, 0 rows affected (0.04 sec)

# 将复制配置指向新的主库(Master1)，这里将上述步骤中获取的 Relay_Master_Log_File 和 Exec_Master_Log_
Pos 的值分别填写到 master_log_file 和 master_log_pos 选项中
mysql> change master to master_host='10.10.30.161', master_user='repl', master_
password='password',master_port=3306,master_log_file='mysql-bin.000005',master_log_pos=
96811786;
Query OK, 0 rows affected, 2 warnings (0.01 sec)
```

在 Slave2 中，启动复制线程，并查看复制状态信息：

```
# 启动复制线程
mysql> start slave;
Query OK, 0 rows affected (0.02 sec)

# 查看复制状态信息
mysql> show slave status\G
*************************** 1. row ***************************
             Slave_IO_State: Waiting for master to send event
```

```
              Master_Host: 10.10.30.161
              Master_User: repl
              Master_Port: 3306
              Connect_Retry: 60
              Master_Log_File: mysql-bin.000007
          Read_Master_Log_Pos: 48269360
               Relay_Log_File: mysql-relay-bin.000007
                Relay_Log_Pos: 48120295
        Relay_Master_Log_File: mysql-bin.000007
             Slave_IO_Running: Yes   # I/O 线程正在运行中
            Slave_SQL_Running: Yes   # SQL 线程正在运行中
......
           Exec_Master_Log_Pos: 48120082
               Relay_Log_Space: 48269827
......
         Seconds_Behind_Master: 1
......
            Retrieved_Gtid_Set:
             Executed_Gtid_Set:
                 Auto_Position: 0
......
```

重新启动 Slave1 的 SQL 线程：

```
# 启动 SQL 线程
mysql> start slave sql_thread;
Query OK, 0 rows affected (0.06 sec)

# 查看复制状态信息
mysql> show slave status\G
*************************** 1. row ***************************
              Slave_IO_State: Waiting for master to send event
              Master_Host: 10.10.30.161
              Master_User: repl
              Master_Port: 3306
              Connect_Retry: 60
              Master_Log_File: mysql-bin.000007
          Read_Master_Log_Pos: 96662017
               Relay_Log_File: mysql-relay-bin.000014
                Relay_Log_Pos: 112656192
        Relay_Master_Log_File: mysql-bin.000005
             Slave_IO_Running: Yes   # I/O 线程正在运行中
            Slave_SQL_Running: Yes   # SQL 线程正在运行中
......
           Exec_Master_Log_Pos: 112655979
               Relay_Log_Space: 1170409313
......
         Seconds_Behind_Master: 2129
......
```

```
            Retrieved_Gtid_Set:
            Executed_Gtid_Set:
            Auto_Position: 0
......
```

最后，主库停止用 sysbench 加压（此为模拟步骤，真实环境中可忽略）。

如果有需要，可以使用 pt-table-checksum 工具对主从库的数据一致性进行校验：

```
[root@localhost ~]# pt-table-checksum h='localhost',u='admin',p='letsg0',P=3306 --databases=
sysbench --nocheck-replication-filters --replicate=test.checksums --no-check- binlog-format
Checking if all tables can be checksummed ...
Starting checksum ...
            TS ERRORS DIFFS ROWS CHUNKS SKIPPED TIME TABLE
07-21T17:35:27  0       0    6189  4      0    0.336 sysbench.sbtest1
......
07-21T17:35:29  0       0    6190  1      0    0.323 sysbench.sbtest8
```

至此，我们就完成了传统复制模式下串行复制变更为并行复制的操作。

18.3　GTID 复制模式下的复制拓扑在线调整

在 GTID 复制模式下，调整复制拓扑对读服务的影响比在传统复制模式下调整的影响小一些（至少有一个从库可以正常提供读服务），但为保险起见，仍然建议在业务低峰期进行操作。

- GTID 复制模式下，并行复制变更为串行复制的关键步骤如图 18-3 所示。

图 18-3

- GTID 复制模式下，串行复制变更为并行复制的关键步骤如图 18-4 所示。

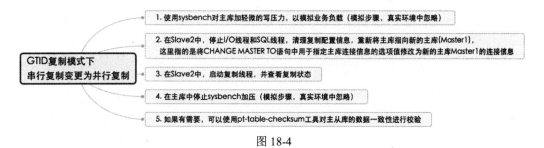

图 18-4

18.3.1 并行复制变更为串行复制

假定已经搭建好了一个一主两从的使用 GTID 复制模式的复制拓扑：Master1→Slave1，Master1→Slave2，现在需要将复制拓扑调整为 Master1→Slave1→Slave2。

首先，使用 sysbench 对主库加轻微的写压力，以模拟业务负载（此为模拟步骤，真实环境中应忽略）。

查看 Slave1 和 Slave2 的复制线程状态信息，确保复制线程正在运行：

```
# 在 Slave1 中查看复制线程状态信息
mysql> show slave status\G
*************************** 1. row ***************************
          Slave_IO_State: Waiting for master to send event
             Master_Host: 10.10.30.161
             Master_User: repl
             Master_Port: 3306
           Connect_Retry: 60
         Master_Log_File: mysql-bin.000004
     Read_Master_Log_Pos: 12066506
          Relay_Log_File: mysql-relay-bin.000002
           Relay_Log_Pos: 4597493
   Relay_Master_Log_File: mysql-bin.000003
        Slave_IO_Running: Yes    # I/O 线程正在运行中
       Slave_SQL_Running: Yes    # SQL 线程正在运行中
......
     Exec_Master_Log_Pos: 4597280
         Relay_Log_Space: 73100277
......
   Seconds_Behind_Master: 124
......
# 在 GTID 复制模式下，从 Retrieved_Gtid_Set 和 Executed_Gtid_Set 字段可以看到 GTID SET 信息，
Retrieved_Gtid_Set 表示 I/O 线程从主库接收的二进制日志包含的 GTID SET，Executed_Gtid_Set 表示从库
SQL 线程当前应用的二进制日志包含的 GTID SET 信息
       Retrieved_Gtid_Set: f3372787-0719-11e8-af1f-0025905b06da:1-32259
        Executed_Gtid_Set: f3372787-0719-11e8-af1f-0025905b06da:1-2211
# 在 GTID 复制模式下，可以启用自动定位功能，Auto_Position 字段值为 1，表示已启用自动定位功能
            Auto_Position: 1

# 在 Slave2 中查看复制线程状态信息
mysql> show slave status\G
*************************** 1. row ***************************
          Slave_IO_State: Waiting for master to send event
             Master_Host: 10.10.30.161
             Master_User: repl
             Master_Port: 3306
           Connect_Retry: 60
         Master_Log_File: mysql-bin.000004
```

```
            Read_Master_Log_Pos: 24525932
              Relay_Log_File: mysql-relay-bin.000002
               Relay_Log_Pos: 35895447
       Relay_Master_Log_File: mysql-bin.000003
            Slave_IO_Running: Yes    # I/O 线程正在运行中
           Slave_SQL_Running: Yes    # SQL 线程正在运行中
......
         Exec_Master_Log_Pos: 35895242
             Relay_Log_Space: 85559687
......
       Seconds_Behind_Master: 91
......
           Retrieved_Gtid_Set: f3372787-0719-11e8-af1f-0025905b06da:1-37757
            Executed_Gtid_Set: f3372787-0719-11e8-af1f-0025905b06da:1-16027
               Auto_Position: 1
......
```

在 Slave2 中，停止 I/O 线程和 SQL 线程，清理复制配置信息，将复制的配置重新指向新的主库（Slave1）：

```
# 停止复制线程
mysql> stop slave;
Query OK, 0 rows affected (0.00 sec)

# 清理复制配置信息
mysql> reset slave all;
Query OK, 0 rows affected (0.15 sec)

# 重新指向新的主库（Slave1）
mysql> change master to master_host='10.10.30.162', master_user='repl', master_
password='password',master_port=3306,master_auto_position=1;
Query OK, 0 rows affected, 2 warnings (0.02 sec)
```

在 Slave2 中，启动复制线程，并查看其状态：

```
# 启动复制线程
mysql> start slave;
Query OK, 0 rows affected (0.02 sec)

# 查看复制线程状态信息，I/O 线程和 SQL 线程都已经正常启动。从这里可以看到，GTID 复制模式下对于复制
拓扑的调整，省略了一大堆手工操作，极其方便快捷
mysql> show slave status\G
*************************** 1. row ***************************
               Slave_IO_State: Waiting for master to send event
                  Master_Host: 10.10.30.162
                  Master_User: repl
                  Master_Port: 3306
                Connect_Retry: 60
              Master_Log_File: mysql-bin.000005
          Read_Master_Log_Pos: 2512208
```

```
                Relay_Log_File: mysql-relay-bin.000002
                 Relay_Log_Pos: 19516417
         Relay_Master_Log_File: mysql-bin.000004
              Slave_IO_Running: Yes          # I/O 线程正在运行中
             Slave_SQL_Running: Yes          # SQL 线程正在运行中
......
            Exec_Master_Log_Pos: 504677831
                Relay_Log_Space: 54222937
......
          Seconds_Behind_Master: 61
......
             Retrieved_Gtid_Set: f3372787-0719-11e8-af1f-0025905b06da:215341-239410
              Executed_Gtid_Set: f3372787-0719-11e8-af1f-0025905b06da:1-224076
                  Auto_Position: 1
......
```

在主库中停止用 sysbench 加压（此为模拟步骤，真实环境中可忽略）。至此，我们就完成了 GTID 复制模式下并行复制变更为串行复制的操作。

提示：GTID SET 中的事务号在一个复制拓扑中允许不一致（因为各个从库的复制可能有不同的延迟），但是 UUID 的数量与字符串内容必须相等（可能存在多个 UUID，UUID 是一个数据库实例的唯一标识，不同的 UUID 和事务号构成了不同的 GTID SET。如果存在多个 GTID SET 时，使用 SHOW MASTER STATUS 语句可以看到用逗号分隔的多个 GTID SET）。当一个复制拓扑中不同数据库实例之间的 UUID 不相等时（例如，在一个实例中查看，只有一个 UUID，而在另一个实例中查看，有两个 UUID），就表示可能存在人为误操作数据的风险，或者可能曾经出现过复制异常，建议在监控系统中增加此监控项。

18.3.2　串行复制变更为并行复制

假定已经搭建好了一个一主两从的使用 GTID 复制模式的复制拓扑：Master1→Slave1→Slave2，现在需要将复制拓扑调整为 Master1→Slave1，Master1→Slave2（在 18.3.1 节的基础上进行调整）。

首先，使用 sysbench 对主库加轻微的写压力，以模拟业务负载（此为模拟步骤，真实环境中可忽略）。

查看 Slave1 和 Slave2 的复制线程状态信息，确保复制线程正在运行：

```
# 在 Slave1 中查看复制线程状态信息
mysql> show slave status\G
*************************** 1. row ***************************
               Slave_IO_State: Queueing master event to the relay log
                  Master_Host: 10.10.30.161
                  Master_User: repl
                  Master_Port: 3306
                Connect_Retry: 60
              Master_Log_File: mysql-bin.000005
```

```
                Read_Master_Log_Pos: 272234
                    Relay_Log_File: mysql-relay-bin.000007
                     Relay_Log_Pos: 267913
             Relay_Master_Log_File: mysql-bin.000005
                  Slave_IO_Running: Yes    # I/O 线程正在运行中
                 Slave_SQL_Running: Yes    # SQL 线程正在运行中
......
                Exec_Master_Log_Pos: 267700
                    Relay_Log_Space: 272741
......
             Seconds_Behind_Master: 0
......
                Retrieved_Gtid_Set: f3372787-0719-11e8-af1f-0025905b06da:1-263983
                 Executed_Gtid_Set: f3372787-0719-11e8-af1f-0025905b06da:1-263983
                    Auto_Position: 1
......

# 在 Slave2 中查看复制线程状态信息
mysql> show slave status\G
*************************** 1. row ***************************
                Slave_IO_State: Waiting for master to send event
                   Master_Host: 10.10.30.162
                   Master_User: repl
                   Master_Port: 3306
                 Connect_Retry: 60
               Master_Log_File: mysql-bin.000005
           Read_Master_Log_Pos: 58420552
                Relay_Log_File: mysql-relay-bin.000004
                 Relay_Log_Pos: 58398225
         Relay_Master_Log_File: mysql-bin.000005
              Slave_IO_Running: Yes    # I/O 线程正在运行中
             Slave_SQL_Running: Yes    # SQL 线程正在运行中
......
           Exec_Master_Log_Pos: 58398012
               Relay_Log_Space: 58421059
......
         Seconds_Behind_Master: 0
......
            Retrieved_Gtid_Set: f3372787-0719-11e8-af1f-0025905b06da:215341-264227
             Executed_Gtid_Set: f3372787-0719-11e8-af1f-0025905b06da:1-264227
                Auto_Position: 1
......
```

在 Slave2 中，停止 I/O 线程和 SQL 线程，清理复制配置信息，将复制的配置重新指向新的主库（Master1）：

```
# 停止复制线程
mysql> stop slave;
Query OK, 0 rows affected (0.01 sec)
```

```
# 清理复制配置信息
mysql> reset slave all;
Query OK, 0 rows affected (0.05 sec)

# 重新指向新的主库(Master1)
mysql> change master to master_host='10.10.30.161', master_user='repl', master_
password='password',master_port=3306,master_auto_position=1;
Query OK, 0 rows affected, 2 warnings (0.02 sec)
```

在 Slave2 中，启动复制线程，并查看其状态：

```
# 启动复制线程
mysql> start slave;
Query OK, 0 rows affected (0.01 sec)

# 查看复制线程状态信息，I/O 线程和 SQL 线程都已经正常启动
mysql> show slave status\G
*************************** 1. row ***************************
             Slave_IO_State: Waiting for master to send event
                Master_Host: 10.10.30.161
                Master_User: repl
                Master_Port: 3306
              Connect_Retry: 60
            Master_Log_File: mysql-bin.000005
        Read_Master_Log_Pos: 138374193
             Relay_Log_File: mysql-relay-bin.000002
              Relay_Log_Pos: 19685209
      Relay_Master_Log_File: mysql-bin.000005
           Slave_IO_Running: Yes
          Slave_SQL_Running: Yes
......
        Exec_Master_Log_Pos: 116304078
            Relay_Log_Space: 41755523
......
      Seconds_Behind_Master: 41
......
         Retrieved_Gtid_Set: f3372787-0719-11e8-af1f-0025905b06da:306504-324928
          Executed_Gtid_Set: f3372787-0719-11e8-af1f-0025905b06da:1-315550
              Auto_Position: 1
```

在主库中停止用 sysbench 加压（此为模拟步骤，真实环境中可忽略）。至此，我们就完成了 GTID 复制模式下串行复制变更为并行复制的操作。

提示：

- 如果复制拓扑中使用了半同步复制，则在在线调整复制拓扑时需要注意相关半同步复制变量的设置，否则当主库发现无任何活跃的从库与自己相连时，主库中新提交

的事务会持续等待从库的半同步复制 ACK 消息，直到超过半同步复制变量设置的超时阈值为止。超过超时阈值之后，如果仍然没有活跃从库与主库相连，则主库将半同步复制降级为异步复制。之后，被阻塞的事务以异步方式继续提交，详情可参考第 15 章 "搭建半同步复制" 文末的注意事项。

- 通过本章中对传统复制和 GTID 复制模式下的复制拓扑调整步骤的比较，相信大家已经见识到了 GTID 复制模式的魔力。所以，建议在允许的情况下，尽可能将 MySQL 升级到最新版本，并切换到 GTID 复制模式。

第 19 章　主从实例的例行切换

主从实例的例行切换，在这里指的是根据业务变更或者运维管理的需要，主动将主库的写访问请求转移到其他数据库实例，这个切换不是由故障触发的。那么，在什么场景下需要进行例行切换呢？通常有如下两种场景：

- 数据库的版本升级：例如，当前版本是 MySQL 5.6，要升级到 MySQL 5.7，我们可以先升级复制拓扑中的所有从库（为什么要先升级从库呢？因为从库的版本必须高于或等于主库的版本，才能兼容主库的二进制日志），然后再升级主库。在升级主库之前，只需要进行一次在线切换，对换主库与从库的角色，然后把原来的主库当作从库再执行一次升级，这样就完成了对复制拓扑中所有实例的升级。
- 主机出现异常：例如，假设主库的操作系统出现严重的内核 bug 或其他异常，需要升级操作系统或者打内核补丁，但这通常需要重启主机，甚至重装操作系统。又或者主库主机硬件出现故障，需要更换硬件，但更换硬件通常要关闭主机。在进行此类操作前，如果条件允许，最保险的办法是先进行一次切换，将主库与从库角色互换，等到原主库恢复正常之后，再以从库的角色加入复制拓扑。

提示：需要进行例行切换的场景可能有很多，在这里无法一一列举，但对于操作切换的人员（或者流程）来说，可以将这些场景归纳为两类。

- 在线切换：即主库保持在线，需要考虑如何处理当前正在访问主库的应用连接的问题（这也是本章的重点，具体的切换过程详见下文）。
- 离线切换：即主库需要关闭主机或者关闭数据库进程，无须考虑应用连接的问题。由于离线切换与故障转移有诸多逻辑重叠，为简单起见，可统一归为故障转移类别。在执行离线切换时，只需要人为停止主库数据库进程，遵循故障转移的逻辑即可（关于故障转移，详情可参考第 20 章"数据库故障转移"）。

19.1　操作环境信息

服务器的 IP 地址信息如下：

- 主库（Master）：10.10.30.163

- 从库（Slave）：10.10.30.164

为了尽可能减少切换过程对应用的影响，需要考虑如何将主库中的读/写业务快速转移到从库。通常，最简单的实现方案是增加一个专用于提供写访问的 VIP 地址（Virtual Internet Protocol Address，虚拟网络协议地址）挂载到主库的主机中，应用使用该 VIP 地址发起写请求。当需要将写业务转移到其他数据库时，只需要将 VIP 地址切换并挂载到其他数据库实例的主机即可。另外，为了便于实现读负载均衡，通常也会增加一个专门提供读访问的 VIP 地址挂载到负载均衡器的主机，应用只需要使用该 VIP 地址发起读请求即可。当读请求到达负载均衡器主机之后，由负载均衡器按照设定好的算法分发读请求到各个提供读服务的从库数据库。

承载写业务的主库服务器需要挂载一个 VIP 地址（且需要挂载到物理网卡上），在主库发生故障转移或进行例行切换时，无须更改应用访问写库的地址，这样可以尽可能降低应用受此类问题的影响（本章使用 10.10.30.171 作为提供写请求的 VIP 地址）。

承载读业务的从库通常有多个，需要配合负载均衡器来实现读业务的负载均衡（例如开源的 LVS）。当有一个从库发生故障时，负载均衡器会将其自动剔除，待到故障从库恢复正常之后，负载均衡器会将其自动加入进来（对于负载均衡器，有兴趣的读者可自行研究，本书不过多阐述）。

首先做如下假设：

- 数据库版本需要从 MySQL 5.6.45 升级到 MySQL 5.7.27。
- 复制拓扑中启用了 GTID。
- 主库提供写服务，从库提供读服务，从库同时作为主库的故障转移备选实例（备主）。

19.2 在线切换

本章以升级数据库版本的场景为例，详细介绍在线切换的整个过程。根据对应用连接的处理，在线切换也可以有两种如下方式：

第一种方式：删除账号。在执行切换前删除账号，使应用无法连接主库，也就无法继续写入新的数据。这种方式效果最好也最彻底，不会导致连接挂死，也不会出现连接数被应用连接占满的问题，但缺点是比较粗暴。这种方式的关键步骤如图 19-1 所示。

第二种方式：修改连接数。在执行切换前，将连接数修改为 1，然后将应用连接全部杀掉，使应用因为连接数被占满而无法连接主库，也就无法继续写入新的数据。但唯一的连接数也可能被应用抢占，而且如果使用了 VIP 地址，则可能会造成应用连接被挂死。这种方式的关键步骤如图 19-2 所示。

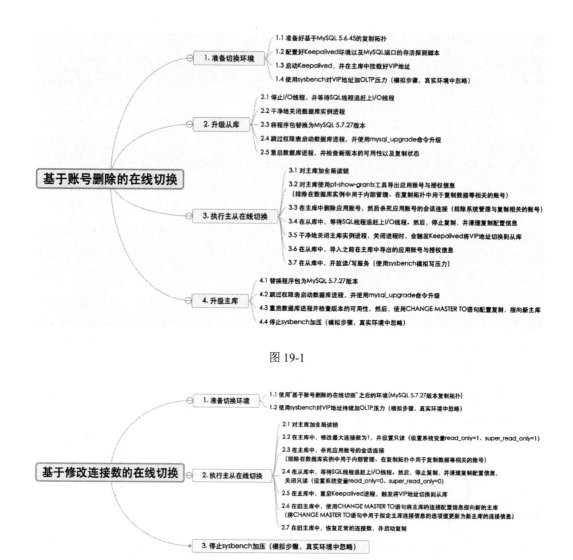

图 19-1

图 19-2

19.2.1 基于账号删除的在线切换

19.2.1.1 切换环境的准备

假设已经准备好 MySQL 5.6.45 版本的一主一从复制拓扑，Keepalived 已配置好且已启动，相关配置如下（这里只给出简单的演示，具体的生产环境中需要怎样的配置请自行评估。但有两点需要注意：第一，不要将 Keepalived 配置在不承接高可用切换的从库服务器上，要确保承接高可用切换的备主在切换前拥有最新的数据；第二，不要与用于读负载均衡的 Keepalived 混淆。由于这些不是本章的重点，有兴趣的读者可自行研究，这里不做过多解释）：

假设主从库都已配置好 MySQL 的检测脚本。这里的是演示用脚本,在生产环境中通常需要更严谨些。例如,在检测脚本中需要探测 3 次,确认失败才返回非零的退出状态。建议使用 SELECT 操作甚至 UPDATE 操作,来实现探测机制
```bash
[root@localhost ~]# cat /usr/local/keepalived/chkmysql.sh
#!/bin/bash
log_file=/usr/local/keepalived/chkmysql.log
process_tag=mysqld_safe
cur_date=`date '+%x %X'`
port=3306

if [ "$port" == "" ]
then
    echo "Usage: $0 <gateway_tab> <port>"
    echo "$cur_date: Usage: $0 <gateway_dir> <port>" >> $log_file
    exit 1;
fi

listen_port=`netstat -plnt|grep ":$port "`
if [ "$listen_port" == "" ]
then
    echo "$cur_date: port $port hasn't been listened." >> $log_file
    exit 1;
else
    echo "$cur_date: port $port is OK." >> $log_file
    exit 0;
fi
```

主库的 Keepalived 配置
```
[root@localhost ~]# cat /etc/keepalived/keepalived.conf
! Configuration File for keepalived

vrrp_script chk_mysql {
  interval 5
 script "/usr/local/keepalived/chkmysql.sh"
}

global_defs {
    router_id ta_103
}

vrrp_instance VI_1 {
    state BACKUP
    interface eth0
    virtual_router_id 103
    priority 100
    advert_int 1
    nopreempt
    authentication {
        auth_type PASS
```

```
        auth_pass 1111
    }

    virtual_ipaddress {
        10.10.30.171 dev eth0 label eth0:1
    }

    track_script {
        chk_mysql
    }
}

# 从库（备主）的 Keepalived 配置
[root@localhost ~]# cat /etc/keepalived/keepalived.conf
! Configuration File for keepalived

vrrp_script chk_mysql {
  interval 5
  script "/usr/local/keepalived/chkmysql.sh"
}

global_defs {
    router_id ta_103
}

vrrp_instance VI_1 {
    state BACKUP
    interface eth0
    virtual_router_id 103
    priority 100
    advert_int 1
    nopreempt
    authentication {
        auth_type PASS
        auth_pass 1111
    }

    virtual_ipaddress {
        10.10.30.171 dev eth0 label eth0:1
    }

    track_script {
        chk_mysql
    }
}
```

先查看主库主机（10.10.30.163）是否已挂载 VIP 地址（10.10.30.171）：

```
[root@localhost ~]# ip addr
```

```
......
2: eth0: <BROADCAST,MULTICAST,UP,LOWER_UP> mtu 1500 qdisc pfifo_fast state UP group
default qlen 1000
    link/ether 52:54:00:c3:37:52 brd ff:ff:ff:ff:ff:ff
    inet 10.10.30.163/24 brd 10.10.30.255 scope global eth0
       valid_lft forever preferred_lft forever
    inet 10.10.30.171/32 scope global eth0:1  # VIP 地址在这里
......
```

使用 sysbench 对 VIP 地址持续施加小压力（注意，这里 sysbench 加压的是 VIP 地址，不是主库的 IP 地址。另外，该步骤为模拟步骤，真实环境中可忽略）：

```
# sysbench 版本为 sysbench-1.0.9
[root@localhost ~]# sysbench --db-driver=mysql --time=99999 --threads=2 --report-
interval=1 --mysql-host=10.10.30.171 --mysql-port=3306 --mysql-user=qbench --mysql-password=
qbench --mysql-db=sbtest --tables=2 --table-size=5000000 oltp_read_write --db-ps-mode=
disable run
```

19.2.1.2 升级从库

登录从库（10.10.30.164）停止 I/O 线程，等待 SQL 线程追赶上 I/O 线程（目的是防止后续的操作导致大事务回滚）：

```
mysql> stop slave io_thread;
Query OK, 0 rows affected (0.00 sec)

# 对于二进制日志文件位置的比较，主要是比较两个条件：Master_Log_File 是否等于 Relay_Master_
Log_File，以及 Read_Master_Log_Pos 是否等于 Exec_Master_Log_Pos
mysql> show slave status\G
*************************** 1. row ***************************
          Slave_IO_State:
             Master_Host: 10.10.30.163
             Master_User: repl
             Master_Port: 3306
           Connect_Retry: 10
         Master_Log_File: mysql-bin.000008  # I/O 线程读取的主库二进制日志文件名称
     Read_Master_Log_Pos: 220178459  # I/O 线程读取的主库二进制日志文件位置
          Relay_Log_File: mysql-relay-bin.000023
           Relay_Log_Pos: 220178669
   Relay_Master_Log_File: mysql-bin.000008  # SQL 线程重放的主库二进制日志文件名称
        Slave_IO_Running: No
       Slave_SQL_Running: Yes
......
      Exec_Master_Log_Pos: 220178459  # SQL 线程重放的主库二进制日志文件位置
         Relay_Log_Space: 220178960
......
    Seconds_Behind_Master: NULL
......
      Retrieved_Gtid_Set: 05c72bee-b4cf-11e9-bde0-525400c33752:1-70626  # I/O 线程
接收的 GTID SET
```

```
                Executed_Gtid_Set: 05c72bee-b4cf-11e9-bde0-525400c33752:1-70626  # SQL 线程重
放的 GTID SET
                Auto_Position: 1
```

如上所述，当发现 SQL 线程追赶至 I/O 线程的位置时，就可以继续往后操作了。判断 SQL 线程是否追赶上 I/O 线程，可以通过两个比较来进行判断，即同时满足：Master_Log_File = Relay_Master_Log_File，Read_Master_Log_Pos = Exec_Master_Log_Pos。

在从库（10.10.30.164）中，干净地关闭数据库实例：

```
# 登录从库数据库实例，设置如下系统变量
mysql> set global innodb_fast_shutdown=0;
Query OK, 0 rows affected (0.00 sec)

# 关闭从库数据库实例进程
[root@localhost ~]# service mysqld stop
Shutting down MySQL.. SUCCESS!
```

在从库中，用 MySQL 5.7.27 版本的程序包替换 MySQL 5.6.45 版本的程序包（注意，数据文件不能动，只是替换程序包）：

```
# 此处省略替换过程（只需要将当前正在使用的程序替换为新的版本即可），替换完之后查看一下版本号
[root@localhost mysql]# mysql --version
mysql Ver 14.14 Distrib 5.7.27, for linux-glibc2.12 (x86_64) using EditLine wrapper
```

在从库中，使用 mysqld_safe 命令启动从库数据库进程，并使用 --skip-grant-tables 选项跳过权限系统表（相较于 MySQL 5.6，在 MySQL 5.7 中与权限相关的数据字典表结构有变更，升级之前必须跳过权限系统表）：

```
[root@localhost mysql]# mysqld_safe --defaults-file=/etc/my.cnf --skip-grant-tables &
```

在从库中，使用 mysql_upgrade 命令升级：

```
[root@localhost mysql]# mysql_upgrade --defaults-file=/etc/my.cnf -uroot -p
Enter password:     # 由于前面的步骤中已经使用 --skip-grant-tables 选项跳过了权限系统表，所以这里
密码为空，直接敲回车键即可
Checking if update is needed.
Checking server version.
Running queries to upgrade MySQL server.
Checking system database.
mysql.columns_priv                                 OK
mysql.db                                           OK
......
sbtest.sbtest1                                     OK
sbtest.sbtest2                                     OK
sys.sys_config                                     OK
Upgrade process completed successfully.   # 看到 "successfully" 就表示升级成功，否则就表示升
级失败
Checking if update is needed.
```

在从库中重启数据库进程：

```
[root@localhost mysql]# service mysqld restart
Shutting down MySQL.. SUCCESS!
```

现在，登录从库数据库，查看升级之后其是否可以正常使用（执行几项简单的查询即可确认）。

```
mysql> show variables like 'log_bin';
+---------------+-------+
| Variable_name | Value |
+---------------+-------+
| log_bin       | ON    |
+---------------+-------+
1 row in set (0.00 sec)

mysql> show slave status\G
*************************** 1. row ***************************
               Slave_IO_State: Queueing master event to the relay log
                  Master_Host: 10.10.30.163
                  Master_User: repl
                  Master_Port: 3306
                Connect_Retry: 10
              Master_Log_File: mysql-bin.000009
          Read_Master_Log_Pos: 526140110
               Relay_Log_File: mysql-relay-bin.000032
                Relay_Log_Pos: 23072361
        Relay_Master_Log_File: mysql-bin.000008
             Slave_IO_Running: Yes
            Slave_SQL_Running: Yes
......
```

提示：

版本升级时有如下注意事项：

在执行升级之前，停止旧版本的数据库时，一定要让其干净地关闭（执行 SET GLOBAL innodb_fast_shutdown=0 语句），否则使用新的程序包无法启动旧版本的数据文件，同时在错误日志中可以看到类似如下的错误信息。

```
2019-07-31T06:56:48.744233Z 0 [ERROR] InnoDB: Upgrade after a crash is not supported.
This redo log was created before MySQL 5.7.9. Please follow the instructions at http://
dev.mysql.com/doc/refman/5.7/en/upgrading.html
2019-07-31T06:56:48.744250Z 0 [ERROR] InnoDB: Plugin initialization aborted with error
Generic error
......
```

版本升级之后，一定要重启数据库，否则 MySQL 的一些功能可能无法正常使用，例如：

```
mysql> show variables like '%log_bin%';
ERROR 1682 (HY000): Native table 'performance_schema'.'session_variables' has the wrong structure
```

19.2.1.3　执行基于账号删除的在线切换

对主库（10.10.30.163）加全局读锁，以防止在获取用户账号信息的过程中，有新的用户被创建（由于后续切换过程中不会修改连接数，所以此举也是为了避免在后续切换过程中，使用新创建的用户连接时被挂死）：

```
mysql> flush table with read lock;
Query OK, 0 rows affected (0.00 sec)
```

现在，假设要对主从实例执行在线切换。先在主库中使用 pt-show-grants 工具捕获用户账号信息（请在--ignore 选项中排除不能删除的或管理系统依赖的账号，以避免删除账号之后发生运维事故）：

```
# pt 工具包的版本为 percona-toolkit-3.0.6
[root@localhost ~]# pt-show-grants --ignore="'mysql.session'@'localhost','mysql.sys'
@'localhost','root'@'localhost','repl'@'%','mysqlxsys'@'localhost','mysql.infoschema'@'
localhost'" -uroot -ppassword --drop --flush > /data/user_info.sql
[root@localhost ~]# cat /data/user_info.sql
......
DROP USER 'program'@'%';
DELETE FROM `mysql`.`user` WHERE `User`='program' AND `Host`='%';
-- Grants for 'program'@'%'
CREATE USER IF NOT EXISTS 'program'@'%';
ALTER USER 'program'@'%' IDENTIFIED WITH 'mysql_native_password' AS '*3B3D7D2FD587C
29C730F36CD52B4BA8CCF4C744F' REQUIRE NONE PASSWORD EXPIRE DEFAULT ACCOUNT UNLOCK;
GRANT CREATE, DELETE, DROP, INSERT, REPLICATION SLAVE, SELECT, UPDATE ON *.* TO
'program'@'%';
DROP USER 'qbench'@'%';
DELETE FROM `mysql`.`user` WHERE `User`='qbench' AND `Host`='%';
-- Grants for 'qbench'@'%'
CREATE USER IF NOT EXISTS 'qbench'@'%';
ALTER USER 'qbench'@'%' IDENTIFIED WITH 'mysql_native_password' AS '*1966B10B87AA6A1F
8E1215A1C81DDD5FBBA6B0D0' REQUIRE NONE PASSWORD EXPIRE DEFAULT ACCOUNT UNLOCK;
GRANT ALL PRIVILEGES ON *.* TO 'qbench'@'%';
FLUSH PRIVILEGES;
```

在主库中删除用户账号及其对应的授权（这里从上述 pt-show-grants 工具打印的文本中过滤出 DROP USER 和 DELETE FROM 语句来执行账号删除操作即可）：

```
[root@localhost ~]# cat /data/user_info.sql |grep -E 'DROP USER|DELETE FROM' >
/data/user_delete.sql
[root@localhost ~]# cat /data/user_delete.sql
DROP USER 'admin'@'%';
DELETE FROM `mysql`.`user` WHERE `User`='admin' AND `Host`='%';
DROP USER 'program'@'%';
DELETE FROM `mysql`.`user` WHERE `User`='program' AND `Host`='%';
DROP USER 'qbench'@'%';
DELETE FROM `mysql`.`user` WHERE `User`='qbench' AND `Host`='%';

# 执行解锁，并在主库中将/data/user_delete.sql 文件导入数据库，删除用户账号，然后加锁（导入前需要解
锁，否则无法执行导入。导入之后，建议立即加全局读锁。为了尽可能避免在此期间创建用户，建议这三个语句一并执行）
mysql> unlock tables; source /data/user_delete.sql; flush table with read lock;
```

```
Query OK, 0 rows affected (0.00 sec)
......
```

```
# 在主库中将 /data/user_info.sql 文件传输到从库，以便完成切换之后，将其导入新主库（现在的从库）
[root@localhost ~]# scp /data/user_info.sql 10.10.30.164:/data/
```

在主库中查询当前用户账号的活跃连接 ID，轮询杀死这些会话：

```
# 查看应用账号的会话（在查询语句的 where 条件中排除了系统账号与管理账号，因此查询结果中不会包含这些账
号的会话）
mysql> select * from information_schema.processlist where user not in ('mysql.session',
'mysql.sys','root','repl','mysqlxsys','mysql.infoschema','admin');
+----+--------+---------+--------+---------+------+------------+------+
| ID | USER   | HOST    | DB     | COMMAND | TIME | STATE      | INFO |
+----+--------+---------+--------+---------+------+------------+------+
| 12 | qbench | 10.10.30.16:61875 | sbtest | Query | 160 | Waiting for global read lock
| UPDATE sbtest2 SET k=k+1 WHERE id=2459059 |
| 13 | qbench | 10.10.30.16:61873 | sbtest | Query | 160 | Waiting for global read lock
| UPDATE sbtest2 SET k=k+1 WHERE id=2514674 |
+----+--------+---------+--------+---------+------+--------+--------+
2 rows in set (0.00 sec)

# 杀死 qbench 用户的会话
mysql> kill 12;
Query OK, 0 rows affected (0.00 sec)

mysql> kill 13;
Query OK, 0 rows affected (0.00 sec)

# 再次查看当前线程状态信息，发现 qbench 用户已经不在了（账号被删除，此时已无法再次连接数据库）。注意，
sysbench 此时会因为连接被杀死而自动退出
mysql> select * from information_schema.processlist where user not in ('mysql.session',
'mysql.sys','root','repl','mysqlxsys','mysql.infoschema','admin');
Empty set (0.00 sec)
```

在从库（10.10.30.164）中，等待 SQL 线程追赶上 I/O 线程，可通过不断轮询查看从库的复制线程状态，并获取 SQL 线程和 I/O 线程的位置进行比较：

```
mysql> show slave status\G
*************************** 1. row ***************************
               Slave_IO_State: Waiting for master to send event
                  Master_Host: 10.10.30.163
                  Master_User: repl
                  Master_Port: 3306
                Connect_Retry: 10
              Master_Log_File: mysql-bin.000008  # I/O 线程读取的主库二进制日志文件名称
          Read_Master_Log_Pos: 207156669  # I/O 线程读取的主库二进制日志文件的位置
               Relay_Log_File: mysql-relay-bin.000023
                Relay_Log_Pos: 207156879
        Relay_Master_Log_File: mysql-bin.000008  # SQL 线程重放的主库二进制日志文件名称
```

```
            Slave_IO_Running: Yes
            Slave_SQL_Running: Yes
......
      Exec_Master_Log_Pos: 207156669    # SQL 线程重放的主库二进制日志文件位置
          Relay_Log_Space: 207157170
......
    Seconds_Behind_Master: 0
......
       Retrieved_Gtid_Set: 05c72bee-b4cf-11e9-bde0-525400c33752:1-63179  # I/O 线程
接收的 GTID SET
        Executed_Gtid_Set: 05c72bee-b4cf-11e9-bde0-525400c33752:1-63179  # SQL 线程
重放的 GTID SET
            Auto_Position: 1
```

提示：如果需要利用 MySQL 5.7 的多线程复制提高复制性能，此时可在配置文件 my.cnf 的 [mysqld] 标签下，添加如下关键变量，然后重启从库（10.10.30.164）数据库进程（由于需要重启，所以从库需要在切换为主库之前添加这些变量，并重启数据库进程。而主库升级完之后就可以立即添加这些变量并重启数据库进程）。

```
[root@localhost ~]# cat my.cnf
[mysqld]
slave_parallel_workers=16
slave_parallel_type=LOGICAL_CLOCK
slave_preserve_commit_order=ON
log_timestamps=SYSTEM
```

在从库中，停止复制线程（SQL 线程和 I/O 线程），并清理复制配置信息（如果未启用 GTID，则这里需要使用 SHOW MASTER STATUS 获取从库的二进制日志文件名和位置信息，以便后面主库作为从库加入复制拓扑时使用）：

```
mysql> stop slave;
Query OK, 0 rows affected (0.01 sec)

mysql> reset slave all;
Query OK, 0 rows affected (0.03 sec)

mysql> show slave status\G
Empty set (0.00 sec)
```

现在，干净地关闭（干净地关闭 MySQL，要求 InnoDB 引擎完成撤销日志清理和插入缓冲合并操作，并将所有的脏页刷新到磁盘中，因此关闭过程较慢，但更安全）主库（10.10.30.163）。当关闭主库数据库进程之后，Keepalived 会自动将 VIP 地址切换到从库：

```
# 设置 innodb_fast_shutdown=0
mysql> set global innodb_fast_shutdown=0;
Query OK, 0 rows affected (0.01 sec)

# 关闭主库数据库实例
[root@localhost ~]# service mysqld stop
```

```
Shutting down MySQL... SUCCESS!

# 在主库中查看 VIP 地址，可以发现已经没有了，而是漂移到了从库上
## 主库服务器
[root@localhost ~]# ip addr
......
2: eth0: <BROADCAST,MULTICAST,UP,LOWER_UP> mtu 1500 qdisc pfifo_fast state UP group
default qlen 1000
    link/ether 52:54:00:c3:37:52 brd ff:ff:ff:ff:ff:ff
    inet 10.10.30.163/24 brd 10.10.30.255 scope global eth0
......

## 从库服务器
[root@localhost ~]# ip addr
......
2: eth0: <BROADCAST,MULTICAST,UP,LOWER_UP> mtu 1500 qdisc pfifo_fast state UP group
default qlen 1000
    link/ether 52:54:00:bd:d1:f2 brd ff:ff:ff:ff:ff:ff
    inet 10.10.30.164/24 brd 10.10.30.255 scope global eth0
       valid_lft forever preferred_lft forever
    inet 10.10.30.171/32 scope global eth0:1
......
```

在从库中导入之前从主库导出的用户信息（这些信息包含了在执行切换操作时删除的账号和权限），然后从库就可以对外开放写服务了：

```
# 登录从库数据库，导入/data/user_info.sql 文件，重新加载用户账号和权限信息
mysql> source /data/user_info.sql;

# 查看用户列表进行确认
mysql> select user from mysql.user;
+---------------+
| user          |
+---------------+
| admin         |
| program       |
| qbench        |
| repl          |
| mysql.session |
| mysql.sys     |
| root          |
+---------------+
7 rows in set (0.00 sec)

# 从库开放读/写服务，提升为新主
mysql> set global read_only=0;
Query OK, 0 rows affected (0.00 sec)
```

这里使用 sysbench 继续对 VIP 地址加压（此为模拟步骤，真实场景中可忽略）：

```
[root@localhost ~]# sysbench --db-driver=mysql --time=99999 --threads=2 --report-
interval=1 --mysql-host=10.10.30.171 --mysql-port=3306 --mysql-user=qbench --mysql-password=
qbench --mysql-db=sbtest --tables=2 --table-size=5000000 oltp_read_write --db-ps-mode=
disable run
```

19.2.1.4 升级主库

在原主库（10.10.30.163）中，用 MySQL 5.7.27 的程序包替换 MySQL 5.6.45 的程序包（注意，数据文件不能动，只替换程序包）：

```
# 此处省略替换过程，替换完之后查看版本号
[root@localhost mysql]# mysql --version
mysql Ver 14.14 Distrib 5.7.27, for linux-glibc2.12 (x86_64) using EditLine wrapper
```

在原主库中，使用 mysqld_safe 命令启动数据库进程，并使用--skip-grant-tables 选项跳过权限系统表：

```
[root@localhost mysql]# mysqld_safe --defaults-file=/etc/my.cnf --skip-grant-tables &
```

在原主库中，使用 mysql_upgrade 命令升级：

```
[root@localhost mysql]# mysql_upgrade --defaults-file=/etc/my.cnf -uroot -p
Enter password:    # 由于前面的步骤中已经使用--skip-grant-tables 选项跳过了权限系统表，所以这里
密码为空，直接敲回车键即可
......
sbtest.sbtest1 OK
sbtest.sbtest2 OK
sys.sys_config OK
Upgrade process completed successfully.  # 看到"successfully"字样就表示升级成功，否则就表
示升级失败
Checking if update is needed.
```

在原主库中，在 my.cnf 的[mysqld]配置组下添加如下优化复制性能的变量，并重启数据库实例：

```
# 添加配置变量
[root@localhost mysql]# cat /etc/my.cnf
[mysqld]
slave_parallel_workers=16
slave_parallel_type=LOGICAL_CLOCK
slave_preserve_commit_order=ON
log_timestamps=SYSTEM

# 重启数据库实例
[root@localhost mysql]# service mysqld restart
```

现在，登录原主库的数据库，查看升级之后其是否可以正常使用：

```
mysql> show variables like 'log_bin';
+---------------+-------+
| Variable_name | Value |
+---------------+-------+
| log_bin       | ON    |
```

```
+----------------+-------+
1 row in set (0.01 sec)
```

在原主库（10.10.30.163）中，使用 CHANGE MASTER TO 语句配置复制，将其指向新主库（10.10.30.164），将原主库作为新的从库加入复制拓扑：

```
# 配置复制
mysql> change master to master_host='10.10.30.164',master_user='repl',master_
password='password',master_auto_position=1;
Query OK, 0 rows affected, 2 warnings (0.01 sec)

# 启动复制
mysql> start slave;

# 查看复制的状态
mysql> show slave status\G
*************************** 1. row ***************************
               Slave_IO_State: Queueing master event to the relay log
                  Master_Host: 10.10.30.164
                  Master_User: repl
                  Master_Port: 3306
                Connect_Retry: 60
              Master_Log_File: mysql-bin.000014
          Read_Master_Log_Pos: 166923651
               Relay_Log_File: mysql-relay-bin.000002
                Relay_Log_Pos: 3777044
        Relay_Master_Log_File: mysql-bin.000014
             Slave_IO_Running: Yes
            Slave_SQL_Running: Yes
......
           Retrieved_Gtid_Set: 0308db53-b4cf-11e9-bde0-525400bdd1f2:1-135177
            Executed_Gtid_Set: 0308db53-b4cf-11e9-bde0-525400bdd1f2:1-3229,
05c72bee-b4cf-11e9-bde0-525400c33752:1-702855
                Auto_Position: 1
         Replicate_Rewrite_DB:
                 Channel_Name:
           Master_TLS_Version:
......
```

在整个复制拓扑中删除之前导出的用户信息文件，避免误操作：

```
# 原主库（新从库）
[root@localhost ~]# rm -f /data/user_info.sql /data/user_delete.sql

# 原从库（新主库）
[root@localhost ~]# rm -f /data/user_info.sql /data/user_delete.sql
```

至此，我们就完成了基于删除账号的在线切换。停止 sysbench 的加压（此为模拟步骤，真实场景中可忽略）。

注意：

- 基于删除账号的方式在线切换主从实例，虽然粗暴，但对于阻止用户创建新连接最有效。

- 当账号被删除之后，之前导出的账号、密码不能丢失，否则会导致应用访问故障，所以建议将 pt 工具导出的账号和权限信息保存在某个管理数据库中，避免丢失。

- 这里以升级数据库版本为例，一并介绍了在线切换的过程。如果在实际应用场景下不需要升级数据库，只是做在线切换，那么将与升级相关的步骤剔除即可。

- 关于 MySQL 数据库版本的升级，由于不是本章的重点，对升级步骤的介绍较为简单，更详细的介绍以及注意事项，详见《千金良方：MySQL 性能优化金字塔法则》的第 2 章 "MySQL 常用的两种升级方法"。

19.2.2 基于修改连接数的在线切换

19.2.2.1 切换环境的准备

这里我们直接使用 19.2.1 节中完成切换与升级的环境（MySQL 5.7.27），但在接下来的步骤中，会将主从角色再次互换（也可以说是还原）。

先查看新主库主机（10.10.30.164）上是否已挂载 VIP 地址（10.10.30.171）：

```
[root@localhost ~]# ip addr
......
2: eth0: <BROADCAST,MULTICAST,UP,LOWER_UP> mtu 1500 qdisc pfifo_fast state UP group default qlen 1000
    link/ether 52:54:00:bd:d1:f2 brd ff:ff:ff:ff:ff:ff
    inet 10.10.30.164/24 brd 10.10.30.255 scope global eth0   # 确认主机 IP 地址
       valid_lft forever preferred_lft forever
    inet 10.10.30.171/32 scope global eth0:1   # VIP 地址在这里
......
```

使用 sysbench 对 VIP 地址持续施加小压力（此为模拟步骤，真实环境中可忽略）：

```
# sysbench 版本为 sysbench-1.0.9
[root@localhost ~]# sysbench --db-driver=mysql --time=99999 --threads=2 --report-interval=1 --mysql-host=10.10.30.171 --mysql-port=3306 --mysql-user=qbench --mysql-password=qbench --mysql-db=sbtest --tables=2 --table-size=5000000 oltp_read_write --db-ps-mode=disable run
```

19.2.2.2 执行基于修改连接数的在线切换

对于应用的账号，通常不允许授予 SUPER 权限，因为这会严重干扰数据库管理系统对数据库的有效管理，甚至导致一些人为故障(这里我们使用 qbench 账号模拟应用的写压力，该账号具有数据库中的所有操作权限，权限过大，必须撤销该账号具有的 SUPER 权限，否则后面在 VIP 地址切换时可能会导致一些莫名其妙的问题。后续会详细介绍因应用账号具备 SUPER 权限而导致切换出现的问题，这里先不过多阐述)：

```
# 在新主库中，先查看具有 SUPER 权限的应用账号
mysql> select user,host,super_priv from mysql.user where super_priv='Y' and user not in
('mysql.session','mysql.sys','root','repl','mysqlxsys','mysql.infoschema','admin');
+--------+------+------------+
| user   | host | super_priv |
+--------+------+------------+
| qbench | %    | Y          |
+--------+------+------------+
1 row in set (0.00 sec)

# 下面我们演示如何撤销 SUPER 权限
mysql> revoke super on *.* from qbench@'%';
Query OK, 0 rows affected (0.01 sec)

mysql> select user,host,super_priv from mysql.user where super_priv='Y' and user not in
('mysql.session','mysql.sys','root','repl','mysqlxsys','mysql.infoschema','admin');
Empty set (0.00 sec)
```

对新主库（10.10.30.164）加全局读锁，让已提交的事务落盘，阻止后续的事务修改数据：

```
mysql> flush table with read lock;
Query OK, 0 rows affected (0.00 sec)
```

在新主库中，将最大连接数修改为 1：

```
# 先查看原有的连接数
mysql> show variables like 'max_connections';
+-----------------+-------+
| Variable_name   | Value |
+-----------------+-------+
| max_connections | 3000  |
+-----------------+-------+
1 rows in set (0.00 sec)

# 设置连接数为 1
mysql> set global max_connections=1;
Query OK, 0 rows affected (0.00 sec)

# 设置只读
mysql> set global read_only=1;
Query OK, 0 rows affected (0.00 sec)
```

在新主库中，查看应用连接会话，使用 ID 列的值来轮询杀死应用账号的会话：

```
# 查看应用账号的会话（这里在查询语句的 where 条件中排除了系统账号与管理账号，因此查询结果中不会包含这些账号的会话）
mysql> select * from information_schema.processlist where user not in ('mysql.session',
'mysql.sys','root','repl','mysqlxsys','mysql.infoschema','admin');
+----+------+------+----+---------+------+-------+------+
| ID | USER | HOST | DB | COMMAND | TIME | STATE | INFO |
```

```
+----+--------+--------------------+--------+---------+------+---------------+-----+
| 27 | qbench | 10.10.30.16:61875  | sbtest | Query   |  160 | Waiting for global read lock
| UPDATE sbtest2 SET k=k+1 WHERE id=2459059 |
| 26 | qbench | 10.10.30.16:61873  | sbtest | Query   |  160 | Waiting for global read lock
| UPDATE sbtest2 SET k=k+1 WHERE id=2514674 |
+----+--------+--------------------+--------+---------+------+---------------+-----+
2 rows in set (0.00 sec)

# 这里我们杀死 ID 为 26 与 27 的会话
mysql> kill 26;
Query OK, 0 rows affected (0.00 sec)

mysql> kill 27;
Query OK, 0 rows affected (0.00 sec)
```

再次查看应用账号的会话信息。注意，sysbench 是单进程多线程的压力测试工具，因此，一个线程在数据库中的会话被杀死，会导致整个 sysbench 进程退出而不再继续在数据库中创建新的会话。但是，应用程序基本都具备短连重连功能，如果在真实的环境中，将应用的会话杀死，执行管理操作的会话也同时断开，那么，唯一允许的连接数就有可能被重连的应用占用，但由于此时数据库中已经加了全局读锁，所有修改数据的操作将会被全局读锁阻塞。这种情况下，如果应用重连的时间恰好在 VIP 地址切换期间，则该连接可能会被挂死

```
mysql> select * from information_schema.processlist where user not in ('mysql.session',
'mysql.sys','root','repl','mysqlxsys','mysql.infoschema','admin');
Empty set (0.00 sec)
```

在原主库（10.10.30.163）中等待 SQL 线程追赶上 I/O 线程，可通过不断轮询查看从库的复制线程状态，并获取 SQL 和 I/O 线程的位置进行比较：

```
mysql> show slave status\G
*************************** 1. row ***************************
               Slave_IO_State: Waiting for master to send event
                  Master_Host: 10.10.30.164
                  Master_User: repl
                  Master_Port: 3306
                Connect_Retry: 60
              Master_Log_File: mysql-bin.000022 # I/O 线程读取的主库二进制日志文件名称
          Read_Master_Log_Pos: 241290517 # I/O 线程读取的主库二进制日志文件位置
               Relay_Log_File: mysql-relay-bin.000026
                Relay_Log_Pos: 241290690
        Relay_Master_Log_File: mysql-bin.000022 # SQL 线程重放的主库二进制日志文件名称
             Slave_IO_Running: Yes
            Slave_SQL_Running: Yes
......
          Exec_Master_Log_Pos: 241290517 # SQL 线程重放的主库二进制日志文件位置
              Relay_Log_Space: 241290984
......
        Seconds_Behind_Master: 0
......
            Retrieved_Gtid_Set: 0308db53-b4cf-11e9-bde0-525400bdd1f2:1-2148900
             Executed_Gtid_Set: 0308db53-b4cf-11e9-bde0-525400bdd1f2:1-2148900,
```

```
         05c72bee-b4cf-11e9-bde0-525400c33752:1-702855
             Auto_Position: 1
        Replicate_Rewrite_DB:
             Channel_Name:
         Master_TLS_Version:
......
```

在原主库中停止复制线程，并清理复制配置信息，并关闭只读系统变量（注意，如果未启用 GTID，则这里需要使用 SHOW MASTER STATUS 获取原主库的二进制日志文件名和位置信息，以便原从库重新加入复制拓扑时使用）：

```
mysql> stop slave;
Query OK, 0 rows affected (0.02 sec)

mysql> reset slave all;
Query OK, 0 rows affected (0.09 sec)

mysql> set global read_only=0;
Query OK, 0 rows affected (0.00 sec)
```

在新主库（10.10.30.164）中，重启 Keepalived（注意，这里没有升级数据库的操作，因为不需要重启数据库，而且正规的在线切换是不允许重启数据库的），让 VIP 地址切换到原主库：

```
# 重启 keepalived
[root@localhost ~]# service keepalived restart
Restarting keepalived (via systemctl): [ OK ]

# 查看 VIP 地址，可以发现在新主库中已经没有 VIP 地址了
[root@localhost ~]# ip addr
......
2: eth0: <BROADCAST,MULTICAST,UP,LOWER_UP> mtu 1500 qdisc pfifo_fast state UP group default qlen 1000
    link/ether 52:54:00:bd:d1:f2 brd ff:ff:ff:ff:ff:ff
    inet 10.10.30.164/24 brd 10.10.30.255 scope global eth0
......
```

登录原主库（10.10.30.163）查看，可以发现 VIP 地址已经切换过来了：

```
[root@localhost ~]# ip addr
......
2: eth0: <BROADCAST,MULTICAST,UP,LOWER_UP> mtu 1500 qdisc pfifo_fast state UP group default qlen 1000
    link/ether 52:54:00:c3:37:52 brd ff:ff:ff:ff:ff:ff
    inet 10.10.30.163/24 brd 10.10.30.255 scope global eth0
       valid_lft forever preferred_lft forever
    inet 10.10.30.171/32 scope global eth0:1  # 发现 VIP 地址
......
```

现在，新主库（10.10.30.164）已经还原为最初的从库，而旧主库也已经还原为最初的

主库。使用 sysbench 重新对 VIP 地址（10.10.30.171）加压（此为模拟步骤，真实环境中可忽略）。

在从库（10.10.30.164）中，使用 CHANGE MASTER TO 语句配置复制，并指向主库（10.10.30.163），以便作为新的从库加入复制拓扑：

```
# 配置复制
mysql> change master to master_host='10.10.30.163', master_user='repl', master_password='password',master_auto_position=1;
Query OK, 0 rows affected, 2 warnings (0.01 sec)
......
```

在从库（10.10.30.164）中恢复正常的连接数，并启动复制线程：

```
# 恢复正常的连接数为 3000
mysql> set global max_connections=3000;
Query OK, 0 rows affected (0.00 sec)

# 启动复制线程
mysql> start slave;
Query OK, 0 rows affected (0.04 sec)

mysql> show slave status\G
*************************** 1. row ***************************
               Slave_IO_State: Queueing master event to the relay log
                  Master_Host: 10.10.30.163
                  Master_User: repl
                  Master_Port: 3306
                Connect_Retry: 60
              Master_Log_File: mysql-bin.000020
          Read_Master_Log_Pos: 299261055
               Relay_Log_File: mysql-relay-bin.000002
                Relay_Log_Pos: 2820272
        Relay_Master_Log_File: mysql-bin.000020
             Slave_IO_Running: Yes
            Slave_SQL_Running: Yes
......
           Exec_Master_Log_Pos: 220479661
               Relay_Log_Space: 81601873
......
        Seconds_Behind_Master: 955
......
           Retrieved_Gtid_Set: 05c72bee-b4cf-11e9-bde0-525400c33752:702856-741274
            Executed_Gtid_Set: 0308db53-b4cf-11e9-bde0-525400bdd1f2:1-2148900,
05c72bee-b4cf-11e9-bde0-525400c33752:1-704336
                Auto_Position: 1
......
```

至此，我们就完成了基于修改连接数的在线切换。停止 sysbench 加压（此为模拟步骤，真实场景中可忽略)。

基于修改连接数的在线切换方式有两个风险点:

- 当连接数被改为 1 之后,在完成 VIP 地址的切换操作之前,操作修改连接数的管理账号会话必须与数据库保持连接,否则一旦断开,这个唯一的连接可能被具有 SUPER 权限的其他用户占用,导致后续的数据库相关操作无法进行。

- 在 VIP 地址切换前的一瞬间,如果有客户端使用 VIP 地址成功连接到主库,则在 VIP 地址切换之后,这些客户端自己必须具备重连机制,否则将会挂死(因为原主库的 VIP 地址已经切换到新主库,无法再通过 VIP 地址将通信数据包返回给客户端。如果客户端程序不具备断线重连机制,就可能会死等主库回包,从而造成客户端挂死的现象)。

提示:

- 如果采用了双主复制的拓扑,且使用了半同步复制,建议在复制拓扑中至少新增一个从库,而不仅仅是只有双主(即需要"两主 + 至少一从"的复制拓扑)。否则,当其中一个主库发生宕机故障时,如果代表 ACK 确认时间的系统变量设置得太大,则在从库提升为主库之后,由于没有任何从库可以为主库回应 ACK 消息,可能会导致新主库中所有写入操作被半同步复制机制阻塞,而发生写业务故障。当然,如果不具备增加从库的条件,可以将系统变量 rpl_semi_sync_master_wait_no_slave 设置为 OFF,然后当主库中没有任何从库连接时,主库立即将半同步复制降级为异步复制,解决该问题。这一点在第 15 章"搭建半同步复制"的注意事项中提到过,可以参考。

- MySQL 5.7 及其之后的版本中,复制延迟的问题基本被解决(后续的 MySQL 版本的复制效率也不断提高),而且 MySQL 5.7 的半同步复制只要使用得当,也能够确保主从库数据的一致性以及数据不丢失。实际上,使用 MySQL 5.7 及其之后的版本搭建双主复制拓扑所带来的麻烦(例如,双主复制拓扑中自增步长岔开的设置可能导致 sysbench 基准测试数据不精确,在升级 MySQL 版本时,需要通过额外的步骤来防止从库升级之后,新版本的二进制日志被同步到旧版本的主库中)比它所能带来的好处更多。所以,除了特殊情况之外,在 MySQL 5.7 及其之后的版本中,笔者不建议再使用双主复制拓扑。

- 整个复制拓扑中,如果有多个从库,且这些从库只用作读负载均衡而不用作主库高可用切换,则其 CHANGE MASTER TO 语句配置复制时可直接指向主库中挂载的写 VIP 地址(10.10.30.171)。这样,在发生主从切换时,只读从库就无须额外地调整复制拓扑,只需要重启复制进程即可正常连接到新主库,从而恢复复制。

- 主从角色的互换(这里指在线切换)在实际工作中使用的频率比较高且较为重要,如操作不当可能导致主从库数据不一致,所以在实际环境中操作时需要谨慎。本章所提及的步骤并不能确保百分之百与你的场景契合,在部署到生产环境之前,请先对功能、性能及健壮性进行严格的测试。

- 在跨地多 IDC（Internet Data Center）架构，甚至是大型业务系统架构的高可用方案中，我们不建议采用 VIP 地址，而是采用支持自动服务发现、主从角色识别及读写请求路由转发的读写分离中间件（例如 maxscale 和 proxysql）结合支持分布式的复制拓扑管理高可用软件 Orchestrator、Consul、replication-manager 来替代 VIP 地址实现业务访问的快速转移。

第 20 章 数据库故障转移

数据库故障转移，在这里指的是由于主库故障而触发的，将主库读/写业务转移到其他数据库实例继续对外提供读/写服务的过程。例如，当探测到主库主机宕机、主库数据库进程不存在、主库数据库无法登录、主库数据库无法执行查询或更新操作时，为尽量减小主库故障对业务造成的影响，需要尽快将读/写访问入口转移到处于正常状态的数据库实例上。

在实际场景中，造成主库无法提供读/写服务的原因多种多样。对于数据库管理系统来说，考虑的因素越多，就会越复杂，可靠性就会越差。所以，通常建议将问题分为可确定故障原因和不可确定故障原因的场景，然后分别采用不同的方式解决。如果混为一谈，可能经常导致误操作转移。

对于可确定故障原因且通过自动故障转移能够恢复业务访问的场景，可以使用数据库管理系统自动进行故障转移。可确定故障原因的场景主要有以下几种：

- 主机宕机。
- 主机存活，但数据库进程不存在（可能从未启动，也可能曾经启动，但之后由于某些原因导致进程崩溃再也无法启动）。
- 主机存活，数据库进程存在，但数据库无法正常登录。
- 主机存活，数据库进程存在，数据库能登录，但执行更新操作时报错。
- 对于无法确定具体故障原因的场景，不建议使用数据库管理系统自动进行故障转移，除非你的数据库业务要优先保证高可用性，才可以在无法确定具体故障原因的场景下执行自动故障转移，而且需要密切关注监控告警系统中的提示（故障原因不明确时，自动故障转移功能可能陷入循环转移的尴尬境地，还可能破坏主从库数据的一致性，甚至带来灾难性的影响），通过监控系统的告警与预警机制，提前规避可能的故障。不可确定故障原因的场景有很多，下面列出几种常见的场景：
- 数据库存在更新数据的大事务，造成其他更新数据的事务被阻塞（需要监控事务表，确认是否存在大事务，且确认该大事务是否为正常操作。如果是正常操作，必须等待其执行完成，无须做额外的操作，但后续需要对大事务进行优化）。
- 网络不稳定（例如，路由器或者交换机设备有故障）。
- 主机磁盘被写满、数据库连接数达到设置的阈值上限、网卡吞吐量达到物理极限、主机 CPU、内存、磁盘负载过高等。

提示：本章不讨论由于一些不规范、不严谨的操作导致出现问题的场景，也不讨论一些有意无意忽略监控预警最终导致出现问题的场景。例如：

- 因为权限规划不规范导致的心跳探测账号被误删或误改，心跳表被误删、误改、误锁定等。
- 在未关闭心跳探测的情况下执行一些需要停机的例行维护。
- 心跳探测的超时时间太短、探测次数不够、探测频率过高等。

20.1 操作环境信息

服务器 IP 地址以及 VIP 地址信息如下：

- 主库（MySQL Master，挂载 VIP 地址，部署 Keepalived）：10.10.30.162
- 从库 1（MySQL Slave 1，作为备主，部署 Keepalived）：10.10.30.163
- 从库 2（MySQL Slave 2，作为只读从库，不部署 Keepalived）：10.10.30.164
- VIP 地址：10.10.30.172

下面我们将针对可确定故障原因中的"数据库进程不存在"因素，介绍数据库故障转移的详细步骤。

提示：

- 关于主从复制拓扑中 Keepalived 挂载 VIP 地址的详细步骤，可参考第 19 章"主从实例的例行切换"。
- 数据库的故障转移，需要考虑数据零丢失的问题。通常，在主从复制拓扑中，要保证数据的零丢失，最简单的办法就是保证主库的二进制日志不丢失。但对于异步复制来说，数据库本身的机制无法保证在主库发生故障时二进制日志不丢失，需要依赖于其他机制（例如，将主库的二进制日志放在一个共享存储中，主库发生故障时，主库的二进制日志可以读取出来，在从库中进行数据补偿）。如果一定要使用数据库本身的机制来保证主库发生故障时二进制日志不丢失，可以使用 MySQL 5.7 及其之后版本的半同步复制。为简单起见，本章的示例中采用半同步复制。关于半同步复制的详细介绍，可参考第 15 章"搭建半同步复制"。
- 如果采用半同步复制拓扑，主库的故障转移到其他实例之后，故障主库恢复时，可能出现比其他数据库实例多出一些数据的情况（是因为主库二进制日志中可能存在一部分已经写入磁盘但还未在存储引擎中提交事务的日志。数据库实例自身的恢复机制，使得故障主库恢复时将这部分事务在存储引擎中重新提交。实际上对于应用程序来说，这些重新提交的事务可能是未正常执行完的事务，在故障转移到其他实例之后，应用程序可能会将这部分事务重新写入新主库。也就是说，故障主库恢复之后必须回滚这部分事务，因为它们可能是重复的数据）。在将故障主库重新加入复制拓扑前，需要将这部分多余的数据回滚。关于回滚工具，详情可参考《千金良方：

MySQL 性能优化金字塔法则》一书中的第 51 章 "MySQL 主流闪回工具详解"。

- 本章示例中的半同步复制配置变量模板如下（将下列配置变量写在 my.cnf 中 [mysqld] 配置组标签下进行持久化）：

```
# 使用 plugin_load 变量在 mysqld 启动时自动加载半同步复制插件，无须人工安装。注意，该变量必须放在所
有半同步复制的配置变量之前，其作用是加载半同步复制插件。而与半同步复制相关的变量是在半同步复制插件中提供支
持的，因此，如果不先加载半同步复制插件，会导致 MySQL Server 因为无法识别半同步复制相关的变量而启动失败
plugin_load = "rpl_semi_sync_master=semisync_master.so;rpl_semi_sync_slave=semisync_slave.so"
# 半同步复制插件主库端的开关变量
rpl_semi_sync_master_enabled=1
# 半同步复制降级为异步复制的超时时间为 24 小时，单位为毫秒
rpl_semi_sync_master_timeout=172800000
# 当主库检测到状态变量 Rpl_semi_sync_master_clients 为 0 时，立即降级为异步复制，不会等待系统变量
rpl_semi_sync_master_timeout 设置的超时时间
rpl_semi_sync_master_wait_no_slave=OFF
# 主库端设置每个事务提交时需要等待多少个从库回复 ACK 消息
rpl_semi_sync_master_wait_for_slave_count=1
# 主库端设置每个事务在提交时，等待 ACK 消息的位置
rpl_semi_sync_master_wait_point=AFTER_SYNC
# 半同步复制插件从库端的开关变量
rpl_semi_sync_slave_enabled=1
```

20.2　主库故障转移的关键步骤

主库故障转移的关键步骤如图 20-1 所示。

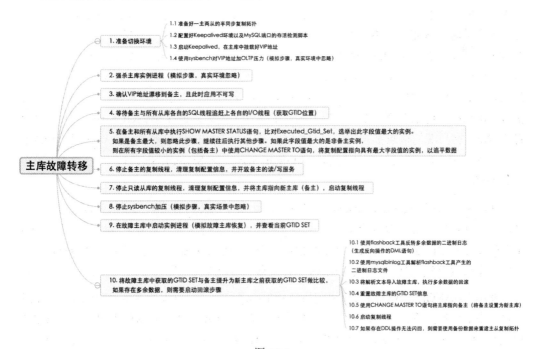

图 20-1

假设故障转移前的复制拓扑如图 20-2 所示（参考 Oracle MySQL 官方资料）。

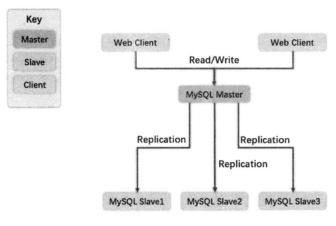

图 20-2

图 20-2 中，两个 Web Client（Web 客户端）对数据库的读取和写入操作都指向单个 MySQL Master（MySQL 主库）。MySQL Master 将数据变更复制到从库中（这里指的是 MySQL Slave1、MySQL Slave2、MySQL Slave3）。图 20-2 中只显示了数据库的读/写请求，并没有显示数据库的只读请求（因为只读请求发生在从库中，发生故障转移时，只读请求并不需要切换服务器）。

在 MySQL 5.7 之前的版本中，所有从库（这里指的是 MySQL Slave1、MySQL Slave2、MySQL Slave3）都必须启用系统变量 log_bin，但可以不启用系统变量 log_slave_updates（未启用该变量时，从库不会将从主库接收的二进制日志记录到自身的二进制日志中。但如果在 MySQL 5.6 中启用 GTID，则必须启用系统变量 log_slave_updates。如果在 MySQL 5.7 中启用系统变量 slave_preserve_commit_order，则系统变量 log_slave_updates 也必须启用）。因此，这里每个从库自身的二进制日志都应该是空的（无数据），当由于某种原因 MySQL Master 变得不可用时，可以选择一个从库提升为新的主库。例如，选择 MySQL Slave1，则应将所有 Web 客户端的请求都重定向到 MySQL Slave1，新的请求产生的数据变更将记录到 MySQL Slave1 的二进制日志中，然后，MySQL Slave2 和 MySQL Slave3 应该将复制连接切换到 MySQL Slave1。

从库中不启用 log_slave_updates，是为了防止发生故障转移时，一个从库被提升为主库之后，其他从库重复读取这个新主库中的二进制日志（而这部分被重复读取的二进制日志来自故障主库中的事件，而且可能在其他从库中已经被读取过），但如果启用了 GTID，则其他从库在读取到重复的二进制日志时，会根据 GTID 自行处理。

在用 CHANGE MASTER TO 语句指向新主库之前，需要确保所有从库都已经处理完中继日志中的所有数据。在所有从库上，先执行 STOP SLAVE IO_THREAD 语句停止 I/O 线程，然后执行 SHOW PROCESSLIST 语句查看 I/O 线程的状态，直到看到 "Slave has read all relay log; waiting for more updates"为止。然后，选举拥有最新数据的从库，将其提升为主库，对其执行 STOP SLAVE 和 RESET MASTER 语句，清理复制配置信息。

当选举出待提升为主库的从库之后，紧接着要在其他从库上执行 STOP SLAVE 和 CHANGE MASTER TO MASTER_HOST ='Slave1'语句（切换主库时，需要确认 MySQL Slave1 上的复制账号和密码、数据库端口等信息），如果启用了 GTID，则在指向新主库时无须指定二进制日志文件位置。然后，执行 START SLAVE 语句，启动其他从库的复制线程。

一旦 MySQL Slave1 被提升为新主库，接下来就需要告诉所有的 Web 客户端，将其请求从 MySQL Master 指向 MySQL Slave1，然后新的请求的数据变更写入 MySQL Slave1 的二进制日志中，MySQL Slave2 和 MySQL Slave3 通过读取 MySQL Slave1 的二进制日志实现新的数据同步。

假设故障转移后的复制拓扑如图 20-3 所示（参考 Oracle MySQL 官方资料）。

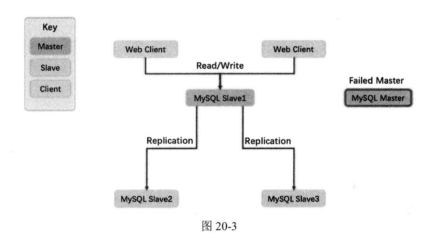

图 20-3

图 20-3 中，当 MySQL Master（MySQL 主库）发生故障，并完成故障转移之后，该故障主库被剔除出复制拓扑。两个 Web Client（Web 客户端）的新读/写请求将指向 MySQL Slave1（即新主库），MySQL Slave2 和 MySQL Slave3 从 MySQL Slave1 中复制数据。

当 MySQL Master 恢复为可用时，应该将其配置为 MySQL Slave1 的从库，并追赶在发生故障期间落后于 MySQL Slave1 的内容（如果落后于 MySQL Slave1 的数据过多而无法直接通过主从复制来追赶数据时，则需要通过一个全新的备份来重建 MySQL Master）。

如果需要将 MySQL Master 再次提升为主库，可以使用上述步骤进行操作。

20.3 主库故障转移的详细过程

20.3.1 环境的准备

先查看主库主机（10.10.30.162）上是否已挂载 VIP 地址（10.10.30.172）：

```
[root@localhost ~]# ip addr
......
2: eth0: <BROADCAST,MULTICAST,UP,LOWER_UP> mtu 1500 qdisc pfifo_fast state UP group default qlen 1000
    link/ether 52:54:00:2a:54:f2 brd ff:ff:ff:ff:ff:ff
    inet 10.10.30.162/24 brd 10.10.30.255 scope global eth0
       valid_lft forever preferred_lft forever
    inet 10.10.30.172/32 scope global eth0:1    # VIP 地址在这里
......
```

使用 sysbench 对 VIP 地址持续施加小压力（此为模拟步骤，真实环境中可忽略）：

```
# sysbench 版本为 sysbench-1.0.9
[root@localhost ~]# sysbench --db-driver=mysql --time=99999 --threads=2 --report-interval=1 --mysql-host=10.10.30.172 --mysql-port=3306 --mysql-user=qbench --mysql-password=qbench --mysql-db=sbtest --tables=2 --table-size=5000000 oltp_read_write --db-ps-mode=disable run
```

查看整个复制拓扑中，所有数据库实例的半同步复制插件状态：

```
# 主库
mysql> select * from performance_schema.global_status where variable_name in
('Rpl_semi_sync_master_status','Rpl_semi_sync_master_no_tx','Rpl_semi_sync_master_no_times','Rpl_semi_sync_master_tx_avg_wait_time');
+----------------------------------------+----------------+
| VARIABLE_NAME                          | VARIABLE_VALUE |
+----------------------------------------+----------------+
| Rpl_semi_sync_master_no_times          | 0              |
| Rpl_semi_sync_master_no_tx             | 0              |
| Rpl_semi_sync_master_status            | ON             | # 主库上该状态变量的值必须为 ON，否则表示主库端半同步复制插件的工作状态不正常
| Rpl_semi_sync_master_tx_avg_wait_time  | 0              |
+----------------------------------------+----------------+
4 rows in set (0.01 sec)

# MySQL Slave 1
mysql> select * from performance_schema.global_status where variable_name in
('Rpl_semi_sync_slave_status');
+----------------------------+----------------+
| VARIABLE_NAME              | VARIABLE_VALUE |
+----------------------------+----------------+
| Rpl_semi_sync_slave_status | ON             | # 从库上该状态变量值必须为 ON，否则表示从库端半同步复制插件的工作状态不正常
+----------------------------+----------------+
1 row in set (0.00 sec)

# MySQL Slave 2
mysql> select * from performance_schema.global_status where variable_name in
('Rpl_semi_sync_slave_status');
+----------------------------+----------------+
```

```
| VARIABLE_NAME             | VARIABLE_VALUE |
+---------------------------+----------------+
| Rpl_semi_sync_slave_status | ON |   # 同上
+---------------------------+----------------+
1 row in set (0.01 sec)
```

检查所有从库的只读变量（为何需要检查只读变量？因为我们的示例中使用了 Keepalived 来探测主库存活状态以及自动切换 VIP 地址，如果不设置只读变量，当主库发生故障时，触发 VIP 地址切换到从库时，将导致应用直接写入从库，而不会做一些必要的数据补偿操作。例如，从库存在延迟时，必须等待 SQL 线程追赶上 I/O 线程，否则就会破坏主从库数据的一致性）：

```
# MySQL Slave1
mysql> show variables like 'read_only';
+---------------+-------+
| Variable_name | Value |
+---------------+-------+
| read_only     | ON    |
+---------------+-------+
1 row in set (0.01 sec)

# MySQL Slave 2
mysql> show variables like 'read_only';
+---------------+-------+
| Variable_name | Value |
+---------------+-------+
| read_only     | ON    |
+---------------+-------+
1 row in set (0.00 sec)
```

在主库中，检查应用账号是否具有 SUPER 权限，因为具有 SUPER 权限的用户在只读状态下仍然可以写入，这在故障转移过程中会破坏主从库数据的一致性（从 MySQL 5.7 开始，支持通过系统变量 super_read_only 将具有 SUPER 权限的用户也设置为只读，但这样做通常会带来管理上的麻烦，也没有太大意义，因为具有 SUPER 权限的用户可自行关闭该变量）：

```
# 这里排除了管理账号与复制账号
mysql> select user,host,super_priv from mysql.user where super_priv='Y' and user not in
('mysql.session','mysql.sys','root','repl','mysqlxsys','mysql.infoschema','admin');
Empty set (0.00 sec)   # 没有发现具有 SUPER 权限的应用账号
```

20.3.2　执行步骤

在主库中，使用 kill -9 命令强杀数据库实例进程（此为模拟步骤，真实环境中可忽略）：

```
[root@localhost ~]# kill -9 `pgrep mysqld`
[root@localhost ~]# pgrep mysqld
[root@localhost ~]#
```

在主库中查看 VIP 地址,可以看到其已经不在该服务器上了:

```
[root@localhost ~]# ip addr
......
2: eth0: <BROADCAST,MULTICAST,UP,LOWER_UP> mtu 1500 qdisc pfifo_fast state UP group default qlen 1000
    link/ether 52:54:00:2a:54:f2 brd ff:ff:ff:ff:ff:ff
    inet 10.10.30.162/24 brd 10.10.30.255 scope global eth0
......
```

在 MySQL Slave1 中查看 VIP 地址,可以看到它已经切换到该服务器(VIP 地址为 10.10.30.172):

```
[root@localhost ~]# ip addr
......
2: eth0: <BROADCAST,MULTICAST,UP,LOWER_UP> mtu 1500 qdisc pfifo_fast state UP group default qlen 1000
    link/ether 52:54:00:c3:37:52 brd ff:ff:ff:ff:ff:ff
    inet 10.10.30.163/24 brd 10.10.30.255 scope global eth0
       valid_lft forever preferred_lft forever
    inet 10.10.30.172/32 scope global eth0:1
......
```

由于 sysbench 是多线程工具,一个线程断开会触发整个进程退出,真实环境中的应用程序可能会立即尝试重连,并重新写入断开连接之前未成功提交的数据。现在,我们重新启动 sysbench 加压,看看会发生什么(此为模拟步骤,真实环境中可忽略):

```
[root@localhost ~]# sysbench --db-driver=mysql --time=99999 --threads=2 --report-interval=1 --mysql-host=10.10.30.172 --mysql-port=3306 --mysql-user=qbench --mysql-password=qbench --mysql-db=sbtest --tables=2 --table-size=5000000 oltp_read_write --db-ps-mode=disable run
......
# 从下面的报错信息中可以看到,只读变量阻止了写入操作。此时不应该允许数据写入,因为并不能确定备主(也就是 MySQL Slave1)的 SQL 线程是否追赶上了 I/O 线程
FATAL: mysql_drv_query() returned error 1290 (The MySQL server is running with the --read-only option so it cannot execute this statement) for query 'UPDATE sbtest2 SET k=k+1 WHERE id=2498724'
FATAL: `thread_run' function failed: /usr/share/sysbench/oltp_common.lua:452: SQL error, errno = 1290, state = 'HY000': The MySQL server is running with the --read-only option so it cannot execute this statement
FATAL: mysql_drv_query() returned error 1290 (The MySQL server is running with the --read-only option so it cannot execute this statement) for query 'UPDATE sbtest2 SET k=k+1 WHERE id=2079678'
FATAL: `thread_run' function failed: /usr/share/sysbench/oltp_common.lua:452: SQL error, errno = 1290, state = 'HY000': The MySQL server is running with the --read-only option so it cannot execute this statement
Error in my_thread_global_end(): 2 threads didn't exit
```

在 MySQL Slave 1 和 MySQL Slave 2 中(如果你的复制拓扑中有多个从库,所有从库都必须执行完此步骤),轮询查看复制线程状态信息,直到各自的 SQL 线程追赶上 I/O 线程:

```
# MySQL Slave1
mysql> show slave status\G
*************************** 1. row ***************************
       Slave_IO_State: Reconnecting after a failed master event read
          Master_Host: 10.10.30.162
          Master_User: repl
          Master_Port: 3306
        Connect_Retry: 10
      Master_Log_File: mysql-bin.000010  # I/O 线程读取的主库二进制日志文件名称
  Read_Master_Log_Pos: 132277813  # I/O 线程读取的主库二进制日志文件位置
       Relay_Log_File: mysql-relay-bin.000029
        Relay_Log_Pos: 132278026
Relay_Master_Log_File: mysql-bin.000010  # SQL 线程重放的主库二进制日志文件名称
     Slave_IO_Running: Connecting
    Slave_SQL_Running: Yes
......
   Exec_Master_Log_Pos: 132277813 # SQL 线程重放的主库二进制日志文件位置
       Relay_Log_Space: 132278320
......
 Seconds_Behind_Master: NULL
Master_SSL_Verify_Server_Cert: No
         Last_IO_Errno: 2003
         Last_IO_Error: error reconnecting to master 'repl@10.10.30.162:3306' -
retry-time: 10 retries: 45
......
 Last_IO_Error_Timestamp: 190805 16:24:02
......
    Retrieved_Gtid_Set: 8809b627-b74d-11e9-8461-5254002a54f2:159517-484826
     Executed_Gtid_Set: 8809b627-b74d-11e9-8461-5254002a54f2:1-484826
         Auto_Position: 1
......

# MySQL Slave2
mysql> show slave status\G
......
```

对比在 MySQL Slave 1 和 MySQL Slave 2（如果复制拓扑中有多个从库，在所有从库中做比较）上 SHOW MASTER STATUS 语句输出信息中的 Executed_Gtid_Set 字段值，如果 MySQL Slave 1 的 GTID SET 大于或者等于 MySQL Slave 2 的 GTID SET，则跳过本步骤进行下一个步骤；反之，则需要先在 MySQL Slave 1 上使用 CHANGE MASTER TO 语句将复制配置指向 MySQL Slave 2，使其追赶上 MySQL Slave 2 的数据。

在 MySQL Slave 1 中，停止复制线程（SQL 线程和 I/O 线程），并清理复制配置信息（注意，如果未启用 GTID，则在这里需要使用 SHOW MASTER STATUS 获取从库的二进制日志文件名和位置信息，以便原主库作为从库加入复制拓扑时使用）：

```
mysql> stop slave;
Query OK, 0 rows affected (0.01 sec)
```

```
mysql> reset slave all;
Query OK, 0 rows affected (0.03 sec)

mysql> show slave status\G
Empty set (0.00 sec)

# 查看当前的 GTID SET，在恢复故障主库时，做闪回要用到它
mysql> show master status\G
*************************** 1. row ***************************
             File: mysql-bin.000010
         Position: 94172559
     Binlog_Do_DB:
 Binlog_Ignore_DB:
Executed_Gtid_Set: 8809b627-b74d-11e9-8461-5254002a54f2:1-484826
1 row in set (0.00 sec)
```

在 MySQL Slave 1 中，关闭只读系统变量，以便对外提供服务：

```
mysql> set global read_only=0;
Query OK, 0 rows affected (0.00 sec)

mysql> show variables like 'read_only';
+---------------+-------+
| Variable_name | Value |
+---------------+-------+
| read_only     | OFF   |
+---------------+-------+
1 row in set (0.00 sec)
```

现在，重新启动 sysbench，对 VIP 地址（10.10.30.172）加压。正常情况下，这时通过 VIP 地址访问已经能将数据写入主库（此为模拟步骤，真实环境中可忽略）。

在 MySQL Slave 2 中停止复制线程，清理复制配置信息，并将复制的配置指向新主库（MySQL Slave 1）：

```
# 停止复制线程
mysql> stop slave;
Query OK, 0 rows affected (0.00 sec)

# 清理复制配置信息
mysql> reset slave all;
Query OK, 0 rows affected (0.00 sec)

# 指向新主库（MySQL Slave1）
mysql> change master to master_host='10.10.30.163', master_port=3306, master_user=
'repl',master_password='password',master_auto_position=1;
Query OK, 0 rows affected, 2 warnings (0.02 sec)

# 启动复制
```

```
mysql> start slave;
Query OK, 0 rows affected (0.04 sec)

# 查看复制线程状态
mysql> show slave status\G
*************************** 1. row ***************************
               Slave_IO_State: Waiting for master to send event
                  Master_Host: 10.10.30.163
                  Master_User: repl
                  Master_Port: 3306
                Connect_Retry: 60
              Master_Log_File: mysql-bin.000010
          Read_Master_Log_Pos: 206349631
               Relay_Log_File: mysql-relay-bin.000002
                Relay_Log_Pos: 3248998
        Relay_Master_Log_File: mysql-bin.000010
             Slave_IO_Running: Yes
            Slave_SQL_Running: Yes
......
          Exec_Master_Log_Pos: 131950160
              Relay_Log_Space: 77648676
......
        Seconds_Behind_Master: 412
......
           Retrieved_Gtid_Set: 9f6feeae-b74d-11e9-8b26-525400c33752:1-37129
            Executed_Gtid_Set: 8809b627-b74d-11e9-8461-5254002a54f2:1-484826,
9f6feeae-b74d-11e9-8b26-525400c33752:1-1603
                Auto_Position: 1
......
```

至此，MySQL Slave 1 就被提升为新主库（即主库已经完成故障转移），停止 sysbench 的加压（此为模拟步骤，真实场景中可忽略）。

现在，我们在故障主库中启动数据库实例进程（模拟主库已经从故障中恢复）：

```
[root@localhost ~]# service mysqld start
Starting MySQL SUCCESS!
```

在故障主库中，查看当前 GTID SET 的位置：

```
mysql> show master status;
+------------------+----------+--------------+------------------+---------------------------------------------------+
| File             | Position | Binlog_Do_DB | Binlog_Ignore_DB | Executed_Gtid_Set                                 |
+------------------+----------+--------------+------------------+---------------------------------------------------+
| mysql-bin.000011 | 194      |              |                  | 8809b627-b74d-11e9-8461-5254002a54f2:1-484827 |   # 从这
里可以看到，GTID SET 中的事务号范围为 1 ~ 484827，而在提升 MySQL Slave 1 为新主库时，GTID SET 事务号
范围为 1 ~ 484826，表示故障主库中多了一个事务，需要执行回滚
+------------------+----------+--------------+------------------+---------------------------------------------------+
1 row in set (0.00 sec)
```

在故障主库中，通过闪回工具反转需要执行回滚的二进制日志：

```
# 注意，这里要指定两个二进制日志文件的原因是，最后一个是故障主库启动时重新产生的二进制日志，里面并不
包含数据，包含数据的是倒数第二个，也就是发生崩溃时的最新的二进制日志文件
[root@localhost binlog]# /root/flashback --binlogFileNames = 'mysql-bin.000010,
mysql-bin.000011' --exclude-gtids='8809b627-b74d-11e9-8461-5254002a54f2:1-484826' --outBinlog
File NameBase='binlog_flashback'

[root@localhost binlog]# ll
total 4705168
# 注意，指定了多少个二进制日志文件，就会产生多少个 binlog_flashback.flashback 文件，其中第一个是
不带数字编号的，后面的文件都带有一个编号。我们在后面执行闪回动作时，从不带编号的 binlog_flashback.
flashback 文件开始，将带编号的文件按照数字编号从小到大依次解析并导入数据库，执行闪回
-rw-r--r-- 1 root root 123 Aug 5 17:07 binlog_flashback.flashback
-rw-r--r-- 1 root root 882 Aug 5 17:07 binlog_flashback.flashback.000001
-rw-r----- 1 mysql mysql 536991157 Aug 5 14:59 mysql-bin.000001
......
-rw-r----- 1 mysql mysql 132279083 Aug 5 16:50 mysql-bin.000010
-rw-r----- 1 mysql mysql 217 Aug 5 17:00 mysql-bin.000011
-rw-r----- 1 mysql mysql 676 Aug 5 17:03 mysql-bin.index
```

使用mysqlbinlog工具解析上述步骤中产生的所有 binlog_flashback.flashback* 文件，并导入数据库，执行数据闪回（回滚）：

```
# 解析所有 binlog_flashback.flashback* 文件
[root@localhost binlog]# mysqlbinlog --skip-gtids --disable-log-bin binlog_flashback.
flashback -vv > a.sql
[root@localhost binlog]# mysqlbinlog --skip-gtids --disable-log-bin binlog_flashback.
flashback.000001 -vv >> a.sql

# 如果文件不大，可以使用 vim 看里边的内容，确认是否有误（如果文件过大，则可能需自行编写一些数据校验程
序代替人工校验）
[root@localhost binlog]# vim a.sql
......
### DELETE FROM `sbtest`.`sbtest2`
### WHERE
###   @1=2270376 /* INT meta=0 nullable=0 is_null=0 */
###   @2=2523812 /* INT meta=0 nullable=0 is_null=0 */
###
@3='31166423918-92136794304-15712636752-67455221609-95913320906-77348420510-49861132528
-08541428325-39113489260-63103968634' /* STRING(360) meta=61032 nullable=0 is_null=0 */
###   @4='02058424893-59656379618-20150720645-13431637194-08124843994' /* STRING(180)
meta=65204 nullable=0 is_null=0 */
......

# 确认无误之后，将 a.sql 文件导入数据库，执行数据闪回
[root@localhost binlog]# mysql -uroot -ppassword < a.sql
```

在故障主库中，登录数据库，确认数据是否已被回滚：

```
# 根据上述 a.sql 文件中的内容，查看表结构，确认第一列的列名，构造 id 值为 2270376 的 WHERE 条件语句进
行查询
mysql> select * from sbtest2 where id=2270376;
```

```
Empty set (0.00 sec)    # 确认数据已被删除
```

在故障主库中重置 GTID SET 信息：

```
# 使用 RESET MASTER 语句清理 GTID SET 信息
mysql> reset master;
Query OK, 0 rows affected (1.01 sec)

mysql> show master status;
+------------------+----------+--------------+------------------+-------------------+
| File             | Position | Binlog_Do_DB | Binlog_Ignore_DB | Executed_Gtid_Set |
+------------------+----------+--------------+------------------+-------------------+
| mysql-bin.000001 | 154      |              |                  |                   |
+------------------+----------+--------------+------------------+-------------------+
1 row in set (0.00 sec)

# 使用 SET GLOBAL GTID_PURGED 语句，设置在 MySQL Slave 1 提升为新主库的时间段内获取的 GTID SET
信息
mysql> set global gtid_purged='8809b627-b74d-11e9-8461-5254002a54f2:1-484826';
Query OK, 0 rows affected (0.01 sec)
```

在故障主库中使用 CHANGE MASTER TO 语句配置复制，将 MySQL Slave 1 作为主库，将故障主库作为从库加入复制拓扑：

```
mysql> change master to master_host='10.10.30.163', master_user='repl', master_password='password',master_port=3306,master_auto_position=1;
Query OK, 0 rows affected, 2 warnings (0.04 sec)
```

在故障主库中启动复制，并查看复制线程的状态：

```
mysql> start slave;
Query OK, 0 rows affected (0.16 sec)

mysql> show slave status\G
*************************** 1. row ***************************
               Slave_IO_State: Queueing master event to the relay log
                  Master_Host: 10.10.30.163
                  Master_User: repl
                  Master_Port: 3306
                Connect_Retry: 60
              Master_Log_File: mysql-bin.000010
          Read_Master_Log_Pos: 208212885
               Relay_Log_File: mysql-relay-bin.000002
                Relay_Log_Pos: 1501962
        Relay_Master_Log_File: mysql-bin.000010
             Slave_IO_Running: Yes
            Slave_SQL_Running: Yes
......
          Exec_Master_Log_Pos: 130203124
              Relay_Log_Space: 79511930
......
```

```
            Seconds_Behind_Master: 3482
......
    Retrieved_Gtid_Set: 9f6feeae-b74d-11e9-8b26-525400c33752:1-38016
    Executed_Gtid_Set: 8809b627-b74d-11e9-8461-5254002a54f2:1-484826,
9f6feeae-b74d-11e9-8b26-525400c33752:1-801
            Auto_Position: 1
......
```

至此，故障主库也已经完成恢复。

提示：文中所使用的闪回工具 flashback 无法回滚 DDL 语句（事实上，目前市面上可见的闪回工具基本都不支持回滚 DDL 语句），如果在故障主库中遇见多余的 DDL 语句，可能需要重新从存活实例中备份一份全量数据进行恢复（清空从库数据目录，并重建主从复制拓扑），或者使用定时备份中的备份数据进行恢复。

若故障主库的主机恢复之后（如果曾经宕机），数据库实例无法正常启动，也许需要使用备份工具重新备份一份数据进行恢复（复制拓扑中如果还有其他从库，则优先使用从库进行备份；如果没有，则需要直接从主库备份，或者使用定时备份做恢复）。数据安全至关重要，任何时候，都要做好数据备份。

常用的备份规则如下：

- 数据的逻辑备份（温备）：最具有代表性的逻辑备份工具就是 MySQL 自带的 mysqldump。如果要用这个备份工具搭建主从复制，还需要使用选项--master-data（默认值为 1），启用该选项后，在备份文件中会包含带有二进制日志记录信息的 CHANGE MASTER TO 语句。将--set-gtid-purged 选项设置为 AUTO（默认值）或 ON，以便在备份文件中包含已执行完的事务的 GTID SET 信息。完成备份之后，在从库服务器中使用 mysql 客户端导入备份文件即可恢复数据。注意，对于 mysqldump，备份的时间点是备份开始的时间，因为数据是在一个大的一致性快照事务中进行备份的（非 InnoDB 存储引擎不受事务控制，如果存在非事务引擎，则在备份时不能使用--single-transaction 选项）。

- 数据的物理备份(热备)：需要使用 Percona 公司的 XtraBackup 或者 MySQL Enterprise Backup 组件中的 mysqlbackup 命令。由于是物理备份，所以备份工具和备份目标数据库必须在同一台服务器中，不能远程备份。对于 InnoDB 表，可以实现热备，但非 InnoDB 表用温备（会加锁）。另外，这些备份工具会在备份过程中持续拷贝新产生的 redo log，因此备份的时间点是备份结束时的时间。

- 拷贝数据的物理文件（冷备）：停止数据库实例进程，将数据目录的内容复制到从库的数据目录，然后重新启动从库实例进程。如果使用此方法，启用 GTID 时需要删除数据目录下的 auto.cnf 文件，以避免 UUID 重复。

- 二进制日志备份：使用 binlog server 从主库服务器备份二进制日志到备份服务器，使用 mysqlbinlog 工具的--read-from-remote-server 和--read-from-remote-master 选项，

指定需要从哪个数据库实例备份二进制日志。

对于半同步复制，可以通过定期采集如下状态变量来检测其运行状态：

- Rpl_semi_sync_master_status：主库半同步复制插件状态变量，其值为 ON 表示生效，为 OFF 则表示不生效。

- Rpl_semi_sync_master_no_tx：主库端检测半同步复制的事务数量，如果持续增长，或者频繁检测到非零值，则说明半同步复制被降级为异步复制，或主从之间的复制网络不稳定。

- Rpl_semi_sync_master_no_times：主库端检测半同步复制插件被关闭的次数，如果次数不断增加，则说明主从库之间的复制网络可能不稳定。

- Rpl_semi_sync_master_tx_avg_wait_time：主库端检测半同步复制事务的平均 ACK 响应时间，单位为毫秒，如果该延迟逐渐增加，则说明主从库之间的复制网络可能不稳定，从库的负载可能在持续增加，导致其 ACK 响应变慢。

- Rpl_semi_sync_slave_status：从库半同步复制插件的状态，其值为 ON 表示生效，为 OFF 则表示不生效。

GTID 复制可以简化数据库故障转移的过程。通常，若无特殊情况，建议在复制拓扑中启用 GTID。

注意：

- 如果采用双主复制拓扑，且使用了半同步复制，建议在复制拓扑中至少配置一个只读从库，而不是仅仅只有双主（即需要两主一从的复制拓扑）。否则，当活跃主库发生宕机故障时，如果 ACK 确认时间被设置得太长，则在备主被提升为新主库之后，由于没有任何从库可以给新主库回应 ACK 消息，可能会导致新主库中所有写入操作被半同步复制机制阻塞（直到 ACK 消息超时，降级为异步复制），从而发生写业务故障。当然，如果无法实现在复制拓扑中至少配置一个从库，就需要将系统变量 rpl_semi_sync_master_wait_no_slave 关闭，当主库没有与任何从库连接时，主库立即将半同步复制降级为异步复制，解决该问题。这一点在第 15 章 "搭建半同步复制" 的注意事项中提到过，可以参考。

- 主库故障转移有一个风险，如果主库中存在大事务的二进制日志未被从库完全接收，从库的 SQL 线程在重放中继日志时，会死等主库发送完整的 event（事件）。用 mysqlbinlog 解析从库最后一个中继日志可以发现如下错误，此时由于从库只知道与主库失联，并不清楚主库是否还能够恢复，所以只能死等。

```
WARNING: The range of printed events ends with a row event or a table map event that does
not have the STMT_END_F flag set. This might be because the last statement was not fully
written to the log, or because you are using a --stop-position or --stop-datetime that refers
to an event in the middle of a statement. The event(s) from the partial statement have not
been written to output.
```

处理方法如下：

- 如果使用异步复制，则无法保证主库已提交事务的二进制日志是否被从库接收。进行故障转移时，不建议等待从库 SQL 线程重放完已读取的中继日志，而是直接停止复制线程，以 SQL 线程应用的二进制日志位置为起点，通过其他方式直接解析主库二进制日志来做数据补偿（所以，主库的二进制日志是否有冗余备份很关键）。

- 如果使用半同步复制，则可以保证已提交事务的二进制日志被从库成功接收，则故障转移时，可以等待从库 SQL 线程重放完。但是，如果发现 SQL 线程重放的位置（SHOW SLAVE STATUS 语句输出中的 Exec_Master_Log_Pos 选项值）长时间未变动，则需要尝试解析 SQL 线程读取的中继日志文件，检查是否有未接收完整 event 的问题。如果有，同时也确认主库短时间内无法恢复正常，则可以停止复制，并清理复制配置信息，然后继续执行后续的故障转移步骤。

第 21 章　搭建多源复制

MySQL 的多源复制（也可以称为"多主复制"），指的是复制拓扑中的从库同时从多个源 Server（主库）接收二进制日志进行重放。多源复制可用于将多个 Server 的数据备份到单个 Server 中（从库），以及在分库分表场景中，将来自多个 Server 的分片表数据合并。从库在应用来自多个主库的二进制日志时，不会执行任何冲突检测或解决冲突，如有需要，则靠应用程序来解决这些问题。在多源复制拓扑中，从库会为每个主库建立一个单独的复制通道（单独的 I/O 线程、协调器线程、Worker 线程），各自重放各自的二进制日志，互不依赖。

关于多源复制的原理与使用场景，详情可参考第 7 章"多源复制"。本章将详细介绍多源复制的搭建步骤。

21.1　操作环境信息

1. Server IP 地址

- 主库 1（MySQL Master1）：10.10.30.162
- 主库 2（MySQL Master 2）：10.10.30.163
- 从库（MySQL Slave）：10.10.30.164

2. MySQL 数据库版本

MySQL 5.7.27

3. sysbench 版本

sysbench 1.0.9

下文将分别详细介绍，在传统复制与 GTID 复制模式下搭建多源复制的步骤。关于复制模式的切换，本章不做过多介绍，详情可参考第 17 章"复制模式的切换"。

提示：

- 在多源复制拓扑中，从库可以使用单线程，也可以使用多线程复制，由系统变量 slave_parallel_workers 控制。该系统变量在 MySQL 5.6 中引入，其值被设置为 0 时，就表示复制为单线程的；大于 1 时（假设为 N），就表示启用了多线程复制（此时，单线程复制中的 SQL 线程被拆分为 1 个协调器线程和 N 个 Worker 线程）。该变量可动态设置，但需要重启复制线程方可生效。
- 在多源复制拓扑中，系统变量 master_info_repository 和 relay_log_info_repository 必须设置为 TABLE。这两个变量可动态设置。
- 本章只提及多源复制、多线程复制相关的配置变量，更多的复制配置变量的设置，可参考 14.4 节"变量模板"。

21.2 基于传统复制的多源复制

在传统复制模式下，搭建多源复制的关键步骤如图 21-1 所示。

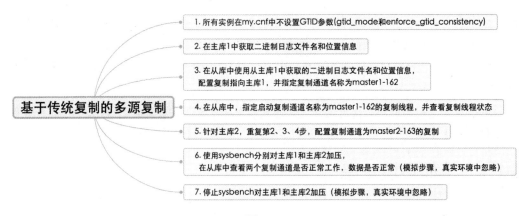

图 21-1

提示：复制拓扑中所有数据库实例都关闭 GTID 变量。

```
# GTID 相关的变量默认关闭，如果之前曾配置过，在 my.cnf 中注释掉即可
# gtid-mode=on
# enforce-gtid-consistency=true
```

21.2.1 传统复制模式下的单线程多源复制

假定主库 1、主库 2、从库这三个实例都已经完成初始化安装，并已启动数据库进程，在所有的数据库实例中确认 GTID 变量已经关闭：

```
mysql> select * from performance_schema.global_variables where variable_name regexp
'enforce_gtid_consistency|gtid_mode';
+--------------------------+----------------+
| VARIABLE_NAME            | VARIABLE_VALUE |
+--------------------------+----------------+
| enforce_gtid_consistency | OFF            |
| gtid_mode                | OFF            |
```

```
+-------------------------+----------------+
2 rows in set (0.01 sec)
```

在从库中,设置复制线程数量为 0,即关闭多线程复制:

```
mysql> set global slave_parallel_workers=0;
Query OK, 0 rows affected (0.00 sec)
```

在主库 1 中,获取当前二进制日志文件名和位置信息:

```
mysql> show master status;
+------------------+----------+--------------+------------------+-------------------+
| File             | Position | Binlog_Do_DB | Binlog_Ignore_DB | Executed_Gtid_Set |
+------------------+----------+--------------+------------------+-------------------+
| mysql-bin.000002 | 154      |              |                  |                   |
+------------------+----------+--------------+------------------+-------------------+
1 row in set (0.00 sec)
```

在从库中,使用在主库 1 中获取的二进制日志文件名和位置信息配置复制,使其指向主库 1,并指定复制通道名称为 master1-162:

```
mysql> change master to master_host='10.10.30.162', master_user='repl', master_password='letsg0',master_port=3306,master_log_file='mysql-bin.000002',master_log_pos=154 for channel 'master1-162';
Query OK, 0 rows affected, 2 warnings (0.02 sec)
```

在从库中,指定启动通道名称为 master1-162 的复制线程:

```
mysql> start slave for channel 'master1-162';
Query OK, 0 rows affected (0.00 sec)

mysql> show slave status for channel 'master1-162'\G
*************************** 1. row ***************************
               Slave_IO_State: Waiting for master to send event
                  Master_Host: 10.10.30.162
                  Master_User: repl
                  Master_Port: 3306
                Connect_Retry: 60
              Master_Log_File: mysql-bin.000002
          Read_Master_Log_Pos: 154
               Relay_Log_File: mysql-relay-bin-master1@002d162.000002
                Relay_Log_Pos: 320
        Relay_Master_Log_File: mysql-bin.000002
             Slave_IO_Running: Yes
            Slave_SQL_Running: Yes
......
          Exec_Master_Log_Pos: 154
              Relay_Log_Space: 543
......
        Seconds_Behind_Master: 0
......
           Retrieved_Gtid_Set:
```

```
            Executed_Gtid_Set:
            Auto_Position: 0
         Replicate_Rewrite_DB:
            Channel_Name: master1-162
            Master_TLS_Version:
1 row in set (0.00 sec)
```

在主库 2 中，获取当前二进制日志文件名和位置信息：

```
mysql> show master status;
+------------------+----------+--------------+------------------+-------------------+
| File             | Position | Binlog_Do_DB | Binlog_Ignore_DB | Executed_Gtid_Set |
+------------------+----------+--------------+------------------+-------------------+
| mysql-bin.000001 | 154      |              |                  |                   |
+------------------+----------+--------------+------------------+-------------------+
1 row in set (0.00 sec)
```

在从库中，使用在主库 2 中获取的二进制日志文件名和位置信息配置复制，使其指向主库 2，并指定复制通道名称为 master2-163：

```
mysql> change master to master_host='10.10.30.163',master_user='repl',master_password=
'letsg0',master_port=3306,master_log_file='mysql-bin.000001',master_log_pos=154       for
channel 'master2-163';
Query OK, 0 rows affected, 2 warnings (0.02 sec)
```

在从库中，指定启动通道名称为 master2-163 的复制线程：

```
mysql> start slave for channel 'master2-163';
Query OK, 0 rows affected (0.00 sec)

mysql> show slave status for channel 'master2-163'\G
*************************** 1. row ***************************
             Slave_IO_State: Waiting for master to send event
                Master_Host: 10.10.30.163
                Master_User: repl
                Master_Port: 3306
              Connect_Retry: 60
            Master_Log_File: mysql-bin.000003
        Read_Master_Log_Pos: 175893342
             Relay_Log_File: mysql-relay-bin-master2@002d163.000002
              Relay_Log_Pos: 104502368
      Relay_Master_Log_File: mysql-bin.000001
           Slave_IO_Running: Yes
          Slave_SQL_Running: Yes
......
         Exec_Master_Log_Pos: 104502202
             Relay_Log_Space: 1250913605
......
       Seconds_Behind_Master: 7018
......
          Retrieved_Gtid_Set:
```

```
          Executed_Gtid_Set:
           Auto_Position: 0
      Replicate_Rewrite_DB:
             Channel_Name: master2-163
       Master_TLS_Version:
1 row in set (0.00 sec)
```

在从库中，查看多源复制相关的状态信息（多源复制拓扑带来的操作语句的变化详见下文）：

```
# 在文件系统层面，可以看到中继日志针对多源复制调整了文件名，中继日志文件名中带有复制通道的名称
[root@localhost ~]# cd /data/mysqldata1/relaylog/
[root@localhost relaylog]# ll
......
## 名字中带有字符master1@002d162的文件，就是名为master1-162的复制通道对应的中继日志文件。也就
是说，这些文件中存放着从主库1中读取的二进制日志
-rw-r----- 1 mysql mysql       223 Aug 6 11:46 mysql-relay-bin-master1@002d162.000001
-rw-r----- 1 mysql mysql       320 Aug 6 11:46 mysql-relay-bin-master1@002d162.000002
-rw-r----- 1 mysql mysql       152 Aug 6 11:46 mysql-relay-bin-master1@002d162.index
## 名字中带有master2@002d163字符的文件，就是名为master2-163的复制通道对应的中继日志文件。也就
是说，这些文件中存放着从主库2中读取的二进制日志
-rw-r----- 1 mysql mysql       270 Aug 6 13:49 mysql-relay-bin-master2@002d163.000022
-rw-r----- 1 mysql mysql 108346556 Aug 6 13:49 mysql-relay-bin-master2@002d163.000023
-rw-r----- 1 mysql mysql       152 Aug 6 13:51 mysql-relay-bin-master2@002d163.index

# 在processlist信息层面，从PERFORMANCE_SCHEMA系统库中可以看到2个I/O线程和2个SQL线程
mysql> select name,processlist_id, processlist_command, processlist_state, processlist_
info,thread_os_id from performance_schema.threads where name regexp 'thread/sql/slave_io|
thread/sql/slave_sql|thread/sql/slave_worker';
+----------------------+----------------+---------------------+-----------------------------------+------------------+--------------+
| NAME                 | PROCESSLIST_ID | PROCESSLIST_COMMAND | PROCESSLIST_STATE                 | PROCESSLIST_INFO | THREAD_OS_ID |
+----------------------+----------------+---------------------+-----------------------------------+------------------+--------------+
| thread/sql/slave_io  | 3              | Connect             | Waiting for master to send event  | NULL             | 23615        |
| thread/sql/slave_sql | 4              | Connect             | Slave has read all relay log; waiting for more updates | NULL | 23616 |
| thread/sql/slave_io  | 5              | Connect             | Waiting for master to send event  | NULL             | 27508        |
| thread/sql/slave_sql | 6              | Connect             | Slave has read all relay log; waiting for more updates | NULL | 27509 |
+----------------------+----------------+---------------------+-----------------------------------+------------------+--------------+
4 rows in set (0.01 sec)
```

使用sysbench分别对主库1和主库2施加压力：

```
# 对主库1加压，主库1中的数据来自名为sbtest的数据库
[root@localhost ~]# sysbench --db-driver=mysql --time=99999 --threads=2 --report-
interval=1 --mysql-host=10.10.30.162 --mysql-port=3306 --mysql-user=qbench --mysql-password=
qbench --mysql-db=sbtest --tables=2 --table-size=5000000 oltp_read_write --db-ps-mode=
disable run
```

```
# 对主库 2 加压，主库 2 中的数据来自名为 sbtest2 的数据库
[root@localhost ~]# sysbench --db-driver=mysql --time=99999 --threads=2 --report-
interval=1  --mysql-host=10.10.30.163 --mysql-port=3306 --mysql-user=qbench  --mysql-
password=qbench  --mysql-db=sbtest2  --tables=2  --table-size=5000000  oltp_read_write
--db-ps-mode=disable run
```

在从库中查看两个通道各自的复制线程状态：

```
# 不指定复制通道时，使用 SHOW SLAVE STATUS 语句可以同时打印所有通道的复制线程状态信息
## 以下内容是名为 master1-162 的通道的复制线程状态信息
mysql> show slave status\G
*************************** 1. row ***************************
              Slave_IO_State: Waiting for master to send event
                 Master_Host: 10.10.30.162
                 Master_User: repl
                 Master_Port: 3306
               Connect_Retry: 60
             Master_Log_File: mysql-bin.000002
         Read_Master_Log_Pos: 94610428
              Relay_Log_File: mysql-relay-bin-master1@002d162.000002
               Relay_Log_Pos: 94608322
       Relay_Master_Log_File: mysql-bin.000002
            Slave_IO_Running: Yes
           Slave_SQL_Running: Yes
......
          Exec_Master_Log_Pos: 94608156
              Relay_Log_Space: 94610817
......
       Seconds_Behind_Master: 0
......
          Retrieved_Gtid_Set:
           Executed_Gtid_Set:
               Auto_Position: 0
        Replicate_Rewrite_DB:
                Channel_Name: master1-162
          Master_TLS_Version:
## 以下内容是名为 master2-163 的通道的复制线程状态信息
*************************** 2. row ***************************
              Slave_IO_State: Waiting for master to send event
                 Master_Host: 10.10.30.163
                 Master_User: repl
                 Master_Port: 3306
               Connect_Retry: 60
             Master_Log_File: mysql-bin.000008
         Read_Master_Log_Pos: 204068726
              Relay_Log_File: mysql-relay-bin-master2@002d163.000023
               Relay_Log_Pos: 204067203
       Relay_Master_Log_File: mysql-bin.000008
            Slave_IO_Running: Yes
```

```
              Slave_SQL_Running: Yes
......
          Exec_Master_Log_Pos: 204066990
              Relay_Log_Space: 204069209
......
        Seconds_Behind_Master: 0
......
            Retrieved_Gtid_Set:
            Executed_Gtid_Set:
                Auto_Position: 0
         Replicate_Rewrite_DB:
                 Channel_Name: master2-163
           Master_TLS_Version:
2 rows in set (0.00 sec)
```

在从库中，查看来自主库 1 和主库 2 中的数据汇总情况：

```
# 来自主库 1 和主库 2 的数据汇总
mysql> show databases;
+--------------------+
| Database           |
+--------------------+
| information_schema |
| mysql              |
| performance_schema |
| sbtest  |    # 来自主库 1 中的 sbtest 数据库
| sbtest2 |    # 来自主库 2 中的 sbtest2 数据库
| sys                |
+--------------------+
6 rows in set (0.00 sec)
```

停止用 sysbench 对主库 1 和主库 2 施加压力（此为模拟步骤，真实环境中可忽略）。

21.2.2　传统复制模式下的多线程多源复制

基于 21.2.1 节中的环境，在从库中修改系统变量 slave_parallel_workers 的值，改为大于 0 的任意数字值（这里设置为 16）：

```
# 先查看系统变量 slave_parallel_workers 的值
mysql> show variables like '%para%';
+------------------------+---------------+
| Variable_name          | Value         |
+------------------------+---------------+
| slave_parallel_type    | LOGICAL_CLOCK |
| slave_parallel_workers | 0             |
+------------------------+---------------+
2 rows in set (0.00 sec)

# 将系统变量 slave_parallel_workers 的值修改为 16
```

```
mysql> set global slave_parallel_workers=16;
Query OK, 0 rows affected (0.00 sec)

# 确认修改值
mysql> show variables like '%para%';
+------------------------+---------------+
| Variable_name          | Value         |
+------------------------+---------------+
| slave_parallel_type    | LOGICAL_CLOCK |
| slave_parallel_workers | 16            |
+------------------------+---------------+
2 rows in set (0.00 sec)

# 此时，对系统变量 slave_parallel_workers 值的修改在复制线程上并没有生效。通过查询 performance_
schema.threads 表中记录的线程数据可以看到，仍然只有 2 个 I/O 线程和 2 个 SQL 线程
mysql> select name,processlist_id, processlist_command, processlist_state, processlist_
info,thread_os_id from performance_schema.threads where name regexp 'thread/sql/slave_io|
thread/sql/slave_sql|thread/sql/slave_worker';
+----------------------+---------------+---------------------+---------------------------+-----------------+--------------+
| NAME                 | PROCESSLIST_ID | PROCESSLIST_COMMAND | PROCESSLIST_STATE         | PROCESSLIST_INFO | THREAD_OS_ID |
+----------------------+---------------+---------------------+---------------------------+-----------------+--------------+
| thread/sql/slave_io  | 3             | Connect             | Waiting for master to send event | NULL | 23615 |
| thread/sql/slave_sql | 4             | Connect             | Slave has read all relay log; waiting for more updates | NULL | 23616 |
| thread/sql/slave_io  | 5             | Connect             | Waiting for master to send event | NULL | 27508 |
| thread/sql/slave_sql | 6             | Connect             | Slave has read all relay log; waiting for more updates | NULL | 27509 |
+----------------------+---------------+---------------------+---------------------------+-----------------+--------------+
4 rows in set (0.00 sec)
```

在从库中，重启所有复制通道的复制线程：

```
# 停止复制线程。不带通道名称停止复制线程，即可同时停止所有通道的复制线程
mysql> stop slave;
Query OK, 0 rows affected (0.02 sec)

# 启动复制线程。不带通道名称启动复制线程，即可同时启动所有通道的复制线程
mysql> start slave;
Query OK, 0 rows affected (0.10 sec)

# 查看所有复制通道的复制线程状态
mysql> show slave status\G
......
```

查看多线程复制的配置是否生效：

```
# 可通过 SHOW PROCESSLIST 语句或者 information_schema.processlist 表查看，也可以通过
performance_schema.threads 表查看（状态信息更详细）。从这些信息中可以看到，新增了 32 个 Worker 线程（2
个复制通道，每个复制通道各 16 个 Worker 线程）
```

```
mysql> select  name,processlist_id, processlist_command, processlist_state, processlist_
info,thread_os_id from performance_schema.threads where name regexp 'thread/sql/slave_
io|thread/sql/slave_sql|thread/sql/slave_worker';
+------------------------+----------------+---------------------+-----------------------------------------------+-----------------+---------------+
| NAME                   | PROCESSLIST_ID | PROCESSLIST_COMMAND | PROCESSLIST_STATE                             | PROCESSLIST_INFO | THREAD_OS_ID |
+------------------------+----------------+---------------------+-----------------------------------------------+-----------------+---------------+
| thread/sql/slave_io    | 8              | Connect             | Waiting for master to send event              | NULL            | 28820         |
| thread/sql/slave_sql   | 9              | Connect             | Slave has read all relay log; waiting for more updates | NULL  | 28821  |
| thread/sql/slave_worker| 10             | Connect             | Waiting for an event from Coordinator         | NULL            | 28822         |
| ......
| thread/sql/slave_worker| 25             | Connect             | Waiting for an event from Coordinator         | NULL            | 28837         |
| thread/sql/slave_io    | 26             | Connect             | Waiting for master to send event              | NULL            | 28838         |
| thread/sql/slave_sql   | 27             | Connect             | System lock                                   | NULL            | 28839         |
| thread/sql/slave_worker| 28             | Connect             | Waiting for an event from Coordinator         | NULL            | 28840         |
| ......
| thread/sql/slave_worker| 43             | Connect             | Waiting for an event from Coordinator         | NULL            | 28855         |
+------------------------+----------------+---------------------+-----------------------------------------------+-----------------+---------------+
36 rows in set (0.00 sec)
```

21.3 基于 GTID 复制的多源复制

在 GTID 复制模式下搭建多源复制的关键步骤如图 21-2 所示。

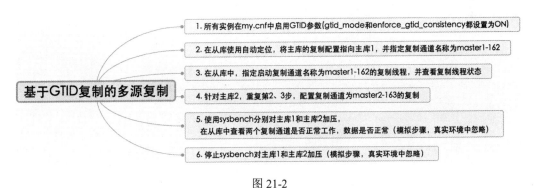

图 21-2

提示：

- 本节基于新初始化的实例介绍搭建多源复制的步骤，而且假定主库 1 和主库 2 中的二进制日志从未被清理过。
- 复制拓扑中所有数据库实例都启用 GTID 变量。

```
# GTID 相关的变量默认是关闭的，如果之前曾配置过，在 my.cnf 中注释掉即可
```

```
gtid-mode=on
enforce-gtid-consistency=true
```

21.3.1　GTID 复制模式下的单线程多源复制

假定主库 1、主库 2 和从库这三个实例都已完成初始化安装,且数据库进程已启动,在所有数据库实例中确认已经启用 GTID 变量:

```
mysql> select * from performance_schema.global_variables where variable_name regexp
'enforce_gtid_consistency|gtid_mode';
+--------------------------+----------------+
| VARIABLE_NAME            | VARIABLE_VALUE |
+--------------------------+----------------+
| enforce_gtid_consistency | ON             |
| gtid_mode                | ON             |
+--------------------------+----------------+
2 rows in set (0.01 sec)
```

在从库中,设置复制线程数量为 0,即关闭多线程复制:

```
mysql> set global slave_parallel_workers=0;
Query OK, 0 rows affected (0.00 sec)
```

在从库中配置复制,指向主库 1,并指定复制通道名称为 master1-162:

```
mysql> change master to master_host='10.10.30.162', master_user='repl', master_
password='letsg0',master_port=3306,master_auto_position=1 for channel 'master1-162';
Query OK, 0 rows affected, 2 warnings (0.02 sec)
```

在从库中,指定启动通道名称为 master1-162 的复制线程:

```
mysql> start slave for channel 'master1-162';
Query OK, 0 rows affected (0.00 sec)

mysql> show slave status for channel 'master1-162'\G
*************************** 1. row ***************************
               Slave_IO_State: Waiting for master to send event
                  Master_Host: 10.10.30.162
                  Master_User: repl
                  Master_Port: 3306
                Connect_Retry: 60
              Master_Log_File: mysql-bin.000002
          Read_Master_Log_Pos: 154
               Relay_Log_File: mysql-relay-bin-master1@002d162.000002
                Relay_Log_Pos: 367
        Relay_Master_Log_File: mysql-bin.000002
             Slave_IO_Running: Yes
            Slave_SQL_Running: Yes
......
           Exec_Master_Log_Pos: 154
              Relay_Log_Space: 590
```

```
......
        Seconds_Behind_Master: 0
......
            Retrieved_Gtid_Set:
             Executed_Gtid_Set:
             Auto_Position: 1
         Replicate_Rewrite_DB:
                 Channel_Name: master1-162
            Master_TLS_Version:
1 row in set (0.00 sec)
```

在从库中配置复制,指向主库 2,并指定复制通道名称为 master2-163:

```
mysql> change master to master_host='10.10.30.163', master_user='repl', master_password='letsg0',master_port=3306,master_auto_position=1 for channel 'master2-163';
Query OK, 0 rows affected, 2 warnings (0.02 sec)
```

在从库中,指定启动通道名称为 master2-163 的复制线程:

```
mysql> start slave for channel 'master2-163';
Query OK, 0 rows affected (0.00 sec)

mysql> show slave status for channel 'master2-163'\G
*************************** 1. row ***************************
               Slave_IO_State: Waiting for master to send event
                  Master_Host: 10.10.30.163
                  Master_User: repl
                  Master_Port: 3306
                Connect_Retry: 60
              Master_Log_File: mysql-bin.000002
          Read_Master_Log_Pos: 154
               Relay_Log_File: mysql-relay-bin-master2@002d163.000002
                Relay_Log_Pos: 367
        Relay_Master_Log_File: mysql-bin.000002
             Slave_IO_Running: Yes
            Slave_SQL_Running: Yes
......
          Exec_Master_Log_Pos: 154
              Relay_Log_Space: 590
......
        Seconds_Behind_Master: 0
......
            Retrieved_Gtid_Set:
             Executed_Gtid_Set:
             Auto_Position: 1
         Replicate_Rewrite_DB:
                 Channel_Name: master2-163
            Master_TLS_Version:
1 row in set (0.00 sec)
```

使用 sysbench 分别对主库 1 和主库 2 施加压力:

```
# 对主库 1 加压,主库 1 中的数据来自名为 sbtest 的数据库
[root@localhost ~]# sysbench --db-driver=mysql --time=99999 --threads=2 --report-
interval=1 --mysql-host=10.10.30.162 --mysql-port=3306 --mysql-user=qbench --mysql-
password=qbench --mysql-db=sbtest --tables=2 --table-size=5000000 oltp_read_write
--db-ps-mode=disable run

# 对主库 2 加压,主库 2 中的数据来自名为 sbtest2 的数据库
[root@localhost ~]# sysbench --db-driver=mysql --time=99999 --threads=2 --report-
interval=1 --mysql-host=10.10.30.163 --mysql-port=3306 --mysql-user=qbench --mysql-
password=qbench --mysql-db=sbtest2 --tables=2 --table-size=5000000 oltp_read_write --db-
ps-mode=disable run
```

在从库中查看两个通道各自的复制线程状态:

```
# 不指定复制通道时,使用 SHOW SLAVE STATUS 语句可以同时打印所有通道的复制状态线程信息
## 以下为名为 master1-162 的通道的复制线程状态信息
mysql> show slave status\G
*************************** 1. row ***************************
               Slave_IO_State: Waiting for master to send event
                  Master_Host: 10.10.30.162
                  Master_User: repl
                  Master_Port: 3306
                Connect_Retry: 60
              Master_Log_File: mysql-bin.000002
          Read_Master_Log_Pos: 31182550
               Relay_Log_File: mysql-relay-bin-master1@002d162.000002
                Relay_Log_Pos: 31182763
        Relay_Master_Log_File: mysql-bin.000002
             Slave_IO_Running: Yes
            Slave_SQL_Running: Yes
......
           Exec_Master_Log_Pos: 31182550
               Relay_Log_Space: 31182986
......
         Seconds_Behind_Master: 0
......
            Retrieved_Gtid_Set: 8809b627-b74d-11e9-8461-5254002a54f2:1-14840
             Executed_Gtid_Set: 8809b627-b74d-11e9-8461-5254002a54f2:1-14840,
9f6feeae-b74d-11e9-8b26-525400c33752:1-14877
                 Auto_Position: 1
          Replicate_Rewrite_DB:
                  Channel_Name: master1-162
            Master_TLS_Version:
## 以下为名为 master2-163 的通道的复制线程状态信息
*************************** 2. row ***************************
               Slave_IO_State: Waiting for master to send event
                  Master_Host: 10.10.30.163
                  Master_User: repl
                  Master_Port: 3306
```

```
        Connect_Retry: 60
        Master_Log_File: mysql-bin.000002
        Read_Master_Log_Pos: 31057001
         Relay_Log_File: mysql-relay-bin-master2@002d163.000002
          Relay_Log_Pos: 31057214
  Relay_Master_Log_File: mysql-bin.000002
       Slave_IO_Running: Yes
      Slave_SQL_Running: Yes
......
     Exec_Master_Log_Pos: 31057001
         Relay_Log_Space: 31057437
......
   Seconds_Behind_Master: 0
......
   Retrieved_Gtid_Set: 9f6feeae-b74d-11e9-8b26-525400c33752:1-14877
    Executed_Gtid_Set: 8809b627-b74d-11e9-8461-5254002a54f2:1-14840,
9f6feeae-b74d-11e9-8b26-525400c33752:1-14877
        Auto_Position: 1
     Replicate_Rewrite_DB:
         Channel_Name: master2-163
     Master_TLS_Version:
2 rows in set (0.00 sec)
```

在从库中，查看来自主库 1 和主库 2 的数据汇总情况：

```
# 来自主库 1 和主库 2 的数据汇总
mysql> show databases;
+--------------------+
| Database           |
+--------------------+
| information_schema |
| mysql              |
| performance_schema |
| sbtest    |  # 来自主库 1 的 sbtest 数据库
| sbtest2   |  # 来自主库 2 的 sbtest2 数据库
| sys       |
+--------------------+
6 rows in set (0.00 sec)
```

停止 sysbench 对主库 1 和主库 2 的施压（此为模拟步骤，真实环境中可忽略）。

21.3.2　GTID 复制模式下的多线程多源复制

基于 21.3.1 节中的环境，在从库中修改系统变量 slave_parallel_workers 的值，改为大于 0 的任意数值（这里设置为 16）：

```
# 先查看系统变量 slave_parallel_workers 的值
mysql> show variables like '%para%';
+------------------------+----------------+
```

```
| Variable_name          | Value         |
+------------------------+---------------+
| slave_parallel_type    | LOGICAL_CLOCK |
| slave_parallel_workers | 0             |
+------------------------+---------------+
2 rows in set (0.00 sec)

# 将系统变量 slave_parallel_workers 的值修改为 16
mysql> set global slave_parallel_workers=16;
Query OK, 0 rows affected (0.00 sec)

# 确认值已被修改
mysql> show variables like '%para%';
+------------------------+---------------+
| Variable_name          | Value         |
+------------------------+---------------+
| slave_parallel_type    | LOGICAL_CLOCK |
| slave_parallel_workers | 16            |
+------------------------+---------------+
2 rows in set (0.00 sec)
```

在从库中重启所有通道的复制线程：

```
# 停止复制线程
mysql> stop slave;
Query OK, 0 rows affected (0.02 sec)

# 启动复制线程
mysql> start slave;
Query OK, 0 rows affected (0.10 sec)

# 查看复制线程状态
mysql> show slave status\G
......
```

查看多线程复制的配置是否生效：

```
# 可通过 SHOW PROCESSLIST 语句或者 information_schema.processlist 表查看，也可以通过
performance_schema.threads 表查看（状态信息更详细）。从这些信息中可以看到，新增了 32 个 Worker 线程（2
个复制通道，每个通道各 16 个 Worker 线程）
mysql> select name,processlist_id,processlist_command,processlist_state,processlist_
info,thread_os_id from performance_schema.threads where name regexp 'thread/sql/
slave_io|thread/sql/slave_sql|thread/sql/slave_worker';
......
```

21.4 多源复制拓扑中复制相关的操作语句变化

1. 在多源复制拓扑中启动复制线程

```
# 启动所有的复制通道，其中 thread_types 参数不是必需的，值为 io_thread，表示操作 I/O 线程；值为
```

sql_thread，则表示操作 SQL 线程。不指定该参数时，表示同时操作 I/O 和 SQL 线程
    START SLAVE thread_types;

    # 启动指定通道的复制线程
    START SLAVE thread_types FOR CHANNEL channel_name;

2. 在多源复制拓扑中停止复制线程

    # 停止所有复制通道，其中 thread_types 参数不是必需的，值为 io_thread，表示操作 I/O 线程；值为
sql_thread，则表示操作 SQL 线程。不指定该参数时，表示同时操作 I/O 和 SQL 线程
    STOP SLAVE thread_types;

    # 停止指定通道的复制线程
    STOP SLAVE thread_types FOR CHANNEL channel_name;

3. 在多源复制拓扑中清理复制配置信息

    # 清理所有通道的复制配置信息
    RESET SLAVE;

    # 清理指定通道的复制配置信息
    RESET SLAVE FOR CHANNEL channel_name;

4. 在多源复制拓扑中查看复制线程的状态

    # 不指定复制通道名称，查看所有通道的复制线程状态信息
    SHOW SLAVE STATUS\G

    # 查看指定通道的复制线程状态信息
    SHOW SLAVE STATUS FOR CHANNEL channel_name;

可以使用 PERFORMANCE_SCHEMA 系统库下的复制状态表查看更多复制通道的状态数据：

    # 查询 replication_connection_status 表，该表中记录着 I/O 线程的状态信息。如果不指定 Channel_Name
    作为查询条件，则表示查询所有复制通道的状态信息；如果需要指定复制通道，则使用 WHERE CHANNEL_NAME='通道
    名称'来限定
    mysql> select * from performance_schema.replication_connection_status\G
    *************************** 1. row ***************************
              CHANNEL_NAME: master1-162
                GROUP_NAME:
               SOURCE_UUID: 8809b627-b74d-11e9-8461-5254002a54f2
                 THREAD_ID: 139
             SERVICE_STATE: ON
    COUNT_RECEIVED_HEARTBEATS: 138
     LAST_HEARTBEAT_TIMESTAMP: 2019-08-06 16:16:22
     RECEIVED_TRANSACTION_SET: 8809b627-b74d-11e9-8461-5254002a54f2:1-112758
            LAST_ERROR_NUMBER: 0
           LAST_ERROR_MESSAGE:
         LAST_ERROR_TIMESTAMP: 0000-00-00 00:00:00
    *************************** 2. row ***************************
              CHANNEL_NAME: master2-163

```
                  GROUP_NAME:
                 SOURCE_UUID: 9f6feeae-b74d-11e9-8b26-525400c33752
                   THREAD_ID: 157
               SERVICE_STATE: ON
     COUNT_RECEIVED_HEARTBEATS: 274
      LAST_HEARTBEAT_TIMESTAMP: 2019-08-06 16:16:23
      RECEIVED_TRANSACTION_SET: 9f6feeae-b74d-11e9-8b26-525400c33752:1-115972
           LAST_ERROR_NUMBER: 0
          LAST_ERROR_MESSAGE:
        LAST_ERROR_TIMESTAMP: 0000-00-00 00:00:00
2 rows in set (0.00 sec)
```

\# 查看 replication_applier_status_by_coordinator 表，该表中记录着协调器线程的状态信息。如果不指定 Channel_Name 作为查询条件，则表示查询所有复制通道的状态信息；如果需要指定复制通道，则使用 WHERE CHANNEL_NAME='通道名称' 来限定

```
mysql> select * from performance_schema.replication_applier_status_by_coordinator;
+--------------+-----------+---------------+-------------------+--------------------+
| CHANNEL_NAME | THREAD_ID | SERVICE_STATE | LAST_ERROR_NUMBER | LAST_ERROR_MESSAGE |
  LAST_ERROR_TIMESTAMP |
+--------------+-----------+---------------+-------+---------------------+
| master1-162  | 140       | ON            | 0     |                     | 0000-00-00 00:00:00 |
| master2-163  | 158       | ON            | 0     |                     | 0000-00-00 00:00:00 |
+--------------+-----------+---------------+-------+---------------------+
2 rows in set (0.00 sec)
```

\# 查看 replication_applier_status_by_worker 表，该表中记录着 Worker 工作线程的状态信息。如果不指定 Channel_Name 作为查询条件，则表示查询所有复制通道的状态信息；如果需要指定复制通道，则使用 WHERE CHANNEL_NAME='通道名称' 进行限定

```
mysql> select * from performance_schema.replication_applier_status_by_worker;
+--------------+-----------+-----------+---------------+---------------------+
| CHANNEL_NAME | WORKER_ID | THREAD_ID | SERVICE_STATE | LAST_SEEN_TRANSACTION |
  LAST_ERROR_NUMBER | LAST_ERROR_MESSAGE | LAST_ERROR_TIMESTAMP |
+--------------+-----+-----+--------+------+------+---------------------+
| master1-162  | 1   | 141 | ON     |      | 0    | 0000-00-00 00:00:00 |
......
| master1-162  | 16  | 156 | ON     |      | 0    | 0000-00-00 00:00:00 |
| master2-163  | 1   | 159 | ON     |      | 0    | 0000-00-00 00:00:00 |
......
| master2-163  | 16  | 174 | ON     |      | 0    | 0000-00-00 00:00:00 |
+--------------+-----+-----+--------+------+------+---------------------+
32 rows in set (0.00 sec)
```

也可以使用 mysql 系统库下的复制状态表查看更多复制通道的状态数据：

\# 查看 slave_master_info 表，该表中记录着 I/O 线程的状态信息。如果不指定 Channel_Name 作为查询条件，则表示查询所有复制通道的状态信息；如果需要指定复制通道，则使用 WHERE CHANNEL_NAME='通道名称' 来限定

```
mysql> select * from mysql.slave_master_info\G
*************************** 1. row ***************************
       Number_of_lines: 25
```

```
          Master_log_name: mysql-bin.000004
           Master_log_pos: 194
                     Host: 10.10.30.162
                User_name: repl
            User_password: letsg0
                     Port: 3306
            Connect_retry: 60
......
                Heartbeat: 5
                     Bind:
        Ignored_server_ids: 0
                     Uuid: 8809b627-b74d-11e9-8461-5254002a54f2
              Retry_count: 86400
                  Ssl_crl:
              Ssl_crlpath:
      Enabled_auto_position: 1
             Channel_name: master1-162
              Tls_version:
*************************** 2. row ***************************
           Number_of_lines: 25
           Master_log_name: mysql-bin.000002
            Master_log_pos: 154
                      Host: 10.10.30.163
                 User_name: repl
             User_password: letsg0
                      Port: 3306
             Connect_retry: 60
......
                 Heartbeat: 5
                      Bind:
         Ignored_server_ids: 0
                      Uuid: 9f6feeae-b74d-11e9-8b26-525400c33752
               Retry_count: 86400
                   Ssl_crl:
               Ssl_crlpath:
     Enabled_auto_position: 1
              Channel_name: master2-163
               Tls_version:
2 rows in set (0.00 sec)
```

查看slave_relay_log_info表，该表中记录着SQL线程的状态信息。如果不指定Channel_Name作为查询条件，则表示查询所有复制通道的状态信息；如果需要指定复制通道，则使用WHERE CHANNEL_NAME='通道名称'来限定

```
mysql> select * from mysql.slave_relay_log_info\G
*************************** 1. row ***************************
     Number_of_lines: 7
      Relay_log_name: /home/mysql/data/mysqldata1/relaylog/mysql-relay-bin-master1@002d162.000003
       Relay_log_pos: 82953890
```

```
    Master_log_name: mysql-bin.000002
     Master_log_pos: 237319608
          Sql_delay: 0
  Number_of_workers: 16
                 Id: 1
       Channel_name: master1-162
*************************** 2. row ***************************
    Number_of_lines: 7
     Relay_log_name: /home/mysql/data/mysqldata1/relaylog/mysql-relay-bin-master2@002d163.
000005
      Relay_log_pos: 454
    Master_log_name: mysql-bin.000002
     Master_log_pos: 242565817
          Sql_delay: 0
  Number_of_workers: 16
                 Id: 1
       Channel_name: master2-163
2 rows in set (0.01 sec)
```

提示：

- 在多源复制拓扑中，对半同步复制的支持目前还不是很友好。在设计复制拓扑时，请谨慎考虑，做好测试。同时，尽可能地使用最新的 GA 版本（GA 是 General Availability 的缩写，GA 版本表示正式发布的稳定版本）。

- 在多源复制拓扑中，如果要启用 GTID，最好在初始化多源复制拓扑时就一并启用，否则会给维护带来一定的麻烦。例如，如果某个主库的二进制日志曾经执行过清理操作，则需要在从库中使用 SET GLOBAL GTID_PURGED 语句设置被清理的 GTID SET 位置，但执行该语句时需要先在从库中执行 RESET MASTER 操作，而该操作会导致从库中所有的二进制日志以及 GTID 信息被清理，会影响其他复制通道。

- 在多源复制拓扑中，GTID 复制模式与传统复制模式不能混搭。若要使用 GTID 复制模式，则复制拓扑中的所有实例都必须启用 GTID；若要使用传统复制模式，则复制拓扑中所有的实例都必须关闭 GTID。否则，未启用 GTID 的复制源主库对应的复制通道将不能正常工作。

- 更多关于 PERFORMANCE_SCHEMA 和 mysql 系统库的介绍，可参考《千金良方：MySQL 性能优化金字塔法则》一书中的相关章节，也可以参考微信公众号"沃趣技术"或"知数堂"中相关的专栏。

第 22 章 MySQL 版本升级

MySQL 的版本并不需要常升级，但如果要使用新版本的某个新特性，或者为了修复旧版本的某个 bug，就不得不进行版本升级。在生产环境中，为避免单点故障，通常都会用多个 MySQL 实例构建一个复制拓扑。为了使升级操作对业务的影响尽可能小，可以先升级复制拓扑中的只读实例（从库），然后再执行一次主从角色切换（会造成应用短暂中断），最后将主库当作从库再做一次升级即可。

本章只阐述在复制拓扑中升级 MySQL 版本的一些注意事项，关于复制拓扑中的 MySQL 版本升级步骤，可参考 19.2 节"在线切换"。欲了解更多与升级相关的步骤，可参考《千金良方：MySQL 性能优化金字塔法则》一书的第 2 章"MySQL 常用的两种升级方法"。

版本升级操作中的注意事项：

- 跨大版本的升级（例如，从 MySQL 5.5.x 升级到 MySQL 5.7.x）不能直接使用 mysql_upgrade 命令，需要使用 mysqldump 命令从旧版本导出纯 SQL 文本数据，将其导入新版本的数据库实例中后，再使用 mysql_upgrade 命令。
- 小版本或大版本逐级升级，可以直接使用 mysql_upgrade 命令，例如，从 MySQL 5.5.x 升级到 MySQL 5.6.x、从 MySQL 5.6.x 升级到 MySQL 5.7.x、从 MySQL 5.7.x 升级到 MySQL 5.7.y。
- 在一个复制拓扑中，主库的版本不得高于从库的，否则主库的新版本特性在从库中可能无法被识别。
- 如果复制拓扑中启用了 GTID，则在使用 mysql_upgrade 命令升级时不能打开 --write-binlog 选项。

22.1　MySQL 版本之间的复制兼容性

MySQL 支持从一个发行版本系列到下一个更高版本系列的复制（即支持不同主版本之间的复制）。例如，可以从 MySQL 5.6 版本的主库复制到 MySQL 5.7 版本的从库，或从 MySQL 5.7 版本的主库复制到 MySQL 8.0 版本的从库，依此类推。但要注意，如果使用旧版本 MySQL 的主库（比从库使用的版本低）在某个修改数据的语句中依赖的某个特定功能或行为，在使用新版本 MySQL 的从库上不再支持，则在从较旧的主库复制到较新的从库时可能会遇到问题。例如，MySQL 8.0 不再支持超过 64 个字符的外键名称。

无论主从库的 MySQL Server 使用什么样的版本，在涉及多个主库的复制拓扑中都不支持使用三个及其以上的不同版本。例如，如果使用链式（串行）或环式复制拓扑，则不能同时使用 MySQL 5.7.22、MySQL 5.7.23、MySQL 5.7.24 这三个版本，但可以同时使用其中的任意两个版本。如果无特殊需求，建议尽量统一整个复制拓扑中的数据库版本。

强烈建议使用给定 MySQL 发行版系列（某个主版本）中的最新 GA 版本，因为 MySQL Server 的功能在不断改进，这样可以避免很多已知的 bug，其性能与稳定性能够得到更好的保障。

提示：将使用新版本 MySQL 的主库数据变更，复制到使用旧版本 MySQL 的从库，从框架实现上来讲是可行的，但是通常不建议这么做，原因有如下几点：

- 新旧版本 MySQL 中的二进制日志格式可能存在差异，二进制日志格式可能在主要版本号之间发生变更（新版本 MySQL 的二进制日志格式可能有改动）。虽然新版本通常都会试图保持向后兼容性，但这并不总是可行的。
- 新旧版本 MySQL 中的 SQL 语句可能不兼容：如果某个语句使用了新版本 MySQL（主库）中的某个 SQL 特性，而这个 SQL 特性无法在旧版本 MySQL（从库）上使用（旧版本 MySQL 不支持该特性），则无法使用基于 statement 格式正确地复制该语句。但是，如果主库和从库都支持基于 row 格式的复制，且不存在 DDL 语句依赖于主库的 SQL 特性，复制到从库之后也不会出现无法找到该 SQL 依赖的特性的情况，那么可以使用 row 格式复制 DML 语句产生的数据变更，此时版本的差异在理论上不会导致主从库之间的数据复制出现问题。

22.2 升级复制的设置

MySQL Server 版本的升级过程及升级之后复制变量的设置，取决于升级前和升级后的版本。随着版本的不断迭代，MySQL 的主从复制沿着保证数据的一致性和复制效率两条路线不断向前演进。每引入一个新特性，就会引入新的复制配置变量对其进行控制。详情可参考《千金良方：MySQL 性能优化金字塔法则》一书中第 18 章"复制技术的演进"。

要将主库从早期的 MySQL 发行版系列升级到 MySQL 5.7 时，首先应确保该主库的所有从库都使用的是相同的 MySQL 5.7.x 版本。否则，应依次将所有从库升级到 MySQL 5.7，然后重启 mysqld 进程和复制线程。从库升级到 MySQL 5.7 之后，将使用该版本创建中继日志。

严格的 SQL 模式（系统变量 sql_mode 的值为 STRICT_TRANS_TABLES 或 STRICT_ALL_TABLES）可能导致从库更新之后，来自主库的一些变更在应用二进制日志时失败。例如，从 MySQL 5.7.2 开始，严格的 SQL 模式限制时间数据类型插入默认值为 0 的值。如果使用基于 statement 格式的二进制日志记录（系统变量 binlog_format = statement），而且在主库升级到新版本之前先升级从库，则在还未升级的主库上为时间数据类型插入零值不会报错，但是将这些语句复制到从库时，可能导致从库的复制进程报错而中止。要避免这个问题，可以采用如下几种方法：

- 升级从库之前，停止主库中的所有写入，并等待从库追赶上主库数据。

- 升级从库之前，在主库中修改系统变量 binlog_format = row，并等待从库追赶上修改二进制日志格式的时间点位置。

- 升级从库之前，先在从库的配置文件中写入旧版本的系统变量 sql_mode 的值。

所有从库升级后，在复制拓扑中执行在线主从切换，提升一个从库为新主库，原主库降级为从库后关闭，升级到与从库相同的 MySQL 5.7.x 版本，然后重新启动 mysqld 进程，原主库作为一个新的从库加入复制拓扑。如果之前修改过主库的系统变量 binlog_format = row，此时根据需要修改该系统变量的值，新版本（MySQL 5.7）的主库能够读取升级前旧版本写入的二进制日志，并将它们发送到 MySQL 5.7 版本的从库中。从库能够正确识别旧的格式并正确处理这些二进制日志。升级后主库创建的二进制日志采用新的 MySQL 5.7 版本的格式。换句话说，在这个示例中主库升级到 MySQL 5.7 时，从库必须已经是 MySQL 5.7 版本。

请注意，从 MySQL 5.7 降级到较旧版本并不简单：你必须确保主从实例之间都已处理完所有 MySQL 5.7 版本的二进制日志或中继日志，并在删除新版本的二进制日志之后才能够继续执行降级操作。

当从一个 MySQL 版本系列迁移到下一个 MySQL 版本系列时，某些升级可能需要删除并重新创建数据库对象。例如，排序规则的变更可能需要重建表的索引。如有必要，建议在主从库上单独执行这些操作，升级操作的步骤如下：

- 停止所有从库，并执行升级（如果需要在线升级，可以依次逐一升级从库）。在重启从库 mysqld 进程时使用--skip-slave-start 选项，以便重启之后不会自动启动复制连接到主库。然后，执行数据字典的升级（使用 mysql_upgrade 命令）。

- 执行主从切换，主库降级为从库，将原主库作为从库进行升级。

如果要将现有复制设置从不支持 GTID 的 MySQL 版本升级到支持 GTID 的版本，则在确保复制设置满足 GTID 复制的所有要求之前，不应在主库或从库上启用 GTID。例如，系统变量 server_uuid 是在 MySQL 5.6 中添加的，必须确保 MySQL Server 支持该系统变量才能使 GTID 复制模式正常运行。关于将非 GTID 复制切换为 GTID 复制，详情可参考第 17 章"复制模式的切换"。

如果 MySQL Server 已经启用了 GTID（系统变量 gtid_mode = ON），在执行 mysql_upgrade 工具升级时，不能一同启用二进制日志记录功能（这里指的是 mysql_upgrade 命令的--write-binlog 选项）。

第 23 章　将不同数据库的数据复制到不同实例

将不同数据库中的数据复制到不同的实例（MySQL Server），在实现上具体指的就是从库将主库的全量二进制日志拉取到本地的中继日志之后，从库 SQL 线程在重放这些二进制日志时，根据自身配置的复制过滤规则，选择需要应用哪些库与表，以及需要忽略哪些库与表。当然，在主库端也支持复制过滤，虽然在主库端配置复制过滤后能够减少二进制日志的传输量，但主库端只支持库级别的过滤规则，而且容易导致主从库数据不一致。通常不建议在主库端配置复制过滤规则，可靠的复制过滤都是在从库端实现的，因为这样才更合理，每个从库根据自己的需要来灵活配置复制过滤规则。

关于复制过滤的原理与流程，本章不做过多阐述，可参考第 13 章 "MySQL Server 复制过滤"。本章将以在从库端配置复制过滤规则为例，详细介绍其操作步骤。

23.1　操作环境信息

1. Server IP 地址

- 主库（MySQL Master）：10.10.30.16
- 从库 1（MySQL Slave 1）：10.10.30.162
- 从库 2（MySQL Slave 2）：10.10.30.163
- 从库 3（MySQL Slave 3）：10.10.30.164

2. MySQL 数据库版本

MySQL 5.7.27

3. sysbench 版本

sysbench 1.0.9

23.2 通过设置复制过滤规则将不同数据库的数据复制到不同实例

在某些情况下，你可能只有一个主库，承载着多个数据库的业务，你希望将该主库中的不同数据库的数据复制到不同从库。例如，你可能希望将不同的销售数据分发到不同的部门，以便在进行数据分析期间分散负载。

如图 23-1 所示（参考 Oracle MySQL 官方资料），MySQL Master 中（MySQL 主库）存在三个数据库：databaseA、databaseB、databaseC。我们希望将 databaseA 仅复制到 MySQL Slave1，将 databaseB 仅复制到 MySQL Slave2，将 databaseC 仅复制到 MySQL Slave3。要实现此目的，MySQL 主库只需要正常配置复制过滤相关的规则即可。

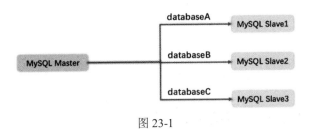

图 23-1

在 MySQL 5.7 及其之后的版本中，对复制过滤的配置支持只读和动态配置两种方式，可分别满足如下两种场景的复制过滤需求：

- 事先能够规划的复制过滤。在全新初始化的复制拓扑中，事先通过规划，在 my.cnf 文件中配置好只读选项，即可完成复制过滤规则的配置，后面配置复制拓扑的步骤与未配置复制过滤规则的完全相同。通过在 my.cnf 文件中配置选项实现复制过滤的方式，是 MySQL 5.7 之前的版本唯一可选的配置方式。但如果要在已运行一段时间的复制拓扑中配置复制过滤规则，采用这种方式就很不方便了。通过只读选项配置复制过滤规则的关键步骤如图 23-2 所示 。

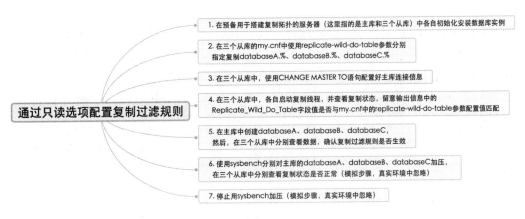

图 23-2

- 事先未规划,后面新增了复制过滤需求。在已运行了一段时间的复制拓扑中,可以利用 MySQL 5.7 及其之后的版本中新增的 CHANGE REPLICATION FILTER 语句,在从库中动态配置复制过滤规则,只需要重启 SQL 线程即可生效,无须重启数据库实例,如果要在 MySQL 5.7 及其之后的版本中配置复制过滤规则,这种方式无疑是首选。通过动态语句配置复制过滤规则的关键步骤如图 23-3 所示 。

图 23-3

下面分别详细介绍这两种配置方式。

23.2.1 通过只读选项配置复制过滤规则

假设我们已经按照 23.1 节"操作环境信息"中的要求完成了数据库实例的全新初始化安装(但未配置复制),现在分别在从库 1、从库 2、从库 3 的服务器上的 my.cnf 中配置如下复制过滤规则:

```
# 从库 1 只复制名为 databaseA 的数据库的数据,则配置如下选项 (.%表示匹配 databaseA 库下的所有表)
replicate-wild-do-table=databaseA.%

# 从库 2 只复制名为 databaseB 的数据库的数据,则配置如下选项
replicate-wild-do-table=databaseB.%

# 从库 3 只复制名为 databaseC 的数据库的数据,则配置如下选项
replicate-wild-do-table=databaseC.%
```

分别登录从库 1、从库 2、从库 3 的数据库,使用 CHANGE MASTER TO 语句配置复制指向主库:

```
# 从库 1(启用 GTID 复制模式)
mysql> change master to master_host='10.10.30.16', master_user='repl', master_password='password',master_port=3306,master_auto_position=1;
Query OK, 0 rows affected, 2 warnings (0.02 sec)

# 在从库 2、从库 3 中分别使用 CHANGE MASTER TO 语句配置复制指向主库(配置步骤与从库 1 的类似,这里不再赘述)
```

分别登录从库 1、从库 2 和从库 3 的数据库,启动复制线程,并查看其状态:

```
# 从库 1
```

```
mysql> start slave;
Query OK, 0 rows affected (0.05 sec)

mysql> show slave status\G
*************************** 1. row ***************************
               Slave_IO_State: Waiting for master to send event
                  Master_Host: 10.10.30.16
                  Master_User: repl
                  Master_Port: 3306
                Connect_Retry: 10
              Master_Log_File: mysql-bin.000002
          Read_Master_Log_Pos: 154
               Relay_Log_File: mysql-relay-bin.000005
                Relay_Log_Pos: 367
        Relay_Master_Log_File: mysql-bin.000002
             Slave_IO_Running: Yes
            Slave_SQL_Running: Yes
              Replicate_Do_DB:
          Replicate_Ignore_DB:
           Replicate_Do_Table:
       Replicate_Ignore_Table:
      Replicate_Wild_Do_Table: databaseA.%   # 这里可以看到，从库 1 中的复制过滤规则为只复制
databaseA.%
  Replicate_Wild_Ignore_Table:
......
           Exec_Master_Log_Pos: 154
              Relay_Log_Space: 787
......
         Seconds_Behind_Master: 0
......
           Retrieved_Gtid_Set:
            Executed_Gtid_Set:
                Auto_Position: 1
......

# 从库 2
mysql> start slave;
Query OK, 0 rows affected (0.06 sec)

mysql> show slave status\G
*************************** 1. row ***************************
               Slave_IO_State: Waiting for master to send event
                  Master_Host: 10.10.30.16
                  Master_User: repl
                  Master_Port: 3306
                Connect_Retry: 10
              Master_Log_File: mysql-bin.000002
          Read_Master_Log_Pos: 154
```

```
            Relay_Log_File: mysql-relay-bin.000005
             Relay_Log_Pos: 367
     Relay_Master_Log_File: mysql-bin.000002
          Slave_IO_Running: Yes
         Slave_SQL_Running: Yes
           Replicate_Do_DB:
       Replicate_Ignore_DB:
        Replicate_Do_Table:
    Replicate_Ignore_Table:
   Replicate_Wild_Do_Table: databaseB.%   # 这里可以看到,从库 2 中的复制过滤规则为只复制
databaseB.%
Replicate_Wild_Ignore_Table:
......
        Exec_Master_Log_Pos: 154
           Relay_Log_Space: 787
......
      Seconds_Behind_Master: 0
......
        Retrieved_Gtid_Set:
         Executed_Gtid_Set:
             Auto_Position: 1
......

# 从库 3
mysql> start slave;
Query OK, 0 rows affected (0.07 sec)

mysql> show slave status\G
*************************** 1. row ***************************
             Slave_IO_State: Waiting for master to send event
                Master_Host: 10.10.30.16
                Master_User: repl
                Master_Port: 3306
              Connect_Retry: 10
            Master_Log_File: mysql-bin.000002
        Read_Master_Log_Pos: 154
             Relay_Log_File: mysql-relay-bin.000005
              Relay_Log_Pos: 367
      Relay_Master_Log_File: mysql-bin.000002
           Slave_IO_Running: Yes
          Slave_SQL_Running: Yes
            Replicate_Do_DB:
        Replicate_Ignore_DB:
         Replicate_Do_Table:
     Replicate_Ignore_Table:
    Replicate_Wild_Do_Table: databaseC.%   # 这里可以看到,从库 3 中的复制过滤规则为只复制
databaseC.%
 Replicate_Wild_Ignore_Table:
```

```
......
          Exec_Master_Log_Pos: 154
            Relay_Log_Space: 787
......
        Seconds_Behind_Master: 0
......
          Retrieved_Gtid_Set:
           Executed_Gtid_Set:
              Auto_Position: 1
......
```

现在，从库 1、从库 2、从库 3 中都没有任何用户数据，我们在主库中分别创建名为 databaseA、databaseB、databaseC 的 3 个数据库，然后在 3 个从库中分别查看：

```
# 在主库中分别创建 databaseA、databaseB、databaseC
mysql> create database databaseA;
Query OK, 1 row affected (0.00 sec)

mysql> create database databaseB;
Query OK, 1 row affected (0.00 sec)

mysql> create database databaseC;
Query OK, 1 row affected (0.00 sec)

# 查看刚才创建的名为 databaseA、databaseB、databaseC 的数据库
mysql> show databases;
+--------------------+
| Database           |
+--------------------+
| information_schema |
| databasea |    # 由于启用了忽略系统变量的大小写，字母 A、B、C 都变成小写字母，所以这里看到的就是名
为 databasea、databaseb、databasec 的 3 个数据库
| databaseb |
| databasec |
| mysql |
| performance_schema |
| sys |
+--------------------+
7 rows in set (0.00 sec)

# 在从库 1 中查看其是否只同步了 databaseA
mysql> show databases;
+--------------------+
| Database           |
+--------------------+
| information_schema |
| databasea |    # 从这里可以看到，从库 1 中只有名为 databasea 的数据库，没有 databaseb 和 databasec
| mysql |
| performance_schema |
```

```
| sys                |
+--------------------+
5 rows in set (0.01 sec)

# 在从库 2 中查看其是否只同步了 databaseB
mysql> show databases;
+--------------------+
| Database           |
+--------------------+
| information_schema |
| databaseb |    # 从这里可以看到，从库 2 中只有名为 databaseb 的数据库，而没有 databasea 和 databasec
| mysql              |
| performance_schema |
| sys                |
+--------------------+
5 rows in set (0.01 sec)

# 在从库 3 中查看其是否只同步了 databaseC
+--------------------+
| Database           |
+--------------------+
| information_schema |
| databasec |    # 从这里可以看到，从库 3 中只有名为 databasec 的数据库，而没有 databasea 和 databaseb
| mysql              |
| performance_schema |
| sys                |
+--------------------+
5 rows in set (0.00 sec)
```

现在，我们使用 sysbench 对主库中的 databaseA（databasea）、databaseB（databaseb）、databaseC（databasec）3 个库分别建表并插入测试数据，然后加 OLTP 压力（此为模拟步骤，真实环境中可忽略）：

```
# 对主库的 databaseA 库进行操作
## sysbench 的 prepare 指令可以在 --mysql-db 选项指定的数据库 databaseA 中新建表，然后插入测试数据
[root@localhost ~]# sysbench --db-driver=mysql --time=99999 --threads=2 --report-interval=1 --mysql-host=10.10.30.16 --mysql-port=3306 --mysql-user=qbench --mysql-password=qbench --mysql-db=databaseA --tables=2 --table-size=5000000 oltp_read_write --db-ps-mode=disable prepare
......
## sysbench 的 run 指令可以对数据库发起 OLTP 模型的并发事务
[root@localhost ~]# sysbench --db-driver=mysql --time=99999 --threads=2 --report-interval=1 --mysql-host=10.10.30.16 --mysql-port=3306 --mysql-user=qbench --mysql-password=qbench --mysql-db=databaseA --tables=2 --table-size=5000000 oltp_read_write --db-ps-mode=disable run

# 另外开启一个服务器终端会话，对主库的 databaseB 库进行操作
[root@localhost ~]# sysbench --db-driver=mysql --time=99999 --threads=2 --report-interval=1 --mysql-host=10.10.30.16 --mysql-port=3306 --mysql-user=qbench --mysql-password=
```

```
qbench --mysql-db=databaseB --tables=2 --table-size=5000000 oltp_read_write --db-ps-
mode=disable prepare
......
    [root@localhost ~]# sysbench --db-driver=mysql --time=99999 --threads=2 --report-
interval=1 --mysql-host=10.10.30.16 --mysql-port=3306 --mysql-user=qbench --mysql-password=
qbench --mysql-db=databaseB --tables=2 --table-size=5000000 oltp_read_write --db-ps-mode=
disable run

# 另外开启一个服务器终端会话，对主库的 databaseC 库进行操作
    [root@localhost ~]# sysbench --db-driver=mysql --time=99999 --threads=2 --report-
interval=1 --mysql-host=10.10.30.16 --mysql-port=3306 --mysql-user=qbench --mysql-password=
qbench --mysql-db=databaseC --tables=2 --table-size=5000000 oltp_read_write --db-ps-mode=
disable prepare
......
    [root@localhost ~]# sysbench --db-driver=mysql --time=99999 --threads=2 --report-
interval=1 --mysql-host=10.10.30.16 --mysql-port=3306 --mysql-user=qbench --mysql-password=
qbench --mysql-db=databaseC --tables=2 --table-size=5000000 oltp_read_write --db-ps-mode=
disable run
```

分别登录从库 1、从库 2、从库 3 的数据库，查看复制线程状态及其数据同步情况：

```
# 从库 1
mysql> show slave status\G
*************************** 1. row ***************************
               Slave_IO_State: Waiting for master to send event
                  Master_Host: 10.10.30.16
                  Master_User: repl
                  Master_Port: 3306
                Connect_Retry: 10
              Master_Log_File: mysql-bin.000023
          Read_Master_Log_Pos: 407892798
               Relay_Log_File: mysql-relay-bin.000068
                Relay_Log_Pos: 383140497
        Relay_Master_Log_File: mysql-bin.000023
             Slave_IO_Running: Yes  # I/O 线程状态正常
            Slave_SQL_Running: Yes  # SQL 线程状态正常
              Replicate_Do_DB:
          Replicate_Ignore_DB:
           Replicate_Do_Table:
       Replicate_Ignore_Table:
      Replicate_Wild_Do_Table: databaseA.%
  Replicate_Wild_Ignore_Table:
                   Last_Errno: 0
                   Last_Error:
                 Skip_Counter: 0
          Exec_Master_Log_Pos: 383140284
              Relay_Log_Space: 407893305
......
        Seconds_Behind_Master: 9
```

```
......
    Retrieved_Gtid_Set: 265af701-b9b0-11e9-b742-0025905b06da:1-68888
    Executed_Gtid_Set: 265af701-b9b0-11e9-b742-0025905b06da:1-53859
        Auto_Position: 1
......

mysql> show databases;
+--------------------+
| Database |
+--------------------+
| information_schema |
| databasea |    # 从库 1 指定复制的 databasea 库在这里，没有多余复制其他数据库（例如 databaseb 和 databasec），表示复制过滤工作正常
| mysql |
| performance_schema |
| sys |
+--------------------+
5 rows in set (0.00 sec)

# 从库 2
mysql> show slave status\G
*************************** 1. row ***************************
               Slave_IO_State: Queueing master event to the relay log
                  Master_Host: 10.10.30.16
                  Master_User: repl
                  Master_Port: 3306
                Connect_Retry: 10
              Master_Log_File: mysql-bin.000024
          Read_Master_Log_Pos: 77049115
               Relay_Log_File: mysql-relay-bin.000068
                Relay_Log_Pos: 519625960
        Relay_Master_Log_File: mysql-bin.000023
             Slave_IO_Running: Yes  # I/O 线程状态正常
            Slave_SQL_Running: Yes  # SQL 线程状态正常
              Replicate_Do_DB:
          Replicate_Ignore_DB:
           Replicate_Do_Table:
       Replicate_Ignore_Table:
      Replicate_Wild_Do_Table: databaseB.%
  Replicate_Wild_Ignore_Table:
                   Last_Errno: 0
                   Last_Error:
                 Skip_Counter: 0
          Exec_Master_Log_Pos: 519625747
              Relay_Log_Space: 613923109
......
        Seconds_Behind_Master: 30
......
```

```
        Retrieved_Gtid_Set: 265af701-b9b0-11e9-b742-0025905b06da:1-189957
        Executed_Gtid_Set: 265af701-b9b0-11e9-b742-0025905b06da:1-136121
            Auto_Position: 1
......

mysql> show databases;
+--------------------+
| Database |
+--------------------+
| information_schema |
| databaseb |    # 从库 2 指定复制的 databaseb 库在这里，没有多余复制其他数据库（例如 databasea 和
databasec），表示复制过滤工作正常
| mysql |
| performance_schema |
| sys |
+--------------------+
5 rows in set (0.01 sec)


# 从库 3
mysql> show slave status\G
*************************** 1. row ***************************
               Slave_IO_State: Waiting for master to send event
                  Master_Host: 10.10.30.16
                  Master_User: repl
                  Master_Port: 3306
                Connect_Retry: 10
              Master_Log_File: mysql-bin.000024
          Read_Master_Log_Pos: 163255754
               Relay_Log_File: mysql-relay-bin.000071
                Relay_Log_Pos: 25089207
        Relay_Master_Log_File: mysql-bin.000024
             Slave_IO_Running: Yes  # I/O 线程状态正常
            Slave_SQL_Running: Yes  # SQL 线程状态正常
              Replicate_Do_DB:
          Replicate_Ignore_DB:
           Replicate_Do_Table:
       Replicate_Ignore_Table:
      Replicate_Wild_Do_Table: databaseC.%
  Replicate_Wild_Ignore_Table:
                   Last_Errno: 0
                   Last_Error:
                 Skip_Counter: 0
          Exec_Master_Log_Pos: 25088994
              Relay_Log_Space: 163256261
......
        Seconds_Behind_Master: 43
......
        Retrieved_Gtid_Set: 265af701-b9b0-11e9-b742-0025905b06da:1-237814
```

```
        Executed_Gtid_Set: 265af701-b9b0-11e9-b742-0025905b06da:1-160588
           Auto_Position: 1
......

mysql > show databases;
+--------------------+
| Database           |
+--------------------+
| information_schema |
| databasec          |    # 从库 3 指定复制的 databasec 库在这里，没有多余复制其他数据库（例如 databasea 和
databaseb），表示复制过滤工作正常
| mysql              |
| performance_schema |
| sys                |
+--------------------+
5 rows in set (0.00 sec)
```

至此，通过只读选项配置复制过滤的步骤已完成，停止 sysbench 的加压（此为模拟步骤，真实环境中忽略）。

提示：

- 在使用基于 statement 的复制和混合复制时（这里指的是系统变量 binlog_format 设置为 statement 或 mixed 值），不要使用库级别复制过滤选项--replicate-do-db。在这种情况下，复制过滤的结果容易与预期不符，但如果使用基于 row 的复制则不会有这个问题。

- 每个从库都会完整接收主库的二进制日志，但是在执行二进制日志时，会根据从库中生效的--replicate-wild-do-table 选项配置的库表进行过滤，只应用与生效的库表相关的二进制日志记录。

23.2.2　通过动态语句配置复制过滤规则

基于 23.2.1 节中的复制拓扑环境，现在我们在从库 1 中使用 CHANGE REPLICATION FILTER 语句指定需要复制 databaseD_change_replication_filter 库下的所有表：

```
# 在从库 1 中执行 CHANGE REPLICATION FILTER 语句，动态修改复制过滤规则。注意，这里需要一并指定
databaseA.%，否则该选项会被覆盖
    mysql> stop slave sql_thread;    # 必须先停止SQL线程,否则后面无法使用CHANGE REPLICATION FILTER
语句动态配置复制过滤规则
    Query OK, 0 rows affected (0.01 sec)

# 查看复制线程状态
    mysql> show slave status\G
    *************************** 1. row ***************************
                Slave_IO_State: Waiting for master to send event
                   Master_Host: 10.10.30.16
                   Master_User: repl
```

```
              Master_Port: 3306
            Connect_Retry: 10
          Master_Log_File: mysql-bin.000032
      Read_Master_Log_Pos: 134758084
           Relay_Log_File: mysql-relay-bin.000095
            Relay_Log_Pos: 134758297
    Relay_Master_Log_File: mysql-bin.000032
         Slave_IO_Running: Yes
        Slave_SQL_Running: No    # 确认 SQL 线程已关闭
          Replicate_Do_DB:
      Replicate_Ignore_DB:
       Replicate_Do_Table:
   Replicate_Ignore_Table:
  Replicate_Wild_Do_Table: databaseA.%
......

# 在从库 1 中，一并指定所有需要复制的库表（多个库之间使用逗号分隔），未指定的库表都会被忽略（未显式指定的库表被视为不需要复制）
mysql> change replication filter replicate_wild_do_table = ('databaseA.%','databaseD_change_replication_filter.%');
Query OK, 0 rows affected (0.00 sec)

# 查看复制的状态
mysql> show slave status\G
*************************** 1. row ***************************
            Slave_IO_State: Waiting for master to send event
               Master_Host: 10.10.30.16
               Master_User: repl
               Master_Port: 3306
             Connect_Retry: 10
           Master_Log_File: mysql-bin.000032
       Read_Master_Log_Pos: 134758084
            Relay_Log_File: mysql-relay-bin.000095
             Relay_Log_Pos: 134758297
     Relay_Master_Log_File: mysql-bin.000032
          Slave_IO_Running: Yes
         Slave_SQL_Running: No    # SQL 线程仍然为关闭状态，此时还未重新启动 SQL 线程
           Replicate_Do_DB:
       Replicate_Ignore_DB:
        Replicate_Do_Table:
    Replicate_Ignore_Table:
   Replicate_Wild_Do_Table: databaseA.%,databaseD_change_replication_filter.%    #
这里可以看到有两条复制过滤规则，一条是 databaseA.%，另一条是 databaseD_change_replication_filter.%

# 启动 SQL 线程
mysql> start slave sql_thread;
Query OK, 0 rows affected (0.05 sec)
```

```
# 查看复制的状态
mysql> show slave status\G
*************************** 1. row ***************************
               Slave_IO_State: Waiting for master to send event
                  Master_Host: 10.10.30.16
                  Master_User: repl
                  Master_Port: 3306
                Connect_Retry: 10
              Master_Log_File: mysql-bin.000032
          Read_Master_Log_Pos: 134758084
               Relay_Log_File: mysql-relay-bin.000095
                Relay_Log_Pos: 134758297
        Relay_Master_Log_File: mysql-bin.000032
             Slave_IO_Running: Yes
            Slave_SQL_Running: Yes    # 确认SQL线程已启动
              Replicate_Do_DB:
          Replicate_Ignore_DB:
           Replicate_Do_Table:
       Replicate_Ignore_Table:
      Replicate_Wild_Do_Table: databaseA.%,databaseD_change_replication_filter.%
......
```

使用相同的方法在从库 2 和从库 3 中配置针对名为 databaseD_change_replication_filter 数据库的复制过滤规则：

```
# 在从库2中执行如下语句（这里只展示与CHANGE REPLICATION FILTER语句相关的步骤，其他步骤省略）
mysql> change replication filter replicate_wild_do_table = ('databaseB.%','databaseD_change_replication_filter.%');
Query OK, 0 rows affected (0.00 sec)

# 在从库3中执行如下语句（这里只展示与CHANGE REPLICATION FILTER语句相关的步骤，其他步骤省略）
mysql> change replication filter replicate_wild_do_table = ('databaseC.%','databaseD_change_replication_filter.%');
Query OK, 0 rows affected (0.00 sec)
```

现在，我们在主库中创建名为 databaseD_change_replication_filter 的数据库，并在从库 1、从库 2 和从库 3 中查看是否其会同步名为 databaseD_change_replication_filter 的数据库的数据：

```
# 在主库中创建名为databaseD_change_replication_filter的数据库
mysql> create database databaseD_change_replication_filter;
Query OK, 1 row affected (0.00 sec)

# 在主库中查看刚才创建的数据库
mysql> show databases;
+-------------------------------------+
| Database                            |
+-------------------------------------+
| information_schema                  |
```

```
| databasea |
| databaseb |
| databasec |
| databased_change_replication_filter |   # 刚才创建的名为 databaseD_change_replication_
filter 的数据库在这里
| mysql |
| performance_schema |
| sys |
+------------------------------------+
8 rows in set (0.01 sec)

# 在从库 1 中查看
mysql> show databases;
+------------------------------------+
| Database |
+------------------------------------+
| information_schema |
| databasea |
| databased_change_replication_filter |   # 已与名为 databaseD_change_replication_filter
的数据库同步
| mysql |
| performance_schema |
| sys |
+------------------------------------+
6 rows in set (0.00 sec)

# 在从库 2 中查看
mysql> show databases;
+------------------------------------+
| Database |
+------------------------------------+
| information_schema |
| databaseb |
| databased_change_replication_filter |   # 已与名为 databaseD_change_replication_filter
的数据库同步
| mysql |
| performance_schema |
| sys |
+------------------------------------+
6 rows in set (0.00 sec)

# 在从库 3 中查看
mysql> show databases;
+------------------------------------+
| Database |
+------------------------------------+
| information_schema |
| databasec |
```

```
| databased_change_replication_filter |  # 已与 databaseD_change_replication_filter 的
数据库同步
| mysql              |
| performance_schema |
| sys                |
+------------------------------------+
6 rows in set (0.01 sec)
```

由于使用 CHANGE REPLICATION FILTER 语句配置的复制过滤规则是临时生效的，若要持久化复制过滤规则，就需要将其配置到从库 1、从库 2 和从库 3 的 my.cnf 中：

```
# 从库 1
[root@localhost ~]# cat /etc/my.cnf
......
[mysqld]
replicate-wild-do-table=databaseA.%
replicate-wild-do-table=databaseD_change_replication_filter.%

# 从库 2
[root@localhost ~]# cat /etc/my.cnf
......
[mysqld]
replicate-wild-do-table=databaseB.%
replicate-wild-do-table=databaseD_change_replication_filter.%

# 从库 3
[root@localhost ~]# cat /etc/my.cnf
......
[mysqld]
replicate-wild-do-table=databaseC.%
replicate-wild-do-table=databaseD_change_replication_filter.%
```

至此，通过动态语句配置复制过滤规则的步骤已完成。

关于 CHANGE REPLICATION FILTER 语句的更多用法，请使用帮助命令查看：

```
mysql> help change replication filter
Name: 'CHANGE REPLICATION FILTER'
Description:
Syntax:
CHANGE REPLICATION FILTER filter[, filter][, ...]

filter:
    REPLICATE_DO_DB = (db_list)
  | REPLICATE_IGNORE_DB = (db_list)
  | REPLICATE_DO_TABLE = (tbl_list)
  | REPLICATE_IGNORE_TABLE = (tbl_list)
  | REPLICATE_WILD_DO_TABLE = (wild_tbl_list)
  | REPLICATE_WILD_IGNORE_TABLE = (wild_tbl_list)
  | REPLICATE_REWRITE_DB = (db_pair_list)
```

```
db_list:
    db_name[, db_name][, ...]
tbl_list:
    db_name.table_name[, db_table_name][, ...]
wild_tbl_list:
    'db_pattern.table_pattern'[, 'db_pattern.table_pattern'][, ...]
db_pair_list:
    (db_pair)[, (db_pair)][, ...]
db_pair:
    from_db, to_db
......
```

提示：

在已经运行了一段时间的复制拓扑中，如果复制过滤需求有变更，可以按照如下几种方法调整：

- 方法一：将主库的所有数据同步到每个从库中，在每个从库中对应删除不需要的数据库表，保留每个从库中各自需要的数据库表。

- 方法二：使用 mysqldump 在主库中为每个从库单独导出一份只包含各从库所需的数据库表，并在每个从库中导入相应的备份文件。

- 方法三：在主库中复制原始的数据文件（只复制每个从库所需的数据库表相关的文件），将其恢复到各个从库中。注意，该方法不适用于 InnoDB 存储引擎的表，除非系统变量 innodb_file_per_table = 1，这样就可以用表空间传输的方式将主库的数据文件复制到从库中进行恢复。

MySQL 8.0 的 PERFORMANCE_SCHEMA 库提供了两张表来查看复制过滤规则：replication_applier_filters（查看按指定复制通道配置的复制过滤规则）、replication_applier_global_filters（如果配置复制过滤规则时未指定复制通道，则默认对全局生效）。同时，CHANGE REPLICATION FILTER 语句也支持多通道语法，如下：

```
# MySQL 8.0中的语法
Syntax:
CHANGE REPLICATION FILTER filter[, filter]
    [, ...] [FOR CHANNEL channel]

# performance_schema.replication_applier_filters表中记录的复制过滤规则示例如下
mysql> select * from replication_applier_filters\G
*************************** 1. row ***************************
  CHANNEL_NAME:    # 使用 CHANGE REPLICATION FILTER 配置复制过滤规则时，如未指定复制通道，默认情况下也会对默认的复制通道生效，默认的通道名称为空字符串
   FILTER_NAME: REPLICATE_WILD_DO_TABLE
   FILTER_RULE: databaseA.%,databaseD_change_replication_filter.%
 CONFIGURED_BY: CHANGE_REPLICATION_FILTER
```

```
   ACTIVE_SINCE: 2019-08-09 19:12:19.972050
         COUNTER: 0
1 row in set (0.01 sec)

# performance_schema.replication_applier_global_filters 表中记录的复制过滤规则示例如下
mysql> select * from replication_applier_global_filters\G
*************************** 1. row ***************************
   FILTER_NAME: REPLICATE_WILD_DO_TABLE
   FILTER_RULE: databaseA.%,databaseD_change_replication_filter.%
 CONFIGURED_BY: CHANGE_REPLICATION_FILTER
  ACTIVE_SINCE: 2019-08-09 19:12:19.972050
1 row in set (0.00 sec)
```

注意：

- 配置了复制过滤规则的从库，由于其中的数据与主库的并不完全相同（此时，从库中缺少被过滤掉的数据），所以不能承载主库的高可用切换。如果有高可用需求，请配置一个完全复制主库（不配置复制过滤）的从库来实现。

- MySQL 数据库的 4 个系统库中，MYSQL 系统库包含了用户账号和权限信息，也许你希望对其所做的修改在整个复制拓扑中保持一致（其他 3 个系统库 INFORMATION_SCHEMA、SYS、PERFORMANCE_SCHEMA 中主要包含每个实例自身的一些状态与统计信息。每个实例各不相同，除非有特殊需求，通常不需要同步），要实现此目的，就必须在 my.cnf 中配置类似 replicate-wild-do-table=mysql.% 的复制过滤规则，或者使用 CHANGE REPLICATION FILTER 语句动态修改复制过滤规则。

- 不要在 MGR（组复制）拓扑中设置复制过滤，这会造成各个数据节点之间的数据不一致（因为 MGR 拓扑支持多节点写，不一致的数据会导致多节点更新时出现问题）。

- CHANGE REPLICATION FILTER 语句在应用复制过滤规则时与对应的复制过滤选项的效果不同（复制过滤选项指的是类似"--replicate-wild-do-table"的配置选项）。

  CHANGE REPLICATION FILTER 语句指定的相同规则只有最后一个规则生效，之前的规则会被忽略。例如：CHANGE REPLICATION FILTER REPLICATE_WILD_DO_TABLE = ('db1.%', 'db2.%'), REPLICATE_WILD_DO_TABLE = ('db3.%','db4.%')，这一个语句指定了两条相同规则，db1.% 和 db2.% 会被忽略，db3.% 和 db4.% 生效。而 CHANGE REPLICATION FILTER REPLICATE_WILD_DO_TABLE = ('db1.%','db2.%') CHANGE REPLICATION FILTER REPLICATE_WILD_DO_TABLE = ('db3.%','db4.%')，使用多个语句指定两条相同规则，最后 db1.% 和 db2.% 被忽略，db3.% 和 db4.% 生效。

  指定多个不同的复制过滤规则时，复制过滤规则会被叠加。

- 在配置了复制过滤规则的复制拓扑中，对主从库数据一致性的校验尤其麻烦，我们常用的 pt-table-checksum 工具在这种情况下已经无法正常工作。所以，除非有特殊需求，否则不建议在复制拓扑中使用复制过滤。

第 24 章　发生数据误操作之后的处理方案

在当今信息大爆炸的时代，数据与信息是一家技术型公司赖以生存的基石。然而，在日常维护存放数据与信息的服务器过程中，对于技术人员来说，难免会有误操作（例如，误修改、误删除、误写入等）。不过，我们可以尽量减少误操作带来的损失。发生误操作之后，可以通过一些方法尽量恢复被误操作的数据。本章将详解介绍 MySQL 中几种常见的简单有效的恢复误删除数据的方法。

在 MySQL 的主从复制拓扑中，根据角色的不同，对数据的误操作可以分为两大类。

第一类是主库上的误操作。通常情况下，主库是承载写业务的（否则也不能称为主库），对数据的误修改会导致被操作的数据同步到复制拓扑的其他所有数据库实例中，致使整个复制拓扑中的数据都不正确，甚至造成数据丢失。针对主库上的误操作，常见的处理方案有如下两种：

- 利用延迟复制的从库找回主库中误删除的数据。
- 利用闪回工具，将误操作的数据反向重新写入主库。

第二类是从库上的误操作。通常情况下，从库是承载读业务的，在从库上发生误修改，只可能导致发生误操作的单个实例无法继续同步数据（当然，也存在这样的情况：从库发生误操作之后，主库的数据变更并未涉及从库中被误操作的数据，在一段时间内，从库端的复制不会出现故障，但实际上此时主从库数据可能不一致。对于这种情况，本章不予讨论），从而导致在该从库上查询不到最新的数据。针对从库上的误操作，常见的处理方式有如下两种：

- 如果是主键冲突或者没有找到记录的错误，可以设置系统变量 slave_exec_mode 为 IDEMPOTENT，跳过错误，让从库中的复制线程正常运行，然后使用 pt-table-checksum 与 pt-table-sync 工具重新同步主从库数据。
- 利用 GTID 的特性，注入空事务，跳过发生报错的事务，让从库中的复制线程正常运行，然后使用 pt-table-checksum 和 pt-table-sync 工具重新同步主从库数据。

提示：
如果误操作的数据量过大，建议采用如下方式处理：

- 如果是主库发生大批量误操作，则可以使用定时备份数据，重新搭建整个主从复制拓扑（该操作影响重大，操作前请先与相关人员沟通具体操作的流程、时间窗口等事宜）。

- 如果是从库发生大批量误操作，则可以重新从主库或其他从库备份全量数据进行恢复，也可以使用定时备份数据进行恢复。

如果采用简单的方法已无法修复数据，则可能需要使用定时备份中的备份数据进行恢复（参考上面的"误操作数据量过大"的处理方法）。更多关于排查复制故障的思路，可参考第 25 章"常用复制故障排除方案"。

24.1 操作环境信息

1. 服务器信息

   - 主库（MySQL Master）：10.10.30.162
   - 从库 1（MySQL Slave 1）：10.10.30.163
   - 从库 2（MySQL Slave 2）：10.10.30.164

2. MySQL 数据库版本

   MySQL 5.7.27

3. percona-toolkit 工具包版本

   percona-toolkit-3.0.12-1.el7.x86_64

启用 GTID 复制模式，然后开始制造测试数据：

```
# 创建名为 misoperation 的数据库
mysql> create database misoperation;
Query OK, 1 row affected (0.00 sec)

# 切换到 misoperation 数据库
mysql> use misoperation
Database changed
# 创建名为 misoperation_table 的表
mysql> create table misoperation_table(id int unsigned not null primary key,xid int,yid int);
Query OK, 0 rows affected (0.03 sec)

# 插入 5 行测试数据
mysql> insert into misoperation_table values(1,1,1),(2,2,2),(3,3,3),(4,4,4),(5,5,5);
Query OK, 5 rows affected (0.01 sec)
Records: 5  Duplicates: 0  Warnings: 0

# 查看插入的 5 行数据
mysql> select * from misoperation_table;
```

```
+----+------+------+
| id | xid  | yid  |
+----+------+------+
|  1 |   1  |   1  |
|  2 |   2  |   2  |
|  3 |   3  |   3  |
|  4 |   4  |   4  |
|  5 |   5  |   5  |
+----+------+------+
5 rows in set (0.01 sec)
```

24.2 主库发生误操作后的数据恢复

关于主库发生误操作的数据恢复，有如下两种方案。

- 通过延迟复制恢复数据，其关键步骤如图 24-1 所示。

图 24-1

- 通过闪回工具恢复数据，其关键步骤如图 24-2 所示。

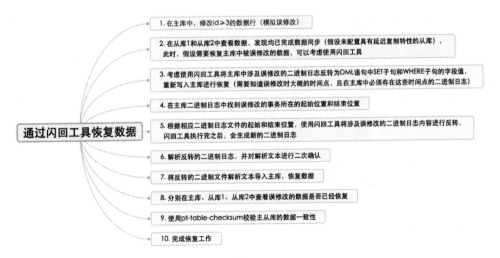

图 24-2

24.2.1 通过延迟复制恢复数据

MySQL 从 5.6 版本开始支持延迟复制，使从库的数据人为刻意滞后于主库一段时间。可以使用 CHANGE MASTER TO 语句的 MASTER_DELAY 选项设置延迟的时间值，默认值为 0。语法如下：

```
mysql> CHANGE MASTER TO ... MASTER_DELAY = N;
```

配置延迟复制之后，从主库接收的事件至少晚 N 秒才开始执行。这里的延迟时间不影响 I/O 线程读取主库二进制日志并写入自己的中继日志，只会影响 SQL 线程重放二进制日志的时间。

延迟复制可用于如下几种目的：

- 防止用户在主库上误操作。DBA 可以利用延迟的从库中的数据来恢复被误删除的数据。

- 测试当从库发生数据滞后时系统的行为方式（表现）。例如，从库负载高时可能导致从库延迟增加。在对应用的测试过程中，模拟从库的负载可能比较困难，使用延迟复制可以轻松模拟出从库复制延迟的情形。

- 要检查数据库中时间比较久远之前的数据，可以利用延迟复制来实现，而不必重新加载备份数据。例如，要查看一周前的数据，就可以配置从库延迟一周的时间进行复制（将 MASTER_DELAY 选项的值配置为一周时间的秒数）。

现在，我们在从库 2 中配置延迟复制：

```
# 配置前需要停止 SQL 线程
mysql> stop slave sql_thread;
Query OK, 0 rows affected (0.01 sec)

# 使用 CHANGE MASTER TO 语句配置延迟复制，时间值设定为 300 s
mysql> change master to MASTER_DELAY=300;
Query OK, 0 rows affected (0.01 sec)
```

在从库 2 中启动 SQL 线程，并查看复制的状态：

```
# 启动 SQL 线程
mysql> start slave sql_thread;
Query OK, 0 rows affected (0.05 sec)

# 查看复制的状态
mysql> show slave status\G
*************************** 1. row ***************************
             Slave_IO_State: Waiting for master to send event
                Master_Host: 10.10.30.162
                Master_User: repl
                Master_Port: 3306
              Connect_Retry: 10
```

```
              Master_Log_File: mysql-bin.000001
          Read_Master_Log_Pos: 961
               Relay_Log_File: mysql-relay-bin.000002
                Relay_Log_Pos: 1174
        Relay_Master_Log_File: mysql-bin.000001
             Slave_IO_Running: Yes
            Slave_SQL_Running: Yes
......
          Exec_Master_Log_Pos: 961
              Relay_Log_Space: 1381
......
        Seconds_Behind_Master: 0
......
                    SQL_Delay: 300    # 这里可以看到配置的复制延迟为 300 秒
          SQL_Remaining_Delay: NULL
      Slave_SQL_Running_State: Slave has read all relay log; waiting for more updates
......
           Retrieved_Gtid_Set: bf320d72-bb2c-11e9-91ab-5254002a54f2:1-3
            Executed_Gtid_Set: bf320d72-bb2c-11e9-91ab-5254002a54f2:1-3
                Auto_Position: 1
......

# 还可以通过 mysql 系统库中的 slave_relay_log_info 表查看延迟复制的配置信息
mysql> select * from mysql.slave_relay_log_info\G
*************************** 1. row ***************************
     Number_of_lines: 7
      Relay_log_name: /home/mysql/data/mysqldata1/relaylog/mysql-relay-bin.000002
       Relay_log_pos: 1174
     Master_log_name: mysql-bin.000001
      Master_log_pos: 961
           Sql_delay: 300    # 这里可以看到配置的复制延迟为 300 秒
   Number_of_workers: 16
                  Id: 1
        Channel_name:

# 还可以通过 PERFORMANCE_SCHEMA 系统库中 replication_applier_configuration 表查看延迟复制的配置信息
mysql> select * from performance_schema.replication_applier_configuration\G
*************************** 1. row ***************************
  CHANNEL_NAME:
 DESIRED_DELAY: 300    # 这里可以看到配置的复制延迟为 300 秒
1 row in set (0.00 sec)
```

在主库中删除一行数据，然后分别到从库 1 和从库 2 中查看数据：

```
# 在主库中删除 id = 5 的数据（模拟误删除操作）
mysql> delete from misoperation_table where id=5;
Query OK, 1 row affected (0.01 sec)
```

```
mysql> select * from misoperation_table;
+----+------+------+
| id | xid | yid |
+----+------+------+
| 1 | 1 | 1 |
| 2 | 2 | 2 |
| 3 | 3 | 3 |
| 4 | 4 | 4 |
+----+------+------+
4 rows in set (0.00 sec)

# 在从库 1 中查看 id = 5 的数据
mysql> select * from misoperation_table where id=5;
Empty set (0.00 sec)   # 已经被删除了

# 在从库 2 中查看 id = 5 的数据，发现数据还在
mysql> select * from misoperation_table where id=5;
+----+------+------+
| id | xid | yid |
+----+------+------+
| 5 | 5 | 5 |
+----+------+------+
1 row in set (0.00 sec)

# 查看一下复制的状态
mysql> show slave status\G
*************************** 1. row ***************************
            Slave_IO_State: Waiting for master to send event
               Master_Host: 10.10.30.162
               Master_User: repl
               Master_Port: 3306
             Connect_Retry: 10
           Master_Log_File: mysql-bin.000001
       Read_Master_Log_Pos: 3121
            Relay_Log_File: mysql-relay-bin.000002
             Relay_Log_Pos: 2969
     Relay_Master_Log_File: mysql-bin.000001
          Slave_IO_Running: Yes
         Slave_SQL_Running: Yes
......
       Exec_Master_Log_Pos: 2756
           Relay_Log_Space: 3541
......
     Seconds_Behind_Master: 76
......
                 SQL_Delay: 300
       SQL_Remaining_Delay: 224    # 从这里可以看到，在主库中被误删除的数据对应的二进制日志（删
除操作产生的数据变更操作在二进制日志中使用 Delete_rows_log_event 记录)，将在 224 秒后在从库 2 中被应用(也
```

就是说，到那时在主库中误删除的数据在从库 2 中也会被删除）
        Slave_SQL_Running_State: Waiting until MASTER_DELAY seconds after master executed event
......
        Retrieved_Gtid_Set: bf320d72-bb2c-11e9-91ab-5254002a54f2:1-10
        Executed_Gtid_Set: bf320d72-bb2c-11e9-91ab-5254002a54f2:1-9
            Auto_Position: 1
......

立即停止 SQL 线程，防止在主库中误删除数据产生的二进制日志被应用到从库 2 中
mysql> **stop slave sql_thread;**
Query OK, 0 rows affected (0.01 sec)

mysql> **show slave status\G**
*************************** 1. row ***************************
               Slave_IO_State: Waiting for master to send event
                  Master_Host: 10.10.30.162
                  Master_User: repl
                  Master_Port: 3306
                Connect_Retry: 10
              Master_Log_File: mysql-bin.000001
          Read_Master_Log_Pos: 3121
               Relay_Log_File: mysql-relay-bin.000002
                Relay_Log_Pos: 2969
        Relay_Master_Log_File: mysql-bin.000001
             Slave_IO_Running: Yes
            Slave_SQL_Running: No
......
          Exec_Master_Log_Pos: 2756
              Relay_Log_Space: 3541
......
        Seconds_Behind_Master: NULL
......
                    SQL_Delay: 300
          SQL_Remaining_Delay: NULL
      Slave_SQL_Running_State:
......
        Retrieved_Gtid_Set: bf320d72-bb2c-11e9-91ab-5254002a54f2:1-10
        Executed_Gtid_Set: bf320d72-bb2c-11e9-91ab-5254002a54f2:1-9
            Auto_Position: 1
......

在从库 2 中用 mysqldump 导出被误删除的数据（我们模拟的误删除数据只有 1 行，但这里的方法适合多行数据被误删除的情况）

用 mysqldump 导出被误删除的数据，这里的--where 选项表示只匹配 id = 5 的数据行，--set-gtid-purged=OFF 表示不生成 SET GLOBAL GTID_PURGED 语句，--no-create-db 表示不生成 CREATE DATABASE 语句，--no-create-info 表示不生成 CREATE TABLE 语句，--replace 表示将 INSERT INTO 语句替换为 REPLACE INTO 语句（这是为了防止在主库中导入恢复的数据之后，由于从库 1 中存在 id = 5 的数据，发生主键冲突而报错），

--complete-insert 表示在 REPLACE INTO 语句中生成字段名

```
[root@localhost ~]# mysqldump -uroot -ppassword misoperation misoperation_table --where 'id=5' --set-gtid-purged=OFF --no-create-db --no-create-info --replace --complete-insert > a.sql

# 查看 a.sql 文件中的内容
[root@localhost ~]# cat a.sql
……
--
-- Dumping data for table `misoperation_table`
--
-- WHERE: id=5

LOCK TABLES `misoperation_table` WRITE;
/*!40000 ALTER TABLE `misoperation_table` DISABLE KEYS */;
REPLACE INTO `misoperation_table` (`id`, `xid`, `yid`) VALUES (5,5,5);   # 我们需要恢复的数据在这里
/*!40000 ALTER TABLE `misoperation_table` ENABLE KEYS */;
UNLOCK TABLES;
……
```

将从库 2 中导出的 a.sql 文件传输到主库，并导入主库数据库：

```
# 在从库 2 中将 a.sql 文件传输到主库
[root@localhost ~]# scp a.sql 10.10.30.162:/tmp/

#将 a.sql 文件导入主库数据库
mysql> source /tmp/a.sql;
……
```

分别在主库、从库 1 和从库 2 中查看 misoperation_table 表中的数据：

```
# 查询主库
mysql> select * from misoperation_table id=5;
+----+------+------+
| id | xid  | yid  |
+----+------+------+
|  5 |   5  |   5  |    # 发现主库中 id = 5 的数据已恢复
+----+------+------+
1 row in set (0.00 sec)

# 查询从库 1
mysql> select * from misoperation_table where id=5;
+----+------+------+
| id | xid  | yid  |
+----+------+------+
|  5 |   5  |   5  |    # 发现从库 1 中 id = 5 的数据已恢复
+----+------+------+
1 row in set (0.00 sec)
```

```
# 查询从库 2
mysql> select * from misoperation_table where id=5;
+----+------+------+
| id | xid  | yid  |
+----+------+------+
| 5  | 5    | 5    |   # 从库 2 中 id = 5 的数据也还在。这是因为从库 2 中 id = 5 的数据本来就没有被删除过，
因为其在更早之前已经停止了 SQL 线程
+----+------+------+
1 row in set (0.00 sec)

# 在从库 2 中启动 SQL 线程，使从库 2 追赶上主库的数据
mysql> start slave sql_thread;
Query OK, 0 rows affected (0.07 sec)

# 在从库 2 中，等待数据同步完成（不断查看复制状态以确认）
mysql> show slave status\G
*************************** 1. row ***************************
               Slave_IO_State: Waiting for master to send event
                  Master_Host: 10.10.30.162
                  Master_User: repl
                  Master_Port: 3306
                Connect_Retry: 10
              Master_Log_File: mysql-bin.000001
          Read_Master_Log_Pos: 3947
               Relay_Log_File: mysql-relay-bin.000002
                Relay_Log_Pos: 2969
        Relay_Master_Log_File: mysql-bin.000001
             Slave_IO_Running: Yes
            Slave_SQL_Running: Yes
......
          Exec_Master_Log_Pos: 2756
              Relay_Log_Space: 4367
......
        Seconds_Behind_Master: 1986
......
                    SQL_Delay: 300
          SQL_Remaining_Delay: 27
      Slave_SQL_Running_State: Waiting until MASTER_DELAY seconds after master executed
event
......
           Retrieved_Gtid_Set: bf320d72-bb2c-11e9-91ab-5254002a54f2:1-13
            Executed_Gtid_Set: bf320d72-bb2c-11e9-91ab-5254002a54f2:1-10    # 从这里可以看到，从库
2 中的 SQL 线程与 I/O 线程相差了 2 个 GTID 事务号，说明此时 SQL 线程并未追赶上 I/O 线程，需要继续等待
                Auto_Position: 1
......

# 在从库 2 中等待 SQL 线程追赶上 I/O 线程之后，再次确认 id = 5 的数据是否还在
mysql> select * from misoperation_table where id=5;
```

```
+----+------+------+
| id | xid  | yid  |
+----+------+------+
| 5  | 5    | 5    |
+----+------+------+
1 row in set (0.00 sec)
```

如有需要,这里可以使用 pt-table-checksum 校验主从库数据的一致性(注意,延迟复制的配置会使 pt-table-checksum 运行时发生等待,如有必要,请先清理延迟复制的配置后再运行该工具):

```
[root@localhost ~]# pt-table-checksum --nocheck-replication-filters --no-check-binlog-
format --replicate=misoperation.checksums h=localhost, u=admin, p=password,P=3306 --databases=
misoperation
Waiting for the --replicate table to replicate to localhost...
......
Replica lag is 300 seconds on localhost. Waiting.
    TS        ERRORS  DIFFS  ROWS  CHUNKS  SKIPPED  TIME     TABLE
0810T14:48:31  0      0      5     1       0        301.035  misoperation.misoperation_table
```

至此,通过延迟复制恢复数据的步骤已完成。

提示:RESET SLAVE 语句将延迟设置的值重置为 0。

SHOW SLAVE STATUS 有三个字段提供与延迟有关的信息:

- SQL_Delay:非负整数,表示从库必须滞后于主库的秒数。

- SQL_Remaining_Delay:当 Slave_SQL_Running_State 字段显示"Waiting until MASTER_DELAY seconds after master executed event"时,SQL_Remaining_Delay 字段的值表示剩余的延迟秒数(即表示在多少秒之后会应用当前被延迟复制时间值阻塞的数据变更记录);否则,在其他情况下,此字段值为 NULL。

- Slave_SQL_Running_State:一个字符串值,显示 SQL 线程的状态(类似于 Slave_IO_State 值)。该值与 SHOW PROCESSLIST 显示的 SQL 线程的 State 值相同。

当从库 SQL 线程正处于延迟复制指定的等待时间内时,SHOW PROCESSLIST 将其状态值显示为"Waiting until MASTER_DELAY seconds after master executed event"。

有关 mysqldump 工具的更多介绍,可参考《千金良方:MySQL 性能优化金字塔法则》一书中的第 48 章"MySQL 主流备份工具之 mysqldump 详解"。

24.2.2 通过闪回工具恢复数据

这里我们基于 24.2.1 节的操作环境继续演示。先在从库 2 中清理延迟复制的配置:

```
# 停止 SQL 线程
mysql> stop slave sql_thread;
Query OK, 0 rows affected (0.00 sec)
```

```
# 关闭延迟复制
mysql> change master to MASTER_DELAY=0;
Query OK, 0 rows affected (0.02 sec)

# 重新启动 SQL 线程
mysql> start slave sql_thread;
Query OK, 0 rows affected (0.04 sec)

# 查看复制的配置信息
mysql> show slave status\G
*************************** 1. row ***************************
               Slave_IO_State: Waiting for master to send event
                  Master_Host: 10.10.30.162
                  Master_User: repl
                  Master_Port: 3306
                Connect_Retry: 10
              Master_Log_File: mysql-bin.000001
          Read_Master_Log_Pos: 13718
               Relay_Log_File: mysql-relay-bin.000002
                Relay_Log_Pos: 13931
        Relay_Master_Log_File: mysql-bin.000001
             Slave_IO_Running: Yes
            Slave_SQL_Running: Yes
......
          Exec_Master_Log_Pos: 13718
              Relay_Log_Space: 14138
......
        Seconds_Behind_Master: 0
......
                    SQL_Delay: 0  # 从这里可以看到，复制延迟的值为 0，表示关闭延迟复制
          SQL_Remaining_Delay: NULL
      Slave_SQL_Running_State: Slave has read all relay log; waiting for more updates
......
           Retrieved_Gtid_Set: bf320d72-bb2c-11e9-91ab-5254002a54f2:1-29
            Executed_Gtid_Set: bf320d72-bb2c-11e9-91ab-5254002a54f2:1-29
                Auto_Position: 1
......
```

登录主库数据库，使用 UPDATE 语句修改 3 行数据（模拟误修改操作）：

```
# 将 id >= 3 的数据行的 yid 列的值增加 10
mysql> update misoperation_table set yid=yid+10 where id >=3;
Query OK, 3 rows affected (0.01 sec)
Rows matched: 3  Changed: 3  Warnings: 0

# 查询修改结果
mysql> select * from misoperation_table;
+----+------+------+
```

```
| id | xid | yid |
+----+-----+-----+
|  1 |  1  |  1  |
|  2 |  2  |  2  |
|  3 |  3  | 13  |   # 修改之后为 13，之前为 3
|  4 |  4  | 14  |   # 修改之后为 14，之前为 4
|  5 |  5  | 15  |   # 修改之后为 15，之前为 5
+----+-----+-----+
5 rows in set (0.00 sec)
```

登录从库 1 和从库 2，查询 misoperation_table 表中的数据：

```
# 登录从库 1，发现修改已同步完成
mysql> select * from misoperation_table where id>=3;
+----+-----+-----+
| id | xid | yid |
+----+-----+-----+
|  3 |  3  | 13  |
|  4 |  4  | 14  |
|  5 |  5  | 15  |
+----+-----+-----+
3 rows in set (0.00 sec)

# 登录从库 2，发现修改已同步完成
mysql> select * from misoperation_table where id>=3;
+----+-----+-----+
| id | xid | yid |
+----+-----+-----+
|  3 |  3  | 13  |
|  4 |  4  | 14  |
|  5 |  5  | 15  |
+----+-----+-----+
3 rows in set (0.00 sec)
```

假设上述修改为误操作，需要将误修改的数据还原为修改之前的状态，此时怎么办呢？没有了延迟复制，从库 1 和从库 2 的数据都已经同步了误修改的数据，这个时候，我们可以解析主库的二进制日志，找到误修改的数据，将 UPDATE 语句 SET 子句中的字段值与 WHERE 子句中的条件字段值互换，然后再写回主库即可完成数据恢复。但这里有两个前提条件：第一，我们需要大概知道误操作的时间点，这样后续可以利用 mysqlbinlog 的 --start-datetime 和 --stop-datetime 选项指定一个时间范围进行解析（如果不知道时间点，就只能从最新的二进制日志文件开始，倒序往前逐个解析进行查找）；第二，主库的二进制日志需要尽可能保留（如果主库有定期清理二进制日志的策略，那么主库必须有定期备份二进制日志的策略），否则被清理之后，就无法利用解析并反转二进制日志的方式恢复数据了。

假设我们知道误操作的时间大约是 2019-08-10 16:17:28~2019-08-10 16:32:25，我们按照这个时间，筛选出符合时间范围的二进制日志文件列表：

```
# 查看主库的二进制日志目录下的磁盘文件列表，留意文件的修改时间
```

```
[root@localhost ~]# cd /data//mysqldata1/binlog/
[root@localhost binlog]# ll
total 20
-rw-r----- 1 mysql mysql 14161 Aug 10 16:30 mysql-bin.000001     # 这里由于是测试环境,只有
一个二进制日志文件 mysql-bin.000001
-rw-r----- 1 mysql mysql    52 Aug 10 13:07 mysql-bin.index
```

在主库中解析 mysql-bin.000001 文件,并查找误操作的 SQL 语句:

```
# 使用 mysqlbinlog 解析 mysql-bin.000001 文件
[root@localhost binlog]# mysqlbinlog -vv mysql-bin.000001 > b.sql

# 使用 vim 命令打开 b.sql 文件,搜索误操作的 SQL 语句文本(注意,如果实际环境中 b.sql 文件的大小超过
10 GB,请先用 grep 等命令找出操作 SQL 语句文本所在的行数,再用 vim 指定行数打开 b.sql 文件)
# at 13687
#190810 16:09:27 server id 33061 end_log_pos 13718 CRC32 0x03ddc158 Xid = 838
COMMIT/*!*/;
# at 13718   # 这里就是我们误操作数据的事务在二进制日志中的第一个 event 的起始位置(13718),记录这
个位置(时间为 190810 16:30:25,这里只需要关注时间点即可)
#190810 16:30:25 server id 33061 end_log_pos 13783 CRC32 0xac005056 GTID last_committed=29
sequence_number=30 rbr_only=yes
/*!50718 SET TRANSACTION ISOLATION LEVEL READ COMMITTED*//*!*/;
SET @@SESSION.GTID_NEXT= 'bf320d72-bb2c-11e9-91ab-5254002a54f2:30'/*!*/;
# at 13783
#190810 16:30:25 server id 33061 end_log_pos 13868 CRC32 0x37490f60 Query thread_id=6
exec_time=0 error_code=0
SET TIMESTAMP=1565425825/*!*/;
SET @@session.sql_mode=1436549152/*!*/;
/*!\C utf8 *//*!*/;
SET
@@session.character_set_client=33,@@session.collation_connection=33,@@session.collation
_server=83/*!*/;
BEGIN
/*!*/;
# at 13868
#190810 16:30:25 server id 33061 end_log_pos 13945 CRC32 0x2525efc8 Rows_query
# 这里就是我们误操作的 SQL 语句。当然,要让二进制日志记录用户的原始 SQL 语句文本,必须事先将系统变量
binlog_rows_query_log_events 设置为 ON,否则不会记录
# update misoperation_table set yid=yid+10 where id >=3
# at 13945
#190810 16:30:25 server id 33061 end_log_pos 14016 CRC32 0x7312826b Table_map:
`misoperation`.`misoperation_table` mapped to number 111
# at 14016
#190810 16:30:25 server id 33061 end_log_pos 14130 CRC32 0xbceee4b3 Update_rows: table
id 111 flags: STMT_END_F

BINLOG '
oYBOXR0lgQAATQAAAHk2AACAADV1cGRhdGUgbWlzb3BlcmF0aW9uX3RhYmxlIHNldCB5aWQ9eWlk
KzEwIHdoZXJlIGlkID49M8jvJSU=
```

```
oYBOXRMlgQAARwAAAMA2AAAAAG8AAAAAAAEADGlpc29wZXJhdGlvbgASbWlzb3BlcmF0aW9uX3Rh
YmxlAAMDAwMABmuCEnM=
oYBOXR8lgQAAcgAAADI3AAAAAG8AAAAAAAEAAgAD///4AwAAAAMAAADAAAA+AMAAAADAAAADQAA
APgEAAAABAAAAAQAAAD4BAAAAAQAAAAOAAAA+AUAAAAFAAAABQAAAPgFAAAABQAAAA8AAACz5O68
'/*!*/;
### UPDATE `misoperation`.`misoperation_table`    # 如果不幸未将系统变量 binlog_rows_
query_log_events 设置为 ON，则只能在这个地方查找了，但它是逐行记录的，查找起来比较费劲
### WHERE
### @1=3 /* INT meta=0 nullable=0 is_null=0 */
### @2=3 /* INT meta=0 nullable=1 is_null=0 */
### @3=3 /* INT meta=0 nullable=1 is_null=0 */
### SET
……
# at 14130
# 这里是我们误操作事务在二进制日志中的最后一个 event 的结束位置(14161)，记录下这个位置(时间为 190810
16:30:25，这里只需要关注一下时间点即可)
#190810 16:30:25 server id 33061 end_log_pos 14161 CRC32 0xa948a12a Xid = 843
COMMIT/*!*/;
SET @@SESSION.GTID_NEXT= 'AUTOMATIC' /* added by mysqlbinlog */ /*!*/;
DELIMITER ;
……
```

找出了误操作数据在主库的哪个二进制日志文件中，也在此文件中找到了相应的位置，那么就可以使用闪回工具反转二进制日志了（这里我们采用美团的闪回工具 flashback）：

```
# 使用 flashback 指定上文中找到的误操作 SQL 语句在二进制日志文件中的起始位置和结束位置
[root@localhost binlog]# /root/flashback --binlogFileNames='mysql-bin.000001' --start-
position=13718 --stop-position=14161 --outBinlogFileNameBase='binlog_flashback'
[root@localhost binlog]# ll binlog_flashback.flashback*
-rw-r--r-- 1 root root 308 Aug 10 17:08 binlog_flashback.flashback

# 使用 mysqlbinlog 工具解析 binlog_flashback.flashback 文件中的内容（在 GTID 复制模式下，这里需
要使用--skip-gtids 选项，否则后续 c.sql 文件导入数据库时，文件中的内容无法被执行）
[root@localhost binlog]# mysqlbinlog -vv binlog_flashback.flashback --skip-gtids > c.sql

# 查看 c.sql 文件中的内容
[root@localhost binlog]# cat c.sql
/*!50530 SET @@SESSION.PSEUDO_SLAVE_MODE=1*/;
/*!50003 SET @OLD_COMPLETION_TYPE=@@COMPLETION_TYPE,COMPLETION_TYPE=0*/;
DELIMITER /*!*/;
# at 4
#190810 13:07:27 server id 33061 end_log_pos 123 CRC32 0x17f26c8e Start: binlog v 4, server
v 5.7.27-log created 190810 13:07:27 at startup
# Warning: this binlog is either in use or was not closed properly.
ROLLBACK/*!*/;
BINLOG '
D1FOXQ8lgQAAdwAAAHsAAAABAAQANS43LjI3LWxvZwAAAAAAAAAAAAAAAAAAAAAAAAAAAAAAAAAAAA
AAAAAAAAAAAAAAAAAAPUU5dEzgNAAgAEgAEBAQEEgAAXwAEGggAAAAICAgCAAAACgoKKKioAEjQA
AY5s8hc=
```

```
'/*!*/;
/*!50616 SET @@SESSION.GTID_NEXT='AUTOMATIC'*//*!*/;
# at 123
#190810 16:30:25 server id 33061  end_log_pos 194  CRC32 0x7312826b  Table_map: `misoperation`.`misoperation_table` mapped to number 111
# at 194
#190810 16:30:25 server id 33061  end_log_pos 308 CRC32 0xbceee4b3 Update_rows: table id 111 flags: STMT_END_F

BINLOG '
oYBOXRMlgQAARwAAAMIAAAAAG8AAAAAAEADG1pc29wZXJhdGlvbgASbWlzb3BlcmF0aW9uX3Rh
YmxlAAMDAwMABmuCEnM=
oYBOXR8lgQAAcgAAADQBAAAAG8AAAAAAEAAgAD///4AwAAAAMAAAANAAAA+AMAAAADAAAAwAA
APgEAAAABAAAAA4AAAD4BAAAAQAAAEAAAA+AUAAAAFAAAADwAAAPgFAAAABQAAAUAAACz5O68
'/*!*/;
### UPDATE `misoperation`.`misoperation_table`
### WHERE
###   @1=3 /* INT meta=0 nullable=0 is_null=0 */
###   @2=3 /* INT meta=0 nullable=1 is_null=0 */
###   @3=13 /* INT meta=0 nullable=1 is_null=0 */
### SET    # 从这里的注释文本可以看到，与原先误修改的 UPDATE 语句相比，SET 子句中的字段值和 WHERE 子句中的条件字段值已经被反转了
###   @1=3 /* INT meta=0 nullable=0 is_null=0 */
###   @2=3 /* INT meta=0 nullable=1 is_null=0 */
###   @3=3 /* INT meta=0 nullable=1 is_null=0 */
### UPDATE `misoperation`.`misoperation_table`
### WHERE
###   @1=4 /* INT meta=0 nullable=0 is_null=0 */
###   @2=4 /* INT meta=0 nullable=1 is_null=0 */
###   @3=14 /* INT meta=0 nullable=1 is_null=0 */
### SET
###   @1=4 /* INT meta=0 nullable=0 is_null=0 */
###   @2=4 /* INT meta=0 nullable=1 is_null=0 */
###   @3=4 /* INT meta=0 nullable=1 is_null=0 */
### UPDATE `misoperation`.`misoperation_table`
### WHERE
###   @1=5 /* INT meta=0 nullable=0 is_null=0 */
###   @2=5 /* INT meta=0 nullable=1 is_null=0 */
###   @3=15 /* INT meta=0 nullable=1 is_null=0 */
### SET
###   @1=5 /* INT meta=0 nullable=0 is_null=0 */
###   @2=5 /* INT meta=0 nullable=1 is_null=0 */
###   @3=5 /* INT meta=0 nullable=1 is_null=0 */
DELIMITER ;
# End of log file
/*!50003 SET COMPLETION_TYPE=@OLD_COMPLETION_TYPE*/;
/*!50530 SET @@SESSION.PSEUDO_SLAVE_MODE=0*/;
```

现在，我们将 c.sql 文件导入主库数据库（注意，导入之前，请先确认是否有多余的其

他数据，如有，则请重新矫正反转的数据，或者删除多余的数据，否则会造成数据被再次误操作）：

```
mysql> source /data/mysqldata1/binlog/c.sql;
......
```

分别在主库、从库 1、从库 2 中校验数据是否已经恢复：

```
# 查询主库数据
mysql> select * from misoperation_table where id>=3;
+----+------+------+
| id | xid  | yid  |
+----+------+------+
|  3 |   3  |   3  |   # 可以看到，数据已恢复为误删除之前的状态了
|  4 |   4  |   4  |
|  5 |   5  |   5  |
+----+------+------+
3 rows in set (0.00 sec)

# 查询从库 1 中的数据
mysql> select * from misoperation_table where id>=3;
+----+------+------+
| id | xid  | yid  |
+----+------+------+
|  3 |   3  |   3  |   # 可以看到，数据也已恢复为误删除之前的状态了
|  4 |   4  |   4  |
|  5 |   5  |   5  |
+----+------+------+
3 rows in set (0.01 sec)

# 查询从库 2 中的数据
mysql> select * from misoperation_table where id>=3;
+----+------+------+
| id | xid  | yid  |
+----+------+------+
|  3 |   3  |   3  |   # 可以看到，数据也已恢复为误删除之前的状态了
|  4 |   4  |   4  |
|  5 |   5  |   5  |
+----+------+------+
3 rows in set (0.00 sec)
```

如有需要，这里可以使用 pt-table-checksum 校验主从库数据的一致性：

```
[root@localhost ~]# pt-table-checksum --nocheck-replication-filters --no-check-binlog-format --replicate=misoperation.checksums h=localhost,u=admin,p=password,P=3306 --databases=misoperation
        TS      ERRORS  DIFFS  ROWS  CHUNKS  SKIPPED  TIME   TABLE
0810T17:24:54      0      0     5       1       0     0.272  misoperation.misoperation_table
```

至此，通过闪回工具恢复数据的步骤已完成。

提示：有关闪回工具的更多介绍，可参考《千金良方：MySQL 性能优化金字塔法则》一书中的第 51 章 "MySQL 主流闪回工具详解"。

24.3 从库发生误操作后的数据恢复

关于从库误操作后的数据恢复，有如下两种方案。

- 通过修改系统变量 slave_exec_mode 恢复从库数据，关键步骤如图 24-3 所示。

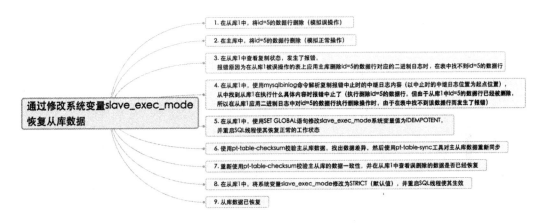

图 24-3

- 通过 GTID 特性注入空事务恢复从库数据，关键步骤如图 24-4 所示。

图 24-4

24.3.1 通过修改系统变量 slave_exec_mode 恢复数据

这里我们基于 24.2.2 节的操作环境继续演示。在从库 1 中，将 id = 5 的数据行删除（模拟误操作）：

```
# 删除前先查询 id = 5 的数据行
mysql> select * from misoperation_table where id=5;
+----+------+------+
| id | xid  | yid  |
+----+------+------+
|  5 |  5   |  5   |
+----+------+------+
1 row in set (0.00 sec)

# 用 DELETE 语句执行删除
mysql> delete from misoperation_table where id=5;
Query OK, 1 row affected (0.01 sec)
```

在主库中，使用 UPDATE 语句对 id = 5 的数据行执行修改：

```
# 修改前先查询 id = 5 的数据行
mysql> select * from misoperation_table where id=5;
+----+------+------+
| id | xid  | yid  |
+----+------+------+
|  5 |  5   |  5   |
+----+------+------+
1 row in set (0.00 sec)

# 用 UPDATE 语句执行修改
mysql> update misoperation_table set yid=yid+10 where id=5;
Query OK, 1 row affected (0.00 sec)
Rows matched: 1  Changed: 1  Warnings: 0

# 再次查询，发现 yid 列的值已被修改为 15
mysql> select * from misoperation_table where id=5;
+----+------+------+
| id | xid  | yid  |
+----+------+------+
|  5 |  5   |  15  |
+----+------+------+
1 row in set (0.00 sec)
```

在从库 1 中查看数据及其复制状态：

```
# 在从库 1 中查询数据，发现不存在 id = 5 的数据
mysql> select * from misoperation_table where id=5;
Empty set (0.00 sec)

# 查看从库 1 的复制状态，发现报错了
mysql> show slave status\G
*************************** 1. row ***************************
           Slave_IO_State: Waiting for master to send event
              Master_Host: 10.10.30.162
              Master_User: repl
```

```
              Master_Port: 3306
            Connect_Retry: 10
          Master_Log_File: mysql-bin.000001
      Read_Master_Log_Pos: 18249
           Relay_Log_File: mysql-relay-bin.000002   # 留意这里的中继日志文件名，后面会用到
            Relay_Log_Pos: 18073 # 留意这里的中继日志位置，后面会用到
    Relay_Master_Log_File: mysql-bin.000001
         Slave_IO_Running: Yes
        Slave_SQL_Running: No # 发现SQL线程已停止，从后面的Last_Error和Last_SQL_Error
字段值上看，发生了错误。这两个字段的值表示应用数据变更时发生了找不到记录的错误，但是这里无法看到具体的错误
发生在哪个表上
......
               Last_Errno: 1032
               Last_Error: Coordinator stopped because there were error(s) in the worker(s).
The most recent failure being: Worker 1 failed executing transaction 'bf320d72-bb2c-
11e9-91ab-5254002a54f2:37' at master log mysql-bin.000001, end_log_pos 18218. See error log
and/or performance_schema.replication_applier_status_by_worker table for more details about
this failure or others, if any.
             Skip_Counter: 0
      Exec_Master_Log_Pos: 17860
          Relay_Log_Space: 18669
......
    Seconds_Behind_Master: NULL
......
           Last_SQL_Errno: 1032
           Last_SQL_Error: Coordinator stopped because there were error(s) in the worker(s).
The most recent failure being: Worker 1 failed executing transaction 'bf320d72-bb2c-
11e9-91ab-5254002a54f2:37' at master log mysql-bin.000001, end_log_pos 18218. See error log
and/or performance_schema.replication_applier_status_by_worker table for more details about
this failure or others, if any.
......
   Last_SQL_Error_Timestamp: 190810 18:01:48
           Master_SSL_Crl:
       Master_SSL_Crlpath:
       Retrieved_Gtid_Set: bf320d72-bb2c-11e9-91ab-5254002a54f2:1-37
        Executed_Gtid_Set: bf320d72-bb2c-11e9-91ab-5254002a54f2:1-36,
c67c9093-bb2c-11e9-8055-525400c33752:1
            Auto_Position: 1
......

# 由于在上述错误代码和错误信息中看不到具体的错误，根据提示，我们在 performance_schema.
replication_applier_status_by_worker表中查看更详细的信息
mysql> select * from performance_schema.replication_applier_status_by_worker where
last_error_message !=''\G
*************************** 1. row ***************************
         CHANNEL_NAME:
            WORKER_ID: 1
            THREAD_ID: NULL
```

```
         SERVICE_STATE: OFF
LAST_SEEN_TRANSACTION: bf320d72-bb2c-11e9-91ab-5254002a54f2:37
    LAST_ERROR_NUMBER: 1032
# 从下面的 LAST_ERROR_MESSAGE 字段信息可以看到，错误发生在 misoperation_table 表，在应用主库的二
进制日志 mysql-bin.000001，end_log_pos 18218 位置的 Update_rows event 时发生了错误，原因为 Can't
find record，即执行 UPDATE 操作时找不到记录
   LAST_ERROR_MESSAGE: Worker 1 failed executing transaction 'bf320d72-bb2c-11e9-
91ab-5254002a54f2:37' at master log mysql-bin.000001, end_log_pos 18218; Could not execute
Update_rows event on table misoperation.misoperation_table; Can't find record in
'misoperation_table', Error_code: 1032; handler error HA_ERR_KEY_NOT_FOUND; the event's
master log mysql-bin.000001, end_log_pos 18218
  LAST_ERROR_TIMESTAMP: 2019-08-10 18:01:48
1 row in set (0.00 sec)
```

在从库 2 中查看数据及其复制状态：

```
# 在从库 2 中查询数据
mysql> select * from misoperation_table where id=5;
+----+------+------+
| id | xid  | yid  |
+----+------+------+
|  5 |   5  |  15  |   # id = 5 的数据行中的 yid 列正常从主库同步为 15
+----+------+------+
1 row in set (0.00 sec)

# 在从库 2 中查看复制状态，正常
mysql> show slave status\G
*************************** 1. row ***************************
               Slave_IO_State: Waiting for master to send event
                  Master_Host: 10.10.30.162
                  Master_User: repl
                  Master_Port: 3306
                Connect_Retry: 10
              Master_Log_File: mysql-bin.000001
          Read_Master_Log_Pos: 18249
               Relay_Log_File: mysql-relay-bin.000002
                Relay_Log_Pos: 18462
        Relay_Master_Log_File: mysql-bin.000001
             Slave_IO_Running: Yes
            Slave_SQL_Running: Yes
......
          Exec_Master_Log_Pos: 18249
              Relay_Log_Space: 18669
......
        Seconds_Behind_Master: 0
......
```

现在，我们需要想办法修复从库 1 中报告的问题。在报错信息中只能看到发生错误原因和位置，无法知道在报错的位置具体是什么数据记录。可以使用从库 1 中 SQL 线程报错

中止时查看到的中继日志位置（上文代码段中提示过留意的地方：Relay_Log_File: mysql-relay-bin.000002 和 Relay_Log_Pos: 18073），解析从库的中继日志，查看在报错位置上是什么数据（我们假设前面对从库 1 的操作为误操作，所以这里就假装不知道从库 1 发生了误操作，需要想办法找出误操作发生前的准确记录）：

```
# 使用mysqlbinlog命令解析从库的中继日志
[root@localhost relaylog]# mysqlbinlog -vv --start-position=18073 mysql-relay-bin.000002 > d.sql

# 使用vim命令打开d.sql文件查看内容
[root@localhost relaylog]# vim d.sql
……
# at 18073   # 这里就是报错的位置 18073
#190810 16:30:25 server id 33061 end_log_pos 17925 CRC32 0x567147f3 GTID last_committed=36 sequence_number=37 rbr_only=yes
/*!50718 SET TRANSACTION ISOLATION LEVEL READ COMMITTED*//*!*/;
SET @@SESSION.GTID_NEXT= 'bf320d72-bb2c-11e9-91ab-5254002a54f2:37'/*!*/;   # 从这里可以看到，报错的位置上是在执行 GTID 为 bf320d72-bb2c-11e9-91ab-5254002a54f2:37 的事务
# at 18138
#190810 16:30:25 server id 33061 end_log_pos 18010 CRC32 0xdb93271a Query thread_id=6 exec_time=5483 error_code=0
SET TIMESTAMP=1565425825/*!*/;
……
BEGIN
/*!*/;
# at 18223
#190810 16:30:25 server id 33061 end_log_pos 18085 CRC32 0x9a7bcd62 Rows_query
# 从这里的 SQL 注释可以看到，报错的位置在执行 UPDATE 操作的 id = 5 的数据行。在前面我们确认过，从库 1 中不存在 id = 5 的数据行，所以二进制日志被应用到这里时就会报错
# update misoperation_table set yid=yid+10 where id=5
# at 18298
#190810 16:30:25 server id 33061 end_log_pos 18156 CRC32 0x7b13002d Table_map: `misoperation`.`misoperation_table` mapped to number 111
# at 18369
#190810 16:30:25 server id 33061 end_log_pos 18218 CRC32 0x231ba9a9 Update_rows: table id 111 flags: STMT_END_F

BINLOG '
oYBOXR01gQAASwAAAKVGAACAADN1cGRhdGUgbWlzb3BlcmF0aW9uX3RhYmxlIHNldCB5aWQ9eWlk
KzEwIHdoZXJlIGlkPTViZXua
oYBOXRM1gQAARwAAAOxGAAAAAG8AAAAAAEADG1pc29wZXJhdGlvbgASbWlzb3BlcmF0aW9uX3Rh
YmxlAAMDAwMABi0AE3s=
oYBOXR81gQAAPgAAACpHAAAAAG8AAAAAAEAAgAD///4BQAAAAUAAAAFAAAA+AUAAAAFAAAADwAA
AKmpGyM=
'/*!*/;
### UPDATE `misoperation`.`misoperation_table`
### WHERE
###   @1=5 /* INT meta=0 nullable=0 is_null=0 */
```

```
### @2=5 /* INT meta=0 nullable=1 is_null=0 */
### @3=5 /* INT meta=0 nullable=1 is_null=0 */
### SET
### @1=5 /* INT meta=0 nullable=0 is_null=0 */
### @2=5 /* INT meta=0 nullable=1 is_null=0 */
### @3=15 /* INT meta=0 nullable=1 is_null=0 */
# at 18431
#190810 16:30:25 server id 33061 end_log_pos 18249 CRC32 0x5f0592c3 Xid = 914
COMMIT/*!*/;
......
```

现在，我们已经找出了报错的原因和位置：从库 1 中不存在 id = 5 的记录，而有一个事务对其执行 UPDATE 操作，所以只要想办法先跳过这个事务，先将复制恢复为正常状态，再使用 pt-table-checksum 和 pt-table-sync 工具重新同步主从库数据即可。

这里以将系统变量 slave_exec_mode 的值改为 IDEMPOTENT 为例进行演示：

```
# 在从库 1 中将 slave_exec_mode 的值改为 IDEMPOTENT
mysql> set global slave_exec_mode=IDEMPOTENT;
Query OK, 0 rows affected (0.00 sec)

# 尝试启动 SQL 线程
mysql> start slave sql_thread;
Query OK, 0 rows affected (0.06 sec)

# 查看复制的状态
mysql> show slave status\G
*************************** 1. row ***************************
               Slave_IO_State: Waiting for master to send event
                  Master_Host: 10.10.30.162
                  Master_User: repl
                  Master_Port: 3306
                Connect_Retry: 10
              Master_Log_File: mysql-bin.000001
          Read_Master_Log_Pos: 18249
               Relay_Log_File: mysql-relay-bin.000002
                Relay_Log_Pos: 18462
        Relay_Master_Log_File: mysql-bin.000001
             Slave_IO_Running: Yes
            Slave_SQL_Running: Yes   # SQL 线程已经正常启动了
......
                   Last_Errno: 0
                   Last_Error:
                 Skip_Counter: 0
          Exec_Master_Log_Pos: 18249
              Relay_Log_Space: 18669
......
        Seconds_Behind_Master: 0
Master_SSL_Verify_Server_Cert: No
                Last_IO_Errno: 0
```

```
        Last_IO_Error:
        Last_SQL_Errno: 0  # 错误代码变为 0
        Last_SQL_Error:    # 错误信息已经消失
......
    Retrieved_Gtid_Set: bf320d72-bb2c-11e9-91ab-5254002a54f2:1-37
    Executed_Gtid_Set: bf320d72-bb2c-11e9-91ab-5254002a54f2:1-37,
c67c9093-bb2c-11e9-8055-525400c33752:1
        Auto_Position: 1
......
```

现在，使用 pt-table-checksum 和 pt-table-sync 工具重新同步主从库数据：

```
# 在主库中使用 pt-table-checksum 先校验主从库数据一致性，找出其中的差异（注意，某些版本的 percona-
toolkit 中的 pt-table-checksum 工具在这里可能无法比对出主从库数据的差异）
[root@localhost binlog]# pt-table-checksum --nocheck-replication-filters --no-check-
binlog-format --replicate=misoperation.checksums h=10.10.30.162,u=admin,p=password,P=3306
--databases=misoperation
    Checking if all tables can be checksummed ...
    Starting checksum ...
# 发现 misoperation.misoperation_table 表有数据差异（DIFFS 列不为 0）
      TS      ERRORS DIFFS ROWS DIFF_ROWS CHUNKS SKIPPED TIME  TABLE
0810T19:01:42 0      1     5    0         1      0       0.329
misoperation.misoperation_table
```

使用 pt-table-sync 工具将 pt-table-checksum 工具找出的主从库之间的差异数据，以主库数据为准重新进行同步，使得主从库之间的数据一致：

```
# 用 pt-table-sync 工具先打印出修复主从库数据有差异的语句（执行 --print）
[root@localhost  binlog]#  pt-table-sync  h=10.10.30.162  -P3306  -uadmin  -ppassword
--replicate=misoperation.checksums --databases=misoperation --print
# 下面就是修复主从库数据不一致的语句，检查是否有误
REPLACE INTO `misoperation`.`misoperation_table`(`id`, `xid`, `yid`) VALUES ('5', '5',
'15') /*percona-toolkit src_db:misoperation src_tbl:misoperation_table src_dsn:P=3306,
h=10.10.30.162,p=...,u=admin dst_db:misoperation dst_tbl:misoperation_table dst_dsn:P=
3306,h=10.10.30.163,p=...,u=admin lock:1 transaction:1 changing_src:misoperation.checksums
replicate:misoperation.checksums bidirectional:0 pid:25741 user:root host:localhost*/;

# 用 pt-table-sync 工具将主从库的差异数据重新同步（执行 --execute）
[root@localhost  binlog]#  pt-table-sync  h=10.10.30.162  -P3306  -uadmin  -ppassword
--replicate=misoperation.checksums --databases=misoperation --execute
```

使用 pt-table-checksum 工具再次校验主从库的数据：

```
[root@localhost binlog]# pt-table-checksum --nocheck-replication-filters --no-check-
binlog-format --replicate=misoperation.checksums h=10.10.30.162,u=admin,p=password,P=3306
--databases=misoperation
    Checking if all tables can be checksummed ...
    Starting checksum ...
# 从这里可以看到，该表的 DIFFS 列已经变更为 0，表示主从库数据已一致
     TS      ERRORS DIFFS ROWS DIFF_ROWS CHUNKS SKIPPED TIME  TABLE
0810T19:14:43 0      0     5    0         1      0       0.346 misoperation.misoperation_table
```

在从库 1 中，查看 misoperation.misoperation_table 表中 id = 5 的数据行是否已经恢复：

```
mysql> select * from misoperation_table where id=5;
+----+------+------+
| id | xid  | yid  |
+----+------+------+
|  5 |  5   |  15  |   # 可以看到，id = 5 的数据行已恢复
+----+------+------+
1 row in set (0.00 sec)
```

在从库 1 中，将系统变量 slave_exec_mode 修改为 STRICT（默认值），并重启 SQL 线程使其生效：

```
mysql> set global slave_exec_mode=STRICT;
Query OK, 0 rows affected (0.00 sec)

mysql> stop slave sql_thread;
Query OK, 0 rows affected (0.01 sec)

mysql> start slave sql_thread;
Query OK, 0 rows affected (0.05 sec)
```

至此，通过修改系统变量 slave_exec_mode 恢复从库数据的步骤已完成。

提示：通过修改系统变量 slave_exec_mode 来跳过复制错误，无法精准地跳过某个事务（会导致在应用二进制日志时所有的主键冲突和缺少记录的错误都被直接跳过），如果不是碰到大量此类问题，不建议通过修改系统变量 slave_exec_mode 来跳过复制错误。

24.3.2　通过 GTID 特性注入空事务恢复数据

这里我们基于 24.3.1 节中的操作环境继续演示。在从库 2 中，将 id 大于或等于 4 的数据行删除（模拟误删除）：

```
# 先查询 id >= 4 的数据
mysql> select * from misoperation_table where id>=4;
+----+------+------+
| id | xid  | yid  |
+----+------+------+
|  4 |  4   |  4   |
|  5 |  5   |  15  |
+----+------+------+
2 rows in set (0.00 sec)

# 执行删除
mysql> delete from misoperation_table where id>=4;
Query OK, 2 rows affected (0.01 sec)

# 再次查看 id >= 4 的数据，发现已不存在
mysql> select * from misoperation_table where id>=4;
```

```
Empty set (0.00 sec)
```

在主库中对 id 大于或等于 4 的数据行执行更新：

```
# 先查看 id >= 4 的数据行
mysql> select * from misoperation_table where id>=4;
+----+------+------+
| id | xid  | yid  |
+----+------+------+
|  4 |   4  |   4  |
|  5 |   5  |  15  |
+----+------+------+
2 rows in set (0.00 sec)


# 执行修改
mysql> update misoperation_table set xid=xid+10 where id>=4;
Query OK, 2 rows affected (0.00 sec)
Rows matched: 2  Changed: 2  Warnings: 0


# 再次查看数据，发现 xid 列已被修改
mysql> select * from misoperation_table where id>=4;
+----+------+------+
| id | xid  | yid  |
+----+------+------+
|  4 |  14  |   4  |
|  5 |  15  |  15  |
+----+------+------+
2 rows in set (0.01 sec)
```

在从库 1 中查看数据和复制状态：

```
# 查看 id >= 4 的数据行，已同步完成
mysql> select * from misoperation_table where id>=4;
+----+------+------+
| id | xid  | yid  |
+----+------+------+
|  4 |  14  |   4  |
|  5 |  15  |  15  |
+----+------+------+
2 rows in set (0.00 sec)


# 查看复制状态，一切正常
mysql> show slave status\G
*************************** 1. row ***************************
               Slave_IO_State: Waiting for master to send event
                  Master_Host: 10.10.30.162
                  Master_User: repl
                  Master_Port: 3306
                Connect_Retry: 10
              Master_Log_File: mysql-bin.000001
```

```
                Read_Master_Log_Pos: 154130
                     Relay_Log_File: mysql-relay-bin.000002
                      Relay_Log_Pos: 154343
              Relay_Master_Log_File: mysql-bin.000001
                   Slave_IO_Running: Yes
                  Slave_SQL_Running: Yes
......
                 Exec_Master_Log_Pos: 154130
                    Relay_Log_Space: 154550
......
              Seconds_Behind_Master: 0
......
                 Retrieved_Gtid_Set: bf320d72-bb2c-11e9-91ab-5254002a54f2:1-232
                  Executed_Gtid_Set: bf320d72-bb2c-11e9-91ab-5254002a54f2:1-232,
c67c9093-bb2c-11e9-8055-525400c33752:1
                      Auto_Position: 1
......
```

在从库 2 中查看数据和复制状态：

```
# 查看 id >= 4 的数据行，发现结果集为空
mysql> select * from misoperation_table where id>=4;
Empty set (0.00 sec)

# 查看复制状态，发现报错了
mysql> show slave status\G
*************************** 1. row ***************************
               Slave_IO_State: Waiting for master to send event
                  Master_Host: 10.10.30.162
                  Master_User: repl
                  Master_Port: 3306
                Connect_Retry: 10
              Master_Log_File: mysql-bin.000001
          Read_Master_Log_Pos: 154130
               Relay_Log_File: mysql-relay-bin.000002    # 留意中继日志文件名，后面会用到
                Relay_Log_Pos: 153927    # 留意中继日志报错而中止的位置，后面会用到
        Relay_Master_Log_File: mysql-bin.000001
             Slave_IO_Running: Yes
            Slave_SQL_Running: No    # SQL 线程已停止
......
                   Last_Errno: 1032    # 发现报 1032 错误
                   Last_Error: Coordinator stopped because there were error(s) in the worker(s).
The most recent failure being: Worker 1 failed executing transaction 'bf320d72-bb2c-
11e9-91ab-5254002a54f2:232' at master log mysql-bin.000001, end_log_pos 154099. See error
log and/or performance_schema.replication_applier_status_by_worker table for more details
about this failure or others, if any.
                 Skip_Counter: 0
          Exec_Master_Log_Pos: 153714
              Relay_Log_Space: 154550
```

```
......
        Seconds_Behind_Master: NULL
Master_SSL_Verify_Server_Cert: No
            Last_IO_Errno: 0
            Last_IO_Error:
           Last_SQL_Errno: 1032
           Last_SQL_Error: Coordinator stopped because there were error(s) in the
worker(s). The most recent failure being: Worker 1 failed executing transaction
'bf320d72-bb2c-11e9-91ab-5254002a54f2:232' at master log mysql-bin.000001, end_log_pos
154099. See error log and/or performance_schema.replication_applier_status_by_worker table
for more details about this failure or others, if any.
......
 Last_SQL_Error_Timestamp: 190810 19:31:52
           Master_SSL_Crl:
       Master_SSL_Crlpath:
       Retrieved_Gtid_Set: bf320d72-bb2c-11e9-91ab-5254002a54f2:1-232
        Executed_Gtid_Set: bf320d72-bb2c-11e9-91ab-5254002a54f2:1-231,
caabe772-bb2c-11e9-9e20-525400bdd1f2:1-2
            Auto_Position: 1
......
```

在从库 2 中，解析 SQL 线程报错中止时读取的中继日志位置，看看在这个位置上的是什么数据：

```
# 使用 mysqlbinlog 命令解析从库 2 的中继日志
[root@localhost relaylog]# mysqlbinlog -vv --start-position=153927 mysql-relay-bin.
000002 > e.sql

# 使用 vim 命令打开 e.sql 查看
[root@localhost relaylog]# vim e.sql
......
# at 153927    # 这里就是 SQL 线程中止的位置
#190810 19:31:52 server id 33061 end_log_pos 153779 CRC32 0x1a02b042 GTID last_
committed=231 sequence_number=232 rbr_only=yes
/*!50718 SET TRANSACTION ISOLATION LEVEL READ COMMITTED*//*!*/;
# 这里是 SQL 线程中止时正在应用的事务所对应的 GTID，记住这个 GTID，后面会用到
SET @@SESSION.GTID_NEXT= 'bf320d72-bb2c-11e9-91ab-5254002a54f2:232'/*!*/;
# at 153992
#190810 19:31:52 server id 33061 end_log_pos 153864 CRC32 0x70bf6784 Query thread_id=37
exec_time=0 error_code=0
SET TIMESTAMP=1565436712/*!*/;
......
BEGIN
/*!*/;
# at 154077
#190810 19:31:52 server id 33061 end_log_pos 153940 CRC32 0x92f34da2 Rows_query
# 这里就是发生错误时事务正在执行的 SQL 语句，它正在使用 UPDATE 语句修改 id >= 4 的记录，而之前我们查
询过，确认在从库 2 中并不存在这两条记录
# update misoperation_table set xid=xid+10 where id>=4
```

```
    # at 154153
    #190810 19:31:52 server id 33061  end_log_pos 154011 CRC32 0x0edc7b4a  Table_map:
`misoperation`.`misoperation_table` mapped to number 111
    # at 154224
    #190810 19:31:52 server id 33061  end_log_pos 154099 CRC32 0x0f459d7d  Update_rows: table
id 111 flags: STMT_END_F
    ......
    COMMIT/*!*/;
    SET @@SESSION.GTID_NEXT= 'AUTOMATIC' /* added by mysqlbinlog */ /*!*/;
    DELIMITER ;
    ......
```

在从库 2 中，使用 GTID 机制的特性，注入一个空事务（使用上述步骤中获取的 GTID，bf320d72-bb2c-11e9-91ab-5254002a54f2:232）：

```
# 使用 SET GTID_NEXT 语句指定在上述步骤中获取的 GTID
mysql> set gtid_next='bf320d72-bb2c-11e9-91ab-5254002a54f2:232';
Query OK, 0 rows affected (0.00 sec)

# 使用连续的 BEGIN 和 COMMIT 语句提交一个空事务（这个空事务会根据 gtid_next 系统变量指定的 GTID 进行
分配，相当于在从库的 GTID SET 中填充这个 GTID，也就表示跳过了该事务）
mysql> begin;commit;
Query OK, 0 rows affected (0.00 sec)
Query OK, 0 rows affected (0.01 sec)

# 将系统变量 GTID_NEXT 修改为默认值 AUTOMATIC
mysql> set gtid_next='AUTOMATIC';
Query OK, 0 rows affected (0.00 sec)

# 重新启动 SQL 线程
mysql> start slave sql_thread;
Query OK, 0 rows affected (0.05 sec)

# 查看复制状态，可以发现已恢复正常
mysql> show slave status\G
*************************** 1. row ***************************
             Slave_IO_State: Waiting for master to send event
                Master_Host: 10.10.30.162
                Master_User: repl
                Master_Port: 3306
              Connect_Retry: 10
            Master_Log_File: mysql-bin.000001
        Read_Master_Log_Pos: 154130
             Relay_Log_File: mysql-relay-bin.000002
              Relay_Log_Pos: 154343
      Relay_Master_Log_File: mysql-bin.000001
           Slave_IO_Running: Yes
          Slave_SQL_Running: Yes    # SQL 线程已启动
......
```

```
                Last_Errno: 0
                Last_Error:    # 报错信息已消失
              Skip_Counter: 0
       Exec_Master_Log_Pos: 154130
           Relay_Log_Space: 154550
......
     Seconds_Behind_Master: 0
Master_SSL_Verify_Server_Cert: No
             Last_IO_Errno: 0
             Last_IO_Error:
            Last_SQL_Errno: 0
            Last_SQL_Error:
......
        Retrieved_Gtid_Set: bf320d72-bb2c-11e9-91ab-5254002a54f2:1-232
         Executed_Gtid_Set: bf320d72-bb2c-11e9-91ab-5254002a54f2:1-232,
caabe772-bb2c-11e9-9e20-525400bdd1f2:1-2
             Auto_Position: 1
......
```

现在，从库 2 中的复制状态虽然恢复正常，但主从库之间的数据不一致，需要使用 pt-table-checksum 和 pt-table-sync 工具重新同步主从库的数据：

```
# 用 pt-table-checksum 工具校验主从库的数据
[root@localhost binlog]# pt-table-checksum --nocheck-replication-filters --no-check-binlog-format --replicate=misoperation.checksums h=10.10.30.162, u=admin, p=password,P=3306 --databases=misoperation
Checking if all tables can be checksummed ...
Starting checksum ...
# 发现该表的 DIFFS 列的值不为 0，表示主从库之间的数据有差异
  TS    ERRORS DIFFS ROWS DIFF_ROWS CHUNKS SKIPPED TIME TABLE
0810T19:48:19 0  1    5      1       0       0    0.334
misoperation.misoperation_table

# 用 pt-table-sync 工具打印如何恢复数据一致性
[root@localhost binlog]# pt-table-sync h=10.10.30.162 -P3306 -uadmin -ppassword --replicate=misoperation.checksums --databases=misoperation --print
REPLACE INTO `misoperation`.`misoperation_table`(`id`, `xid`, `yid`) VALUES ('4', '14', '4') /*percona-toolkit src_db:misoperation src_tbl:misoperation_table src_dsn:P=3306,h=10.10.30.162,p=...,u=admin dst_db:misoperation dst_tbl:misoperation_table dst_dsn:P=3306,h=10.10.30.164,p=...,u=admin lock:1 transaction:1 changing_src:misoperation.checksums replicate:misoperation.checksums bidirectional:0 pid:28106 user:root host:localhost*/;
REPLACE INTO `misoperation`.`misoperation_table`(`id`, `xid`, `yid`) VALUES ('5', '15', '15') /*percona-toolkit src_db:misoperation src_tbl:misoperation_table src_dsn:P=3306,h=10.10.30.162,p=...,u=admin  dst_db:misoperation  dst_tbl:misoperation_table  dst_dsn:P=3306,h=10.10.30.164,p=...,u=admin lock:1 transaction:1 changing_src:misoperation.checksums replicate:misoperation.checksums bidirectional:0 pid:28106 user:root host:localhost*/;

# 用 pt-table-sync 工具执行主从库数据的同步
[root@localhost binlog]# pt-table-sync h=10.10.30.162 -P3306 -uadmin -ppassword
```

```
--replicate=misoperation.checksums --databases=misoperation --execute
```

使用 pt-table-checksum 工具重新校验主从库数据的一致性：

```
[root@localhost binlog]# pt-table-checksum --nocheck-replication-filters --no-check-binlog-format --replicate=misoperation.checksums h =10.10.30.162, u=admin, p=password, P=3306 --databases=misoperation
Checking if all tables can be checksummed ...
Starting checksum ...
# 该表的 DIFFS 列值已变为 0，校验结果表示主从库数据已一致
    TS     ERRORS DIFFS ROWS DIFF_ROWS CHUNKS SKIPPED TIME  TABLE
0810T19:50:17 0   0     5    0         1      0       0.333
misoperation.misoperation_table
```

在从库 2 中，查询 id 大于或等于 4 的数据行是否已经恢复：

```
mysql> select * from misoperation_table where id>=4;
+----+------+------+
| id | xid  | yid  |
+----+------+------+
| 4  | 14   | 4    |  # 发现已经可以正常查询 id >= 4 的数据
| 5  | 15   | 15   |
+----+------+------+
2 rows in set (0.00 sec)
```

至此，通过 GTID 特性注入空事务恢复从库数据的步骤已完成。

提示：注入空事务是一个临时解决方案，后面需要尽快恢复主从库数据的一致性，否则，可能就需要清理记录了空事务的二进制日志，以防止空事务在主从库之间流转。

使用 pt-table-checksum 工具校验主从数据一致性时，需要注意如下问题（这些问题都可以通过参数做适当的适配进行优化，有兴趣的读者可自行研究，这里不做介绍）。

- 如果数据库中的数据总量较大（比如超过 100 GB），则执行校验的时间可能比较长。
- 如果从库存在较大延迟，则可能导致 pt-table-checksum 工具一直在等待从库追赶上主库数据，直到从库的复制延迟降低到一个合理的范围时才会执行。
- 如果主库并发访问负载较高，可能导致 pt-table-checksum 工具一直在等待主库负载降低，直到达到合理的负载时才会执行。

本章主要讲解的是数据行级别的误操作与恢复，如果是表级别或者是文件系统级别的误删除，则可以考虑使用 extundelete 或 undrop-for-innodb 等工具进行恢复，有兴趣的读者可以自行研究，本章不做介绍。

第 25 章 常用复制故障排除方案

作为 MySQL 数据库管理人员，我们在日常的工作中或多或少都会碰到 MySQL 复制相关的问题，有些问题可能很快就解决了，有些问题具有一定的迷惑性，可能需要排查很久才能找到具体的原因。对于后者，我们大概率会在日后的工作中再次遭遇它们。再次遇到的时候，你是否有似曾相识但怎么也想不起来具体细节的感觉呢？

为了避免这种尴尬，建议在处理完故障之后，立即以文档形式总结故障现象、其复现与排查过程、解决方案和规避方法，予以留存。

另外，故障虽然多种多样，但其处理思路与流程是具有共性的。本章以处理 MySQL 复制相关的故障为主题，详细介绍一个排除复制故障的通用方案（注意，本章只讲解思路与流程，不介绍细节）。

25.1 确认故障现象

当出现复制故障时，数据库管理人员是如何知道的呢？在不大可能出问题的时候，碰巧我们查看了复制的状态，然后发现了问题（当然，由误操作导致的问题除外。关于 DML 语句误操作而导致的问题的修复方法，可参考第 24 章"发生数据误操作之后的处理方案"）？其实大多数时候，要么是监控告警系统发现了问题，要么是应用端反馈查询数据出现异常，这时，作为数据库管理人员的我们，需要迅速确认两件事情：

- 确认告警系统报告的问题或者应用端反馈的查询异常是否真实存在？如果是，那么在数据库中故障的表现形式是怎样的？例如，在报错的从库中查看具体的报错信息，然后将这些信息收集起来，以备后续排查时使用。
- 粗略评估该故障对应用造成的影响。例如，具体影响到哪些应用系统，对它们造成的影响有多严重等。

25.2 信息收集与故障排查

在确认故障的真实性和表现之后，就需要围绕它进行推理，列出一些可能造成该故障的原因，并收集可能导致该故障的因素的相关信息（用于逐个排除可能导致故障的不确定因素。有了这些信息，就可以将不确定的因素变成确定的，是否是该因素造成的问题就一

清二楚了)。那么,需要收集哪些信息呢?这里我们将这些信息大致分为如下几个类别:

1. 负载类信息
- 实时收集故障从库的主机和数据库的负载信息,判断是否是由于高负载而导致的问题。
- 从监控系统中采集故障从库的主机和数据库的历史负载信息。

2. 环境类信息
- 与相关人员确认,出故障之前,是否有人对系统或者 MySQL Server 做过什么操作或者变更?如果有,具体是如何操作的。
- 收集报错的复制拓扑中所有数据库的变量模板,判断问题是否是由于某个变量设置不当导致的。
- 收集操作系统内核参数的配置、操作系统版本、数据库版本等。

3. 状态类信息
- 实时从故障从库中获取复制状态信息、事务信息、锁信息、线程状态信息(使用 SHOW PROCESSLIST 语句查看),判断是否有大事务、锁等待、不正常的连接等。
- 从监控系统中获取故障从库的历史复制状态信息、事务信息、锁信息、线程状态信息(使用 SHOW PROCESSLIST 语句查看)。
- 收集故障从库的 MySQL 错误日志、操作系统常规日志、操作系统安全日志,判断是否有某些错误或者 bug 被触发。

收集上述信息之后,我们还需要确认复制拓扑是全新创建的,还是已经正常运行了一段时间。

(1)对于全新搭建的且无法从相关日志中找到有效信息的复制拓扑,可按照如下粒度更细的方法排查问题:

- 根据 SHOW MASTER STATUS 语句的输出信息来检查主库是否启用了二进制日志记录。如果已启用,则使用该语句能够查询到二进制日志的位置信息;如果未启用,则查询结果集为空。要启用二进制日志记录,需要在主库上启用--log-bin 选项(或将系统变量 log_bin 添加到 my.cnf 配置文件的[mysqld]配置组标签下)。
- 检查主库和从库是否都设置了--server-id 选项(或在 my.cnf 中设置系统变量 server_id),以及在整个复制拓扑中该 Server ID 是否是唯一的。
- 检查从库复制线程是否正在运行。使用 SHOW SLAVE STATUS 检查 Slave_IO_Running 和 Slave_SQL_Running 的值是否均为 Yes。如果不是,请检查启动从库时使用的选项。例如,如果使用了--skip-slave-start 选项,会禁止自动启动从库复制线程,此时需要手动执行 START SLAVE 语句,重新启动复制线程。

(2)对于已经正常运行了一段时间的复制拓扑,请检查故障从库与主库的连接是否正常。用 SHOW PROCESSLIST 语句找到 I/O 和 SQL 线程并检查 State 列，查看它们的状态。如果 I/O 线程的状态显示为"Connecting to master"，则做以下检查：

- 检查在主库上用于复制的用户是否具有 REPLICATION SLAVE 权限。
- 检查从库的复制配置中主库的主机名是否正确（如果配置的不是 IP 地址，则需要考虑该主机名是否可以解析为正确的 IP）、数据库端口是否正确（通常默认为 3306）。
- 检查主库或从库上是否禁用了网络。在配置文件中查找系统变量 skip-networking，如果存在，请将其注释掉或删除并重启 mysqld 进程。
- 如果主库启用了 IPTABLES 防火墙或者 IP 过滤配置，请确保未过滤 MySQL 使用的网络端口。
- 使用 ping 或 traceroute/tracert 等工具检查从库是否可以访问主库主机网络。

提示：从库复制正常运行一段时间之后意外中止，通常是因为从库在应用主库上的某个语句时遇到问题而报错中止了。如果搭建从库时获取了主库正确的快照数据，而且从库除了通过复制线程从主库同步数据之外，从未有数据库管理员或应用程序在从库上通过任何形式修改过数据，则不可能发生这种情况，以下情况除外：

- 遇到了 bug。
- 遇到了在主从复制技术实现上一些已知的限制因素。

25.3　复制故障的修复

如果问题不是非常棘手，通过对 25.2 节中收集的信息，逐一分析与筛选之后，应该能找出问题的原因，甚至找到解决问题的方法。现在，目标已经明确，按照解决方法修复复制故障即可。

如果遇到在主库上执行成功，但在从库上通过复制线程应用失败的语句，而且不能随意通过"清除"从库的数据来重新搭建从库，则可以尝试如下方法临时解决从库数据的同步异常问题：

- 确定从库上受影响的表是否与主库中的表结构不同。如果是，则试着了解这是如何发生的。然后，找出主从库中表的结构差异，将这些差异手工同步到从库的表中，并执行 START SLAVE 语句启动复制。
- 如果上面的方法不起作用或不适用，可以了解一下在从库手动更新被拒绝执行的语句是否安全（如果需要的话），如果确认是安全的，则可以忽略报错的语句，让复制线程继续运行（关于如何跳过报错事务的步骤，可参考 24.3 节"从库发生误操作后的数据恢复"）。

如果问题仍然无法解决，而且确定主库中的二进制日志完全同步到了从库，复制线程之外的其他用户并没有修改从库数据，那么可能是遇到了 bug。如果你的数据库版本比较旧，可以尝试升级到最新版本；如果版本比较新，则需要向 MySQL 官方提交 bug 信息。

25.4 无法解决的问题

如果确定用户操作没有错误，而复制仍然无法正常进行或者不稳定时，你可能需要花一些时间和精力准备一份好的 bug 报告，尽可能详细地记录故障现象的特征，以便 MySQL 官方能够追踪该 bug。

如果你有该 bug 的复现过程，可以使用 MySQL 官网介绍的方法将该 bug 记录到官方的 bug 数据库中，但如果是"幻象"问题（偶现问题），则可以按照如下步骤抓取一些有效信息：

（1）确认没有用户操作错误。例如，如果用户人工登录到从库，执行了某些数据修改，导致与从库复制线程发生冲突（发生主键冲突或者找不到记录等），那么可以通过人工介入处理。这不是复制机制本身的问题，而是外部干扰导致的复制失败，与 bug 无关。

（2）从库启用 --log-slave-updates 和 --log-bin 选项（或在 my.cnf 中启用系统变量 log_slave_updates 和 log_bin）。这些选项使从库可以将从主库中接收的二进制日志记录到其自己的二进制日志中。

（3）在重置复制状态之前保存所有证据。如果没有足够的信息，很难甚至不可能追查到问题。所以，你需要收集如下证据：

- 主库的所有二进制日志文件。
- 从库的所有二进制日志文件。
- 当发现问题时，主库上 SHOW MASTER STATUS 语句的输出信息。
- 当发现问题时，从库上 SHOW SLAVE STATUS 语句的输出信息。
- 主库和从库的错误日志文件。

（4）使用 mysqlbinlog 解析二进制文件并进行检查：解析的目标二进制日志文件和目标位置，可以通过在从库中执行 SHOW SLAVE STATUS 语句获取（从库 SQL 线程报错中止时的 Master_Log_File 和 Read_Master_Log_Pos 选项值），解析结果可以重定向到一个文本文件中，待到解析完成之后，可通过文本编辑器找到报错位置的二进制日志记录的内容。例如：

```
mysqlbinlog --start-position = log_pos log_file > file.txt
```

在收集证据后，请先尝试复现故障现象，复现过程中尽可能记录详细的复现步骤，并收集更多的信息，然后将复现步骤和收集的信息一并提交到 MySQL 官方的 bug 数据库。

参 考 篇

第 26 章 二进制日志文件的基本组成

在使用 MySQL 数据库的平台上，很多关键的应用场景都是基于二进制日志实现的，例如主从复制（这也是本书的主题，前面用大量的篇幅介绍了复制的原理与使用案例）、基于时间点的备份与恢复、误操作数据的回滚、供数（解析二进制日志文件，并将得到的文本数据传输到另一个平台，如数据仓库、Kafka 等）等，但是很少有人详细了解过二进制日志。本章将从二进制日志事件类型的角度对二进制日志文件中的内容进行详细的介绍。

26.1 什么是二进制日志

二进制日志（binary log，简称 binlog）是 MySQL 的一种日志，记录了 MySQL 中的数据变更操作（INSERT、UPDATE、ALTER 等），而且它仅记录变更操作，像 SELECT、SHOW 等操作都不会被记录到二进制日志中。对于 MySQL 而言，二进制日志是很重要的日志文件，主要有以下用途：

1．复制

二进制日志是 MySQL 复制的关键，主库将变更操作记录在其二进制日志中，并发送给从库以中继日志的形式保存，从库通过解析中继日志获知主库的变更操作，在应用到自身。

2．数据恢复

利用"备份 + 二进制日志恢复"的方式，可以将数据恢复到某一具体时刻或者某一具体位置。先利用备份完成数据恢复，在备份点之后的数据变更操作也都记录在二进制日志中，所以可以利用二进制日志将数据恢复到备份点之后的某一时刻或者某一具体位置。

3．日志审计

用户可以通过解析二进制日志获得 MySQL 执行的所有变更操作，可以对解析的结果进行日志审计。

26.2 二进制日志的组成

二进制日志由日志文件（记录变更信息）和索引文件（记录二进制日志文件列表）组成。例如：

```
[root@localhost ~]# ls
mysql-bin.index    mysql-bin.000001    mysql-bin.000002    mysql-bin.000003
```

其中，mysql-bin.index 是二进制日志中的索引文件，记录数据库中尚未被清理掉的二进制日志文件名：

```
[root@localhost ~]# cat mysql-bin.index
/home/mysql/data/mysqldata1/binlog/mysql-bin.000001
/home/mysql/data/mysqldata1/binlog/mysql-bin.000002
/home/mysql/data/mysqldata1/binlog/mysql-bin.000003
```

mysql-bin.N 是二进制日志的磁盘文件，记录数据库的变更信息。一般我们说二进制日志的时候，都是指二进制日志的磁盘文件。

- 每个二进制日志文件以 4 字节的魔术数开始，后面包含各种用于表示 MySQL 数据变更的事件（event）。
- 4 字节的魔术数固定为 0xfe 0x62 0x69 0x6e，即 0xfe b i n。
- 每个二进制日志文件中第一个事件是 Format_description_event 类型的事件，记录二进制日志的版本、数据库版本、文件的创建时间等基本信息。Format_description_event 类型的事件具体包含的信息及格式可以参考附录 A "二进制日志事件详解"。
- 每个二进制日志文件中的最后一个事件是 Rotate_event 类型的事件，记录下一个二进制日志的文件名（存在二进制日志文件不以 Rotate_event 结尾的情况。MySQL 意外宕机时，数据库重启之后会生成新的二进制日志文件，宕机前的最后一个二进制日志文件不是以 Rotate_event 事件结尾的）。Rotate_event 事件具体包含的信息及格式可以参考附录 A "二进制日志事件详解"。
- 介于 Format_description_event 和 Rotate_event 之间的是各种用于表示 MySQL 数据变更的事件。
- 每个事件包含头部数据部分，记录事件类型、创建时间等信息。
- 每个事件包含数据部分，记录每种类型事件的具体信息，例如数据变更信息。

26.3　二进制日志内容解析

正常情况下，二进制日志是以二进制格式存储的，打开后看到的是乱码。MySQL 提供了一个名为 mysqlbinlog 的工具，用于解析二进制日志，将二进制格式的数据解析成我们可读的信息。接下来，我们就来看如何使用 mysqlbinlog 解析日志文件以及如何解读解析的结果。本节主要介绍二进制日志文件的内容解析，出于篇幅的考虑，不会过多介绍 mysqlbinlog 工具的使用，如有需要可以参考附录 B "mysqlbinlog 命令选项"。

在第 1 章中，我们介绍过 MySQL 支持两种核心的复制格式：基于 row 的复制与基于 statement 的复制。这两种不同的复制格式中，二进制日志文件记录信息的格式也不一样。接下来，我们分别对基于 row 及基于 statement 的复制中的二进制日志文件内容进行解析。

提示：由于本章涉及非常多的细节，为帮助理解，建议初学者跟随下文的步骤同步操作。

26.3.1 基于 row 的复制的二进制日志内容解析

26.3.1.1 数据库操作记录

本节使用的数据库版本为 MySQL 5.7.22，二进制日志为 row 格式。我们对数据库进行创建表、插入数据、修改数据、删除数据等操作，解析常见操作产生的二进制日志的内容。

```
mysql> select version()\G
*************************** 1. row ***************************
version(): 5.7.22-log
1 row in set (0.00 sec)

mysql> show variables like '%binlog_format%'\G
*************************** 1. row ***************************
Variable_name: binlog_format
        Value: ROW
1 row in set (0.00 sec)
## 通过 FLUSH LOGS 语句切换到新的二进制日志文件
mysql> flush logs;
Query OK, 0 rows affected (0.01 sec)

mysql> create table test(id int primary key auto_increment,name varchar(20));
Query OK, 0 rows affected (0.02 sec)

mysql> insert into test(`name`) values('gangshen_row');
Query OK, 1 row affected (0.00 sec)

mysql> udpate test set  name = 'gangshen_row_new' where name = 'gangshen_row';
Query OK, 1 row affected (0.01 sec)
Rows matched: 1 Changed: 1 Warnings: 0

mysql> delete from test where name = 'gangshen_row_new';
Query OK, 1 row affected (0.01 sec)
## 切换到新的二进制日志文件
mysql> flush logs;
Query OK, 0 rows affected (0.01 sec)
```

26.3.1.2 使用 mysqlbinlog 工具解析

在 26.3.1.1 节，我们对数据库执行了一系列变更操作，由于开启了二进制日志，这些变更操作都被记录到二进制日志中。下面我们使用 mysqlbinlog 工具对产生的二进制日志进行解析，命令如下：

```
mysqlbinlog -vv mysql-bin.N
```

mysqlbinlog 工具的使用方式很简单，只需要在命令后面接 "-v" 选项以及要解析的二进制日志文件名即可（若同时解析多个文件，需要用空格分隔文件名）。这里使用了"-vv"选项，它比"-v"选项解析的结果更详细。感兴趣的读者，可以自行操作，验证两者的区别。

由于篇幅有限，而且 26.3.1.3 节会对解析结果逐一解释，所以这里没有列出 mysqlbinlog 工具解析的详细结果。建议读者在阅读 26.3.1.3 节之前，根据 26.3.1.1 节以及本节中的命令自行操作，以便对二进制日志有更完整的认识。

26.3.1.3 对解析结果的解释

第 1 部分

```
/*!50530 SET @@SESSION.PSEUDO_SLAVE_MODE=1*/;
/*!50003 SET @OLD_COMPLETION_TYPE=@@COMPLETION_TYPE,COMPLETION_TYPE=0*/;
DELIMITER /*!*/;
```

在解析的结果中，以上内容是由 mysqlbinlog 工具添加的，并非二进制日志文件中存储的信息：

- "SET @@SESSION.PSEUDO_SLAVE_MODE=1"设置系统变量值为 1，供 MySQL Server 内部使用，具体作用请自行研究。

- "SET @OLD_COMPLETION_TYPE = @@COMPLETION_TYPE, COMPLETION_TYPE = 0"设置系统变量值为 0，表示关闭链式事务，在遇到 COMMIT 或者 ROLLBACK 语句时不会断开连接，当使用二进制日志文件进行数据恢复时，该系统变量的设置很重要。

- "DELIMITER /*!*/;"设置结束符为";"，告诉 MySQL 当遇到";"时表示当前命令已结束。

第 2 部分

```
1. # at 4
2. #190519 10:07:35 server id 330657 end_log_pos 123 Start: binlog v 4, server v 5.7.22-log created 190519 10:07:35
3. BINLOG '
4. J2PhXA+hCwUAdwAAAHsAAAAAAAQANS43LjIyLWxvZwAAAAAAAAAAAAAAAAAAAAAAAAAAAAAAA
5. AAAAAAAAAAAAAAAAAAAAAAAEzgNAAgAEgAEBAQEEgAAXwAEGggAAAAICAgCAAAACgoKKioAEjQA
6. ADMbJo0=
7. '/*!*/;
```

上面的内容为每个二进制日志文件中的第一个事件，Format_description_event 类型，用于记录 MySQL 的一些基本信息（这里为了方便对照阅读代码段及其对其的解释，特意标记了行号。如无特殊说明，默认代码段不添加行号）。

第 1 行"#at 4"为该事件在二进制日志文件中的起始偏移量，单位为字节，表示该事件从二进制日志文件的多少字节处开始。前面提到，每个二进制日志文件是以 4 字节的魔术数（0xfe 0x62 0x69 0x6e）开始的，所以二进制日志文件中第一个事件的起始偏移量为 4。后面的每个事件开始都会有这么一行表示事件起始位置，不再赘述。

第 2 行中：

- "190519 10:09:35"表示写入该事件的时间,每个事件都会包含该部分内容。下文解释其他类型的事件时,不再赘述。

- "server id 330657"表示产生该事件的 MySQL Server ID 为 330657。每个事件都会包含该部分内容,下文解释其他类型的事件时,不再赘述。

- "end_log_pos 123"表示该事件的结束偏移量为 123,即下一个事件的起始偏移量为 123。每个事件都会包含该部分内容,下文解释其他类型的事件时,不再赘述。

- "Start"表示该事件的类型为 Format_description_event,在主从复制时,从库会忽略该事件。

- "binlog v4"表示二进制日志的结构版本为 v4(结构版本定义了二进制日志的存储格式)。二进制日志一共经历过 v1、v2 和 v4 三个版本的迭代,在 MySQL 5.0 以上版本中,一直是 v4 版本。

- "server v 5.7.22-log"表示产生二进制日志的 MySQL 版本为 5.7.22-log。

- "created 190519 10:07:35"表示该二进制日志文件的创建时间,因为 MySQL 在切换新的二进制日志文件时,会写入 Format_description_event 事件,所以二进制日志文件创建的时间与此类型事件的写入时间一致。

第 3~6 行是 Format_description_event 事件的原始二进制数据,若希望 mysqlbinlog 工具解析的结果中不包含该部分,可以在使用 mysqlbinlog 工具时加上 "--base64-output=decode-rows" 选项。

第 7 行表示该事件的结束。

第 3 部分

```
1. # at 123
2. #190519 10:07:35 server id 330657 end_log_pos 190 Previous-GTIDs
3. # 16b154ef-7a12-11e9-b850-0800276d49d5:1-17
```

该部分为 Previous_gtid_log_event 类型的事件,在 MySQL 切换新的二进制日志文件时,会先写入一个 Format_description_event 事件,但是如果开启了 GTID 模式,则会接着写入一个 Previous_gtid_log_event 事件,用于记录创建该日志文件之前执行的全局事务 ID 集合。

第 1 行:请参考本小节"第 2 部分"中 Format_description_event 事件对应部分的解释。

第 2 行中:

- 对"190519 10:07:35 server id 330657 end_log_pos 190"的解释,请参考本小节"第 2 部分"中 Format_description_event 事件对应部分的解释。

- "Previous-GTIDs"表示该事件的类型为 Previous_gtid_log_event,这个事件记录了创建该二进制日志文件之前执行的全局事务 ID 集合。在主从复制时,从库会忽略该事件。

第 3 行：创建该二进制日志文件之前执行的全局事务 ID 集合。

第 4 部分

```
1. # at 190
2. #190519 10:07:42 server id 330657  end_log_pos 251   GTID    last_committed=0    sequence_number=1  rbr_only=no
3. SET @@SESSION.GTID_NEXT= '16b154ef-7a12-11e9-b850-0800276d49d5:18'/*!*/;
```

该部分为 Gtid_log_event 类型的事件。开启 GTID 复制模式后，MySQL 为每一个事务都会分配一个 GTID，事务的 GTID 记录在 Gtid_log_event 事件中。

第 1 行：请参照本小节"第 2 部分"中 Format_description_event 事件对应部分的解释。

第 2 行中：

- 关于"190519 10:07:42 server id 330657 end_log_pos 251"的含义，请参考本小节"第 2 部分"中 Format_description_event 事件对应部分的解释。
- "GTID"表示该事件类型为 Gtid_log_event，该事件记录了事务的 GTID。
- "last_committed = 0"表示提交一个事务时，取该二进制日志文件中最大的已提交事务的 sequence_number，作为将要提交的这个事务的 last_committed 值。当有多个事务同时提交时，这些事务获取的最大的已提交事务的 sequence_number 可能是一样的，所以多个事务的 last_committed 值可能相同。事务的 sequence_number 以及 last_committed 都是为多线程复制而设计的，由于篇幅有限，这里不展开介绍。
- "sequence_number = 1"：单个二进制日志文件中，事务的 sequence_number 是一个有序递增的数字，表示每个事务的提交顺序，每个事务递增 1，所以每个事务的 sequence_number 都是不一样的。在所有二进制日志文件中，sequence_number 都是从 1 开始的。
- rbr_only 表示该事务中是否含有 SBR（基于 statement 的复制）模式的语句，rbr_only = no 表示该事务中包含 SBR 模式的语句或者该事务是一个空事务。rbr_only = yes 表示该事务中不包含 SBR 模式的语句，在此情况下，使用 mysqlbinlog 工具解析二进制日志时，会在该事务的解析结果中增加一个改变事务隔离级别为 READ-COMMIITED 的语句，下面遇到 rbr_only=yes 时会讲到。

第 3 行：在会话级别设置 GTID_NEXT 变量的语句。可以看到该事务的 GTID 为 16b154ef-7a12-11e9-b850-0800276d49d5:18。

第 5 部分

```
1. # at 251
2. #190519 10:07:42 server id 330657  end_log_pos 403  Query  thread_id=3960   exec_time=0  error_code=0
3. use `gangshen`/*!*/;
4. SET TIMESTAMP=1558274862/*!*/;
5. SET @@session.pseudo_thread_id=3960/*!*/;
```

```
  6. SET @@session.foreign_key_checks=1, @@session.sql_auto_is_null=0, @@session.unique_
checks=1, @@session.autocommit=1/*!*/;
  7. SET @@session.sql_mode=1073741824/*!*/;
  8. SET @@session.auto_increment_increment=2, @@session.auto_increment_offset=1/*!*/;
  9. /*!\C utf8 *//*!*/;
  10. SET @@session.character_set_client=33,@@session.collation_connection=33,@@session.
collation_server=83/*!*/;
  11. SET @@session.lc_time_names=0/*!*/;
  12. SET @@session.collation_database=DEFAULT/*!*/;
  13. CREATE TABLE test(id INT PRIMARY KEY AUTO_INCREMENT,name VARCHAR(20))
  14. /*!*/;
```

该部分为 Query_log_event 类型的事件。

第 1 行：请参考本小节"第 2 部分"中 Format_description_event 事件对应部分的解释。

第 2 行中：

- 关于"190519 10:07:42 server id 330657 end_log_pos 403"的含义，请参考本小节"第 2 部分"中 Format_description_event 事件对应部分的解释。

- "Query"表示该事件类型为 Quert_log_event。这个事件记录了执行的语句，在 RBR（基于 row 的复制）模式下，一般为 DDL 语句。

- "thread_id = 3960"表示执行该语句的线程 ID 为 3960，该 thread_id 与 SHOW PROCESSLIST 语句输出结果中的 ID 字段一致。

- "exec_time = 0"表示该语句的执行时间，单位为 s。

- "error_code = 0"表示该语句执行后返回的错误编号。

第 3 行：将当前操作的数据库改为 gangshen。mysqlbinlog 工具在解析二进制日志时，遇到 Query_log_event 以及 Load_log_event 事件，会将事件中记录的数据库名保存下来。在解析到下一个 Query_log_event 或者 Load_log_event 时，会将之前记录的数据库名与事件中记录的进行比较，如果两者不一样，则在解析结果中添加 USE xxxx 语句，这是为了在使用二进制日志进行数据恢复时，保证数据的正确性。

第 4 行的 SET TIMESTAMP 语句将当前时间戳修改为事件记录的时间。添加该语句是为了在使用二进制日志进行数据恢复时，保证数据的正确性，因为如果 Query_log_event 中记录的语句执行的是与时间相关的内容，不设置时间戳可能会导致恢复的数据不正确。

第 5~12 行：该部分在会话级别设置一些变量值。主库执行记录时将这些变量值记录下来，当从库获得主库二进制日志并重放时，就能保证数据的准确性。

- "SET @@session.pseudo_thread_id=3906"设置 pseudo_thread_id 变量在会话级别的值。官方文档称该变量为供 MySQL Server 内部使用，具体作用可自行研究。

- "SET @@session.foreign_key_checks = 1, @@session.sql_auto_is_null = 0, @@session.unique_checks = 1, @@session.autocommit = 1"设置变量 forrign_key_ checks、sql_auto_is_null、unique_checks、autocommit 在会话级别的值，保证从库重放语句时，不会因为主从数据库配置文件对这些变量的设置不一致而发生重放失败或数据不一致的情况。

- "SET @@session.sql_mode=1073741824" 设置 sql_mode 变量在会话级别的值，保证从库在重放语句时，与主库在相同的 sql_mode 限制下，以防止在从库 sql_mode 限制比主库严格的情况下，某些语句在从库上重放失败。这里看到 sql_mode 变量的值是以整数表示的，与我们平时设置 sql_mode 的方式不同，这是因为 sql_mode 变量实际是以 uint64_t 类型（64 位无符号整数，8 字节）表示的，每一位表示一种模式的开和关。mysqlbinlog 在解析时为了方便直接使用了 sql_mode 所代表的整数值，以 MySQL-5.7.22 的代码为例，sql_mode 中的各种模式定义在 sql/sql_class.h 头文件中。

- "SET @@session.auto_increment_increment = 2, @@session.auto_increment_offset = 1" 设置 auto_increment_increment、auto_increment_offset 变量在会话级别的值，保证从库重放语句时，不会因为自增配置变量设置与主库不一致而导致主从库数据不一致。

- "SET @@session.character_set_client = 33, @@session.collation_connection = 33, @@session.collation_server = 83" 设置 character_set_client、collation_connection、collation_server 变量在会话级别的值，保证从库重放语句时与主库保持一样的字符集，与 sql_mode 值一样，该值采用整数直接表示。

- "SET @@session.lc_time_names=0" 设置 lc_time_names 变量在会话级别的值，保证从库重放的语句中包含 DATE_FORMAT()、DAYNAME()、MONTHNAME()函数时，不会因为该系统变量不一致导致主从库数据不一致。

- "SET @@session.collation_database = DEFAULT"设置 collation_database 变量在会话级别的值，保证主从之间默认数据库的校验规则的一致性。

- 以上这些会话级别的变量设置语句会在 mysqlbinlog 工具解析到第一个 Query_log_event 事件时被打印出来。后面解析到其他 Query_log_event 事件时，只有当上述某一变量的值发生改变，才会将对应的值打印出来，否则表示这些变量值未发生变化。

第 13~14 行：Query_log_event 事件中记录的具体执行语句。

第 6 部分

```
1. # at 403
2. #190519 10:07:49 server id 330657  end_log_pos 464    GTID    last_committed=1 sequence_number=2   rbr_only=yes
3. /*!50718 SET TRANSACTION ISOLATION LEVEL READ COMMITTED*//*!*/;
4. SET @@SESSION.GTID_NEXT= '16b154ef-7a12-11e9-b850-0800276d49d5:19'/*!*/;
```

该部分为 Gtid_log_event 类型的事件。

第 1 行：请参考本小节"第 2 部分"中 Format_description_event 事件对应部分的解释。

第 2 行中：

- 关于"190519 10:07:49 server id 330657 end_log_pos 464"的含义，请参考本小节"第 2 部分"中 Format_description_event 事件对应部分的解释。
- "GTID"表示该事件类型为 Gtid_log_event，该事件记录了事务 ID。
- "last_committed=1"表示一个事务提交时，取该二进制日志文件中最大的已提交事务的 sequence_number，作为将要提交的该事务的 last_committed 值。当有多个事务同时提交时，这些事务获取到的最大已提交事务的 sequence_number 可能是一样的，所以多个事务的 last_committed 值可能相同。事务 sequence_number 以及 last_committed 都是为多线程复制而设计的，由于篇幅有限，这里不展开介绍。
- 单个二进制日志文件中事务的 sequence_number 是一个有序递增的数字，表示每个事务的提交顺序，每个事务递增 1。在所有二进制日志文件中，sequence_number 都是从 1 开始的。
- "rbr_only = yes"表示该事务中不包含 SBR（基于 statement 的复制）模式的语句。在此种情况下，使用 mysqlbinlog 工具解析二进制日志时，会在该事务的解析结果中增加一个改变事务隔离级别为 READ-COMMIITED 的语句，因为对于不包含 SBR 模式语句的事务，从库在重放时设置为 READ-COMMITTED 隔离级别是可以的。并且设置为 READ-COMMITTED 隔离级别可以避免一些死锁问题，因为在 READ-COMMITTED 隔离级别下不会使用间隙锁（GAP Lock）。

第 3 行：将当前事务隔离级别改为 READ-COMMIITED，由 mysqlbinlog 添加。

第 4 行：在会话级别设置 GTID_NEXT 变量的语句。可以看到该事务的 GTID 为 16b154ef-7a12-11e9-b850-0800276d49d5:19。

第 7 部分

```
1. # at 464
2. #190519 10:07:49 server id 330657 end_log_pos 541 Query thread_id=3960  exec_time=0 error_code=0
3. SET TIMESTAMP=1558274869/*!*/;
4. BEGIN
5. /*!*/;
```

该部分为 Query_log_event 类型的事件。

第 1 行：请参考本小节"第 2 部分"中 Format_description_event 事件对应部分的解释。

第 2 行中：

- 关于"190519 10:07:49 server id 330657 end_log_pos 541"的含义，请参考上文中"第 2 部分"Format_description_event 事件对应部分的解释。
- "Query"表示该事件类型为 Quert_log_event，这个事件记录了执行的语句，在 RBR

（基于 row 格式复制）模式下，一般为 DDL 语句。

- "thread_id = 3960"表示执行该语句的线程 ID 为 3960，该 thread_id 与 SHOW PROCESSLIST 语句结果中的 ID 字段一致。
- "exec_time=0"表示该语句的执行时间。
- "error_code=0"表示该语句执行后返回的错误编号。

第 3 行中"SET TIMESTAMP=1558274869"语句将当前时间戳修改为事件记录的时间。

第 4~5 行为 Query_log_event 中记录的具体执行语句。

第 8 部分

```
1. # at 541
2. #190519 10:07:49 server id 330657  end_log_pos 608 Rows_query
3. # insert into test(`name`) values('gangshen_row')
```

该部分为 Rows_query_log_event 类型的事件。在正常的 RBR 模式下，对于数据库记录的变更操作都以 row 格式记录，无法看到变更操作的原始语句，但可以通过设置系统变量 binlog_rows_query_log_events 为 1，在二进制日志中记录原始语句，这些语句会被记录到 Rows_query_log_event 类型的事件中。

第 1 行：请参考本小节"第 2 部分"中 Format_description_event 事件对应部分的解释。

第 2 行中：

- 关于"190519 10:07:49 server id 330657 end_log_pos 608"的含义，请参考本小节"第 2 部分"中 Format_description_event 事件对应部分的解释。
- Rows_query 表示该事件的类型为 Rows_query_logs_event，记录数据库变更操作的原始语句。

第 3 行为数据库变更操作的原始语句内容，可以看到语句是以"#"注释的，在使用二进制日志进行数据恢复时，不会执行该语句，防止重复应用。

第 9 部分

```
1. # at 608
2. #190519 10:07:49 server id 330657  end_log_pos 658 Table_map: `gangshen`.`test` mapped
to number 111
```

该部分为 Table_map_event 类型的事件，是 RBR 模式下独有的事件。在 RBR 模式下，一个事务在二进制日志中由 Table_map_event、Write_rows_log_event（对应于 INSERT 操作）、Update_rows_log_event（对应于 UPDATE 操作）、Delete_rows_log_event（对应于 DELETE 操作），以及 Xid_event 等事件组成。Table_map_event 事件记录了发生变更操作的表的结构，以便后续的 Write_rows_log_event/Update_rows_log_event/Delete_rows_log_event 事件应用变更数据时使用。

第 1 行：请参考本小节"第 2 部分"中 Format_description_event 事件对应部分的解释。

第 2 行中：

- 关于"190519 10:07:49 server id 330657 end_log_pos 658"的含义，请参考本小节"第 2 部分"中 Format_description_event 事件对应部分的解释。
- "Table_map"表示该事件类型为 Table_map_event，记录变更表的表结构信息。
- "gangshen.test mapped to number 111"表示发生变更操作的表为 gangshen 库下的 test 表，其对应的 table_id 为 111，在后面的 Write_rows_log_event、Update_rows_log_event 和 Delete_rows_log_event 事件中就是通过 table_id 寻找对应的操作表的。

注意：实际上 Table_map_event 事件中记录的信息不限于上述内容，它还会记录数据库名、表名、字段数量、字段类型以及字段的一些元信息（如精度、是否允许为 NULL 等）。但是 mysqlbinlog 工具的解析结果中，这些信息并没有展示在 Table_map_event 部分，而是体现在后面的 Write_rows_log_event、Update_rows_log_event、Delete_rows_log_event 中，后续会有说明。

第 10 部分

```
1. # at 658
2. #190519 10:07:49 server id 330657 end_log_pos 707 Write_rows: table id 111 flags: STMT_END_F
3. BINLOG '
4. NWPhXB2hCwUAQwAAAGACAACAAC9pbnNlcnQgaW50byBOZXN0KGBuYWllYCkgdmFsdWVzKCdnYW5n
5. c2hlbl9yb3cnKQ==
6. NWPhXBOhCwUAMgAAAJICAAAAAG8AAAAAAEACGdhbmdzaGVuAAR0ZXN0AAIDDwI8AAI=
7. NWPhXB6hCwUAMQAAAMMCAAAAG8AAAAAAEAAgAC//wBAAAADGdhbmdzaGVuAVUX3Jvdw==
8. '/*!*/;
9. ### INSERT INTO `gangshen`.`test`
10. ### SET
11. ###   @1=1 /* INT meta=0 nullable=0 is_null=0 */
12. ###   @2='gangshen_row' /* VARSTRING(60) meta=60 nullable=1 is_null=0 */
```

该部分为 Write_rows_log_event 类型的事件。在 RBR 模式下，对数据库的 INSERT 操作都会以 Write_rows_log_event 事件记录其插入的实际数据。

第 1 行：请参本小节"第 2 部分"中 Format_description_event 事件对应部分的解释。

第 2 行中：

- 关于"190519 10:07:49 server id 330657 end_log_pos 707"的含义，请参考本小节"第 2 部分"中 Format_description_event 事件对应部分的解释。
- "Write_rows"表示该事件类型为 Write_rows_log_event，记录 INSERT 操作的实际数据。
- "table id 111"表示 INSERT 插入的表的 table_id 为 111，Write_rows_log_event 事件通过 table_id 与上一个在二进制日志文件中出现的 table_id 为 111 的 Table_map_event 事件进行关联，获取 INSERT 操作的表结构信息。

- "flags: STMT_END_F"表示该事件是 INSERT 语句中的最后一个 Write_rows_log_event 事件。当 INSERT 语句插入多行记录时，MySQL 会根据大小将多行记录的变更拆分成多个 Write_rows_log_event 事件，STMT_END_F 标志就表示当前 Write_rows_log_event 事件已经为拆分的事件中的最后一个。

第 3~8 行是 Write_rows_log_event 事件的原始二进制数据，若希望 mysqlbinlog 工具解析的结果中不包含该部分，可以在使用 mysqlbinlog 工具时加上"--base64-output=decode-rows"选项。

第 9~12 行记录的是 INSERT 语句插入的具体数据，在二进制日志中使用"@1, @2 ... @n"表示字段，因为在 Table_map_event 事件中没有记录字段的具体名称，仅记录了表中字段的数量以及依次记录了字段类型，所以在解析 Write_rows_log_event 事件时，也是按照字段顺序依次进行的。可以通过字段序号与 information_schema.columns 表中 ordinal_position 字段对应的方式，查找字段名称，前提是在从该事件写入二进制日志到查询 information_schem.columns 表期间，没有使用 DDL 语句变更过该表的字段顺序。

- "### @1=1 /* INT meta=0 nullable=0 is_null=0 */"表示 INSERT 插入值时，对 gangshen.test 表第一个字段插入的值为 1。/*...*/注释部分的内容为第一个字段的元信息，INT 表示字段类型为 INT，meta = 0 表示字段元数据为 0（INT 类型字段没有元数据），nullable = 0 表示该字段不允许为 NULL，is_null = 0 表示该字段插入的值不为 NULL。注释部分的内容实际记录在 Table_map_event 事件中，mysqlbinlog 工具解析时将其写到 Write_rows_log_event 事件部分，是为了便于查看。

- "### @2='gangshen_row' /* VARSTRING(60) meta=60 nullable=1 is_null=0*/"表示 INSERT 插入值时，对 gangshen.test 表第二个字段插入的值为 gangshen_row。/*...*/注释部分的内容为第二个字段的元信息，VARSTING(60) 表示字段类型为 VARSTRING（对应 varchar 类型），且长度为 60；meta=60 表示字段元数据为 60（varchar 类型字段元数据记录的是该字段最终占用的字节数）；nullable = 1 表示该字段允许为 NULL；is_nul = 0 表示该字段插入的值不为 NULL。

第 11 部分

```
1. # at 707
2. #190519 10:07:49 server id 330657 end_log_pos 734 Xid = 106
3. COMMIT/*!*/;
```

该部分为 Xid_event 类型的事件，Xid_event 表示事务提交。

第 1 行：请参本小节"第 2 部分"中 Format_description_event 事件对应部分的解释。

第 2 行中：

- 关于"190519 10:07:49 server id 330657 end_log_pos 734"的含义，请参考本小节"第 2 部分"中 Format_description_event 事件对应部分的解释。

- Xid 表示该事件类型为 Xid_event，106 表示该事务的 xid 为 106。在 MySQL 中，每

一个语句都会被分配一个 query_id，query_id 是全局自增的，并且 MySQL 实例重启时会清零。一个事务中第一个语句的 query_id 会被作为该事务的 xid。在 MySQL 异常恢复阶段，MySQL 会解析 redo 日志中 prepare 状态的事务，得到 xid，然后根据 xid 在二进制日志中查找，如果找到匹配该 xid 的 Xid_event，则对事务重新持久化，否则回滚事务。

第 3 行的 COMMIT 语句表示提交事务。

第 12 部分

```
# at 734
#190519 10:07:55 server id 330657 end_log_pos 795 GTID last_committed=2  sequence_number=3    rbr_only=yes
/*!50718 SET TRANSACTION ISOLATION LEVEL READ COMMITTED*//*!*/;
SET @@SESSION.GTID_NEXT= '16b154ef-7a12-11e9-b850-0800276d49d5:20'/*!*/;
```

该部分为 Gtid_log_event 类型的事件，请参考本小节"第 4 部分"的解释。

第 13 部分

```
# at 795
#190519 10:07:55 server id 330657 end_log_pos 872 Query thread_id=3960  exec_time=0 error_code=0
SET TIMESTAMP=1558274875/*!*/;
BEGIN
```

该部分为 Query_log_event 类型的事件，参考本小节"第 5 部分"的解释。

第 14 部分

```
/*!*/;
# at 872
#190519 10:07:55 server id 330657 end_log_pos 961 Rows_query
# update test set name = 'gangshen_row_new' where name = 'gangshen_row'
```

该部分为 Rows_query_log_event 类型的事件，参考本小节"第 8 部分"的解释。

第 15 部分

```
# at 961
#190519 10:07:55 server id 330657 end_log_pos 1011 Table_map: `gangshen`.`test` mapped to number 111
```

该部分为 Table_map_event 类型的事件，参考本小节"第 5 部分"的解释。

```
1. # at 1011
2. #190519 10:07:55 server id 330657 end_log_pos 1083 Update_rows: table id 111 flags: STMT_END_F
3. BINLOG '
4. O2PhXB2hCwUAWQAAAMEDAACAAEV1cGRhdGUgdGVzdCBzZXQgbmFtZSA9ICdnYW5nc2hlbl9yb3df
5. bmV3JyB3aGVyZSBuYW1lID0gJ2dhbmdzaGVuX3Jvdyc=
6. O2PhXBOhCwUAMgAAAPMDAAAAAG8AAAAAAAEACGdhbmdzaGVuAAR0ZXN0AAIDDwI8AAI=
7. O2PhXB+hCwUASAAADsEAAAAG8AAAAAAAEAAgAC///8AQAAAAxnYW5nc2hlbl9yb3b3f8AQAAABBn
```

```
8.  YW5nc2hlbl9yb3dfbmV3
9.  '/*!*/;
10. ### UPDATE `gangshen`.`test`
11. ### WHERE
12. ###   @1=1 /* INT meta=0 nullable=0 is_null=0 */
13. ###   @2='gangshen_row' /* VARSTRING(60) meta=60 nullable=1 is_null=0 */
14. ### SET
15. ###   @1=1 /* INT meta=0 nullable=0 is_null=0 */
16. ###   @2='gangshen_row_new' /* VARSTRING(60) meta=60 nullable=1 is_null=0 */
```

该部分为 Update_rows_log_event 类型的事件。在 RBR 模式下，对数据库的 UPDATE 操作都会用 Update_rows_log_event 事件记录下其更新的实际数据。

第 1 行：请参考本小节"第 2 部分"中 Format_description_event 事件对应部分的解释。

第 2 行中：

- 关于"190519 10:07:55 server id 330657 end_log_pos 1083"的含义，请参照本小节"第 2 部分"中 Format_description_event 事件对应部分的解释。

- "Update_rows"表示该事件的类型为 Update_rows_log_event，记录 UPDATE 操作的实际数据。

- "table id 111"表示 UPDATE 插入的表的 table_id 为 111，Update_rows_log_event 事件通过 table_id 与上一个在二进制日志中出现的 table_id 为 111 的 Table_map_event 事件进行关联，获取 UPDATE 操作的表结构信息。

- "flags: STMT_END_F"表示该事件是 UPDATE 语句中的最后一个 Update_rows_log_event 事件。当 UPDATE 语句更新多行记录时，MySQL 会根据一定大小将多行记录的变更拆分成多个 Update_rows_log_event 事件，STMT_END_F 标志表示当前 Update_rows_log_event 事件已经为拆分的事件中的最后一个。

第 3～9 行为 Update_rows_log_event 事件的原始二进制数据。若希望 mysqlbinlog 工具解析的结果中不包含该部分，可以在使用 mysqlbinlog 工具时加上"--base64-output=decode-rows"选项。

第 10~16 行记录的是 UPDATE 语句更新的具体数据，紧接着 WHERE 的为更新前的数据，紧接着 SET 的为更新后的数据。复制时，MySQL 根据更新前的数据到数据库中查找，如找到匹配的行，则用更新后的数据覆盖。WHERE 部分和 SET 部分的数据，与 Write_rows_log_event 事件中数据的表示方法一致。

第 16 部分

```
# at 1083
#190519 10:07:55 server id 330657 end_log_pos 1110 Xid = 107
COMMIT/*!*/;
```

该部分为 Xid_event 类型的事件，请参考本小节"第 11 部分"的解释。

第 17 部分

```
# at 1110
#190519  10:08:12  server  id  330657  end_log_pos  1171  GTID  last_committed=3
sequence_number=4  rbr_only=yes
/*!50718 SET TRANSACTION ISOLATION LEVEL READ COMMITTED*//*!*/;
SET @@SESSION.GTID_NEXT= '16b154ef-7a12-11e9-b850-0800276d49d5:21'/*!*/;
```

该部分为 Gtid_log_event 类型的事件,请参考本小节"第 4 部分"的解释。

第 18 部分

```
# at 1171
#190519 10:08:12 server id 330657 end_log_pos 1248 Query    thread_id=3960  exec_time=0
error_code=0
SET TIMESTAMP=1558274892/*!*/;
BEGIN
/*!*/;
```

该部分为 Query_log_event 类型的事件,请参考本小节中"第 5 部分"的解释。

第 19 部分

```
# at 1248
#190519 10:08:12 server id 330657 end_log_pos 1316 Rows_query
# DELETE FROM test WHERE name = 'gangshen_row_new'
```

该部分为 Rows_query_log_event 类型的事件,请参考本小节"第 8 部分"中的解释。

第 20 部分

```
# at 1316
#190519 10:08:12 server id 330657 end_log_pos 1366 Table_map: `gangshen`.`test` mapped
to number 111
```

该部分为 Table_map_event 类型的事件,请参考本小节"第 9 部分"中的解释。

第 21 部分

```
1. # at 1366
2. #190519 10:08:12 server id 330657 end_log_pos 1419 Delete_rows: table id 111 flags:
STMT_END_F
3. BINLOG '
4. TGPhXB2hCwUARAAAACQFAACAADBkZWxldGUgZnJvbSB0ZXN0IHdoZXJlIG5hbWUgPSAnZ2FuZ3No
5. ZW5fcm93X25ldyc=
6. TGPhXBOhCwUAMgAAAFYFAAAAG8AAAAAAAEACGdhbmdzaGVuAAR0ZXN0AAIDDwI8AAI=
7. TGPhXCChCwUANQAAAIsFAAAAAG8AAAAAAAEAAgAC//wBAAAAAEGdhbmdzaGVuAAX3Jvd19uZXXc=
8. '/*!*/;
9. ### DELETE FROM `gangshen`.`test`
10. ### WHERE
11. ###   @1=1 /* INT meta=0 nullable=0 is_null=0 */
12. ###   @2='gangshen_row_new' /* VARSTRING(60) meta=60 nullable=1 is_null=0 */
```

该部分为 Delete_rows_log_event 类型的事件。在 RBR 模式下,对数据库的 DELETE 操作都会以 Delete_rows_log_event 事件记录下其删除的实际数据。

第 1 行：请参考本小节"第 2 部分"中 Format_description_event 事件对应部分的解释。

第 2 行中：

- 关于"190519 10:08:12 server id 330657 end_log_pos 1366"的含义，请参本小节"第 2 部分"中 Format_description_event 事件对应部分的解释。

- "Delete_rows"表示该事件的类型为 Delete_rows_log_event，记录 DELETE 操作的实际数据。

- "table id 111"表示 DELETE 删除的表的 table_id 为 111，Delete_rows_log_event 通过 table_id 与最近的 table_id 为 111 的 Table_map_event 进行关联，获取 DELETE 操作的表的结构信息。

- "flags: STMT_END_F"表示该事件是 DELETE 语句中的最后一个 Delete_rows_log_event 事件。当 DELETE 语句删除多行记录时，MySQL 会根据一定大小将多行记录的删除拆分成多个 Delete_rows_log_event 事件，STMT_END_F 标志表示当前 Delete_rows_log_event 事件已经为拆分的事件中的最后一个。

第 3～8 行为 Delete_rows_log_event 事件的原始二进制数据。若希望 mysqlbinlog 工具解析的结果中不包含该部分，可以在使用 mysqlbinlog 工具时加上"--base64-output=decode-rows"选项。

第 9~12 行记录的是 DELETE 语句删除的具体数据，紧接着 WHERE 的部分为删除的记录行的数据。复制时，MySQL 根据记录的行数据到数据库中查找，若找到匹配的行，则将其删除。WHERE 部分的数据与 Write_rows_log_event 事件中数据的表示方法一致。

第 22 部分

```
# at 1419
#190519 10:08:12 server id 330657 end_log_pos 1446 Xid = 108
COMMIT/*!*/;
```

该部分为 Xid_event 事件，请参考本小节"第 11 部分"中的解释。

第 23 部分

```
1. # at 1446
2. #190519 10:08:18 server id 330657 end_log_pos 1489 Rotate to mysql-bin.000009 pos:
4
```

该部分为 Rotate_event 类型的事件，表示二进制日志已经到了结尾。

第 1 行：请参考本小节"第 2 部分"中 Format_description_event 事件对应部分的解释。

第 2 行中：

- 关于"190519 10:08:18 server id 330657 end_log_pos 1489"的含义，请参照本小节"第 2 部分"中 Format_description_event 事件对应部分的解释。

- "Rotate"表示该事件的类型为 Rotate_event，表示二进制日志的切换，是当前二进制日志文件的最后一个事件。

- "to mysql-bin.000009 pos:4"表示下一个二进制日志文件名为 mysql-bin.000009，并且下一个二进制日志文件中第一个事件的起始位置为 4。

第 24 部分

```
SET @@SESSION.GTID_NEXT= 'AUTOMATIC' /* added by mysqlbinlog */ /*!*/;
DELIMITER ;
# End of log file
/*!50003 SET COMPLETION_TYPE=@OLD_COMPLETION_TYPE*/;
/*!50530 SET @@SESSION.PSEUDO_SLAVE_MODE=0*/;
```

该部分内容为 mysqlbinlog 工具在解析时添加的，并非二进制日志文件中存储的信息。

- "SET @@SESSION.GTID_NEXT= 'AUTOMATIC' /* added by mysqlbinlog */ /*!*/;"表示将生成 GTID 的方式修改为自动，防止用户手动更改 gtid_next 变量后影响后面生成 GTID。
- "DELIMITER ;"设置结束符为";"。
- "# End of log file"表示已经是二进制日志文件的末尾。
- "/*!50003 SET COMPLETION_TYPE=@OLD_COMPLETION_TYPE*/;"将 completion_type 变量修改为最初的值，与起始部分对应。
- "/*!50530 SET @@SESSION.PSEUDO_SLAVE_MODE=0*/;"恢复 pseudo_slave_mode 变量的值。

26.3.2 基于 statement 的复制的二进制日志内容解析

26.3.2.1 数据库操作记录

参考 26.3.1.1 节对数据库的操作，不同之处在于将系统变量 binlog_format 的值设置为 statement。

注意：MySQL 在 READ-COMMITTED 隔离级别下，无法将系统变量 binlog_format 设置为 statement。

更准确地说，在此种隔离级别下，将系统变量 binlog_format 设置为 statement 后，执行 INSERT、UPDATE、DELETE 操作会报错：

```
  ERROR 1665 (HY000): Cannot execute statement: impossible to write to binary log since
BINLOG_FORMAT = STATEMENT and at least one table uses a storage engine limited to row-based
logging. InnoDB is limited to row-logging when transaction isolation level is READ COMMITTED
or READ UNCOMMITTED.
  Error (Code 1665): Cannot execute statement: impossible to write to binary log since
BINLOG_FORMAT = STATEMENT and at least one table uses a storage engine limited to row-based
logging. InnoDB is limited to row-logging when transaction isolation level is READ COMMITTED
or READ UNCOMMITTED.
```

```
Error (Code 1015): Can't lock file (errno: 170 - It is not possible to log this statement)
```
但 DDL 语句不受限制，因而不会报错。

所以，对数据库进行操作之前，建议通过系统变量 transaction_isolation 将事务的隔离级别修改为 REPEATABLE-READ。

26.3.2.2 使用 mysqlbinlog 工具解析

与 26.3.1.2 节中一致，使用相同的 mysqlbinlog 命令对二进制日志进行解析，只是解析的二进制日志文件名根据实际情况而定：

```
mysqlbinlog -vv mysql-bin.N
```

由于篇幅有限，而且 26.3.2.3 节会逐一解释解析的结果，所以本节没有列出 mysqlbinlog 工具解析的详细结果。建议读者在阅读 26.3.2.3 节之前，根据 26.3.2.1 节以及本节的介绍，亲自操作，以便对二进制日志有更完整的理解。

26.3.2.3 解析结果解释

第 1 部分

```
/*!50530 SET @@SESSION.PSEUDO_SLAVE_MODE=1*/;
/*!50003 SET @OLD_COMPLETION_TYPE=@@COMPLETION_TYPE,COMPLETION_TYPE=0*/;
DELIMITER /*!*/;
```

在解析的结果中，以上内容是由 mysqlbinlog 工具添加的，并非二进制日志文件中存储的信息。

- "SET @@session.pseudo_slave_mode=1" 设置系统变量值为 1，官方文档称该变量供 MySQL Server 内部使用，具体作用请自行研究。

- "SET @old_completion_type=@@completion_type,completion_type=0" 设置系统变量值为 0，表示关闭链式事务，在遇到 COMMIT 或者 ROLLBACK 语句时不会断开连接。当使用二进制日志文件进行数据恢复时，该系统变量的设置很重要。

- "DELIMITER /*!*/;" 设置结束符为 ";"，告诉 MySQL 当遇到 ";" 时，表示当前命令已结束。

第 2 部分

```
# at 4
#190525 12:22:06 server id 330657 end_log_pos 123 Start: binlog v 4, server v 5.7.22-log created 190525 12:22:06
BINLOG '
rmvpXA+hCwUAdwAAAHsAAAAAAAQANS43LjIyLWxvZwAAAAAAAAAAAAAAAAAAAAAAAAAAAAAA
AAAAAAAAAAAAAAAAAAAAAAEzgNAAgAEgAEBAQEEgAAXwAEGggAAAAICAgCAAAACgoKKioAEjQA
AGD0LHA=
'/*!*/;
```

该部分为 Format_description_event 类型的事件，与基于 row 的复制中的解释无区别，可参考 26.3.1.3 节的内容。

第 3 部分

```
# at 123
#190525 12:22:06 server id 330657 end_log_pos 190 Previous-GTIDs
# 16b154ef-7a12-11e9-b850-0800276d49d5:1-34
```

该部分为 Previous_gtid_log_event 类型的事件，与基于 row 的复制中的解释无区别，可参考 26.3.1.3 节的内容。

第 4 部分

```
# at 190
#190525 12:22:13 server id 330657 end_log_pos 251 GTID last_committed=0  sequence_number=1 rbr_only=no
SET @@SESSION.GTID_NEXT= '16b154ef-7a12-11e9-b850-0800276d49d5:35'/*!*/;
```

该部分为 Gtid_log_event 类型的事件，与基于 row 的复制中的解释无区别，可参考 26.3.1.3 节的内容。

第 5 部分

```
# at 251
#190525 12:22:13 server id 330657 end_log_pos 403 Query thread_id=8 exec_time=0 error_code=0
use `gangshen`/*!*/;
SET TIMESTAMP=1558801333/*!*/;
SET @@session.pseudo_thread_id=8/*!*/;
SET @@session.foreign_key_checks=1, @@session.sql_auto_is_null=0, @@session.unique_checks=1, @@session.autocommit=1/*!*/;
SET @@session.sql_mode=1073741824/*!*/;
SET @@session.auto_increment_increment=2, @@session.auto_increment_offset=1/*!*/;
/*!\C utf8 *//*!*/;
SET  @@session.character_set_client=33,@@session.collation_connection=33,@@session.collation_server=83/*!*/;
SET @@session.lc_time_names=0/*!*/;
SET @@session.collation_database=DEFAULT/*!*/;
create table test(id int primary key auto_increment,name varchar(20))
/*!*/;
```

该部分为 Query_log_event 类型的事件，与基于 row 的复制中的解释无区别，可参考 26.3.1.3 节的内容。

第 6 部分

```
# at 403
#190525 12:22:19 server id 330657 end_log_pos 464 GTID  last_committed=1 sequence_number=2 rbr_only=no
SET @@SESSION.GTID_NEXT= '16b154ef-7a12-11e9-b850-0800276d49d5:36'/*!*/;
```

该部分为 Gtid_log_event 类型的事件，与基于 row 的复制中的解释无区别，可参考 26.3.1.3 节的内容。

注意：上述 Gtid_log_event 事件与 26.3.1.3 节中 INSERT 语句对应的有一点不同，后者 rbr_only 的值为 yes，而这里 rbr_only 的值是 no。

第 7 部分

```
# at 464
#190525 12:22:19 server id 330657 end_log_pos 552 Query thread_id=8 exec_time=0 error_code=0
SET TIMESTAMP=1558801339/*!*/;
BEGIN
/*!*/;
```

该部分为 Query_log_event 类型的事件，与基于 row 的复制中的解释无区别，可参考 26.3.1.3 节的内容。

第 8 部分

```
1. # at 552
2. # at 580
3. #190525 12:22:19 server id 330657 end_log_pos 580 Intvar
4. SET INSERT_ID=1/*!*/;
5. #190525 12:22:19 server id 330657 end_log_pos 710 Query thread_id=8 exec_time=0 error_code=0
6. SET TIMESTAMP=1558801339/*!*/;
7. INSERT INTO test(`name`) VALUES('gangshen_row')
8. /*!*/;
```

该部分为 Intvar_event 和 Query_log_event 类型的事件。前者只会在基于 statement 的复制模式下出现。在两种情况下会出现 Intvar_event：一种情况是 INSERT 语句插入数据时，如果表中有带 AUTO_INCREMENT 属性的字段且 INSERT 语句没有指定该字段值，MySQL 会生成一个自增值，这个自增值会以 INSERT_ID 形式被记录在 Intvar_event 中；另一种情况是 DML 操作中使用了 LAST_INSERT_ID()函数，MySQL 会把该函数的值以 LAST_INSERT_ID 形式记录在 Intvar_event 中。

设计 Intvar_event 的原因是为了保证基于 statement 的复制模式下数据的正确性，如果没有记录 INSERT_ID 或者 LAST_INSERT_ID，当主库和从库采用不同的自增配置时，会导致数据不正确。在基于 row 的复制模式下，因为二进制日志中记录的是整行的变化，可以不依靠 Intvar_event 事件保证数据正确性。

第 1 行和第 2 行：由于每个事件中该部分信息代表的含义相同，对于该部分的解释，请参考 26.3.1.3 节中 Format_description_event 事件的对应部分。

第 3 行中：

- 关于 "190525 12:22:19 server id 330657 end_log_pos 580" 的解释，请参考 26.3.1.3 节中 Format_description_event 事件的对应部分。
- "Intvar" 表示该事件的类型为 Intvar_event，记录 INSERT_ID 或 LAST_INSERT_ID 值，这里记录的是 INSERT_ID 值。

- 第 4 行设置 INSERT_ID 的值为 1。

第 5 行中:

- 关于"190525 12:22:19 server id 330657 end_log_pos 710"的解释,请参考 26.3.1.3 节中 Format_description_event 事件的对应部分。
- "Query"表示该事件类型为 Quert_log_event,该事件记录了执行的语句。
- "thread_id = 3960"表示执行该语句的线程 ID 为 8,该 thread_id 与 SHOW PROCESSLIST 语句输出结果中的 ID 字段一致。
- "exec_time=0"表示该语句的执行时间。
- "error_code=0"表示该语句执行后返回的错误编号。
- 第 6 行的 SET TIMESTAMP=xxxx 语句将当前时间戳修改为事件记录的时间。

第 7 行为 Query_log_event 事件中记录的具体执行语句。

第 9 部分

```
# at 710
#190525 12:22:19 server id 330657 end_log_pos 737 Xid = 99
COMMIT/*!*/;
```

该部分为 Xid_event 类型的事件,与基于 row 的复制中的解释无区别,可参考 26.3.1.3 节的内容。

第 10 部分

```
# at 737
#190525 12:22:25 server id 330657 end_log_pos 798 GTID last_committed=2    sequence_number=3    rbr_only=no
SET @@SESSION.GTID_NEXT= '16b154ef-7a12-11e9-b850-0800276d49d5:37'/*!*/;
```

该部分为 Gtid_log_event 类型的事件,与基于 row 的复制中的解释无区别,可参考 26.3.1.3 节的内容。

第 11 部分

```
# at 798
#190525 12:22:25 server id 330657 end_log_pos 886 Query thread_id=8 exec_time=0 error_code=0
SET TIMESTAMP=1558801345/*!*/;
BEGIN
/*!*/;
```

该部分为 Query_log_event 类型的事件,与基于 row 的复制中的解释无区别,可参考 26.3.1.3 节的内容。

第 12 部分

```
# at 886
#190525 12:22:25 server id 330657 end_log_pos 1038 Query    thread_id=8 exec_time=0
```

```
error_code=0
   SET TIMESTAMP=1558801345/*!*/;
   update test set name = 'gangshen_row_new' where name = 'gangshen_row'
   /*!*/;
```

该部分为 Query_log_event 类型的事件,与基于 row 的复制中的解释无区别,可参考 26.3.1.3 节的内容。

第 13 部分

```
# at 1038
#190525 12:22:25 server id 330657 end_log_pos 1065  Xid = 100
COMMIT/*!*/;
```

该部分为 Xid_event 类型的事件,与基于 row 的复制中的解释无区别,可参考 26.3.1.3 节的内容。

第 14 部分

```
# at 1065
#190525 12:22:34 server id 330657 end_log_pos 1126 GTID last_committed=3    sequence_
number=4    rbr_only=no
SET @@SESSION.GTID_NEXT= '16b154ef-7a12-11e9-b850-0800276d49d5:38'/*!*/;
```

该部分为 Gtid_log_event 类型的事件,与基于 row 的复制中的解释无区别,可参考 26.3.1.3 节的内容。

第 15 部分

```
# at 1126
#190525 12:22:34 server id 330657 end_log_pos 1214   Query    thread_id=8 exec_time=0
error_code=0
SET TIMESTAMP=1558801354/*!*/;
BEGIN
/*!*/;
```

该部分为 Query_log_event 类型的事件,与基于 row 的复制中的解释无区别,可参考 26.3.1.3 节的内容。

第 16 部分

```
# at 1214
#190525 12:22:34 server id 330657 end_log_pos 1345   Query    thread_id=8 exec_time=0
error_code=0
   SET TIMESTAMP=1558801354/*!*/;
   delete from test where name = 'gangshen_row_new'
   /*!*/;
```

该部分为 Query_log_event 类型的事件,与基于 row 的复制中的解释无区别,可参考 26.3.1.3 节的内容。

第 17 部分

```
# at 1345
```

```
#190525 12:22:34 server id 330657  end_log_pos 1372 Xid = 101
COMMIT/*!*/;
```

该部分为 Xid_event 类型的事件，与基于 row 的复制中的解释无区别，可参考 26.3.1.3 节的内容。

第 18 部分

```
1. # at 1372
2. #190525 12:22:42 server id 330657  end_log_pos 1415 Rotate to mysql-bin.000019 pos: 4
```

该部分为 Rotate_event 类型的事件，表示二进制日志已经到了结尾。

第 1 行：请参考 26.3.1.3 节中 Format_description_event 事件的对应部分。

第 2 行中：

- 关于 "190525 12:22:42 server id 330657 end_log_pos 1415" 的解释，请参考 26.3.1.3 节中 Format_description_event 事件的对应部分。

- "Rotate" 表示该事件类型为 Rotate_event，表示二进制日志的切换，是当前二进制日志文件的最后一个事件。

- "to mysql-bin.000019 pos:4" 表示下一个二进制日志文件名为 mysql-bin.000019，并且下一个二进制日志文件中第一个事件的起始位置为 4。

第 19 部分

```
SET @@SESSION.GTID_NEXT= 'AUTOMATIC' /* added by mysqlbinlog */ /*!*/;
DELIMITER ;
# End of log file
/*!50003 SET COMPLETION_TYPE=@OLD_COMPLETION_TYPE*/;
/*!50530 SET @@SESSION.PSEUDO_SLAVE_MODE=0*/;
```

该部分内容为 mysqlbinlog 工具在解析时添加的，并非二进制日志文件中存储的信息。

- "SET @@SESSION.GTID_NEXT= 'AUTOMATIC' /* added by mysqlbinlog */ /*!*/;" 将 GTID 的生成方式改为自动生成，防止用户手动更改 gtid_next 变量后影响后续生成 GTID。

- "DELIMITER ;" 设置结束符为 ";"。

- "# End of log file" 表示已经是二进制日志文件的末尾。

- "/*!50003 SET COMPLETION_TYPE=@OLD_COMPLETION_TYPE*/;" 将 completion_type 变量值修改为最初的值，与起始部分对应。

- "/*!50530 SET @@SESSION.PSEUDO_SLAVE_MODE=0*/;" 恢复 pseudo_slave_mode 变量的值。

26.4 小结

通过 26.3.1 和 26.3.2 节，我们了解了常见的 DDL、INSERT、UPDATE、DELETE 语句在基于 row 以及基于 statement 的复制格式下是如何保存在二进制日志中的。下面对本章内容进行总结：

- 在基于 row 的复制格式下，一个事务由 Table_map_event、Write_rows_log_event/Update_rows_log_event/Delete_rows_log_event、Xid_event 组成。
- 在基于 statement 的复制格式下，一个事务由 Query_log_event（当遇到自增字段或使用 LAST_INSERT_ID()时包含 Intvar_event）、Xid_event 组成。
- 在基于 row 的复制格式下，可以设置系统变量 binlog_rows_query_log_event 等于 1，将原始的 DML 语句记录到二进制日志中。
- 所有类型的事件都会记录该事件被写入二进制日志时的时间、生成事件的 MySQL 实例的 Server ID，以及事件在文件中的起始和结束位置。
- 相较而言，基于 statement 的复制格式生成的二进制日志文件占用的磁盘空间更少，复制时占用的网络带宽更小，因为它记录的是主库上执行的原始语句，而基于 row 的复制格式记录的是每一行发生的更改的信息。
- 基于 row 的复制格式比基于 statement 的复制格式更能保证主从库数据一致性。
- MySQL 通过在会话级别修改一些变量的值，保证了 Query_log_event 事件中存储的语句在从库重放后，主从库数据的一致性。

mysqlbinlog 工具解析二进制日志时，只展示一些必要的信息，实际上很多事件中存储的信息（比如事件的 checksum）并没有完全展示出来，但是在通过二进制日志进行数据恢复时，这些未显示的信息并不影响数据恢复的准确性。感兴趣的读者可以参考附录 A "二进制日志事件详解"。

第 27 章　常规 DDL 操作解析

MySQL 中的操作基本上可以分三大类：

第一类是 DML（Data Manipulation Language）操作，如 INSERT、UPDATE 和 DELETE 语句，有时候 SELECT 也被认为属于 DML 操作，因为 SELECT... FOR UPDATE 与 INSERT、UPDATE、DELETE 在加锁方面是一致的。

第二类是 DDL（Data Definition Language）操作，例如 CREATE、ALTER、DROP 和 TRUNCATE 语句等都属于 DDL 操作。

第三类是 DCL（Data Control Language）操作，例如 GRANT、REVOKE 等语句用于控制权限的语句都属于 DCL 操作。

第 26 章已经介绍了 MySQL 中常见 DML 操作（语句）在二进制日志中的存储形式，以及 MySQL 如何保证复制过程中 DML 操作的正确性，本章会介绍 MySQL 中常规 DDL 操作在二进制日志中的存储形式以及如何保证其复制时的正确性。

一些常规的 DDL 操作在二进制日志中的记录形式是类似的，所以本章把它们归在一起来说明，它们是：

- CREATE/DROP DATABASE
- CREATE/DROP TABLE
- ALTER TABLE ADD/DROP/MODIFY COLUMN
- ALTER TABLE ADD/DROP INDEX
- TRUNCATE TABLE

27.1　操作环境信息

1. 操作系统版本

   Red Hat Enterprise Linux Server release 7.4（Maipo）

2. 数据库版本

   5.7.22-log MySQL Community Server（GPL）

3. 目的

演示在不同事务隔离级别、系统变量 binlog_format 的不同设置下，常规 DDL 操作如何被记录到二进制日志文件中，如何保证常规 DDL 操作在复制时的正确性。

4. 数据库系统变量

gtid_mode=on enforce_gtid_consistency = true

27.2 常规 DDL 操作示例

为了确保结果的完备性，笔者分别在 READ-UNCOMMITTED、READ-COMMITTED、REPEATABLE-READ 以及 SERIALIZABLE 隔离级别下，将系统变量 binlog_format 值修改为 row、statement、mixed（后续章节中如无特别说明，均表示在以上 4 种隔离级别以及 3 种二进制日志格式下操作过），执行一组相同的 DDL 操作，并对产生的二进制日志文件进行解析，分析 DDL 操作在不同场景下产生的二进制日志是否有区别。

由于都是重复性操作，所以这里只给出隔离级别为 READ-UNCOMMITTED，binlog_format = row 的场景下的操作过程。想动手操作的读者，可以根据下面给出的语句，自行跟随操作。而 DDL 操作在二进制日志中的记录形式类似，此处仅以常见 DDL 操作作为示例：

```
create database gangshen;
use gangshen;
create table user(id int primary key auto_increment,name varchar(20));
insert into user(name) values('san.zhang'),('si.li'),('wu.wang'),('liu.zhao');
alter table user add column address varchar(50);
alter table user modify column address varchar(100);
alter table user drop column address;
truncate table user;
drop table user;
drop database gangshen;
flush logs;
```

以下是隔离级别为 READ-UNCOMMITTED，binlog_format = row 的场景下的操作过程。

（1）在会话级别将数据库隔离级别改为 READ-UNCOMMITTED，将系统变量 binlog_format 修改为 row，并刷新二进制日志：

```
mysql> set transaction_isolation = 'read-uncommitted';
Query OK, 0 rows affected (0.00 sec)

mysql> set binlog_format = 'row';
Query OK, 0 rows affected (0.00 sec)

mysql> show variables where variable_name in ('transaction_isolation','binlog_format')\G
*************************** 1. row ***************************
Variable_name: binlog_format
```

```
        Value: ROW
*************************** 2. row ***************************
Variable_name: transaction_isolation
        Value: READ-UNCOMMITTED
2 rows in set (0.00 sec)

mysql> flush logs;
Query OK, 0 rows affected (0.02 sec)
```

（2）在数据库中执行常规 DDL 操作：

```
mysql> create database gangshen;
Query OK, 1 row affected (0.01 sec)

mysql> use gangshen;
Database changed

mysql> create table user(id int priamry key auto_increment,name varchar(20));
Query OK, 0 rows affected (0.01 sec)

mysql> insert into user(name) values('san.zhang'),('si.li'),('wu.wang'),('liu.zhao');
Query OK, 4 rows affected (0.01 sec)
Records: 4  Duplicates: 0  Warnings: 0

mysql> alter table user add column address varchar(50);
Query OK, 0 rows affected (0.05 sec)
Records: 0  Duplicates: 0  Warnings: 0

mysql> alter table user modify column address varchar(100);
Query OK, 4 rows affected (0.03 sec)
Records: 4  Duplicates: 0  Warnings: 0

mysql> alter table user drop column address;
Query OK, 0 rows affected (0.06 sec)
Records: 0  Duplicates: 0  Warnings: 0

mysql> truncate table user;
Query OK, 0 rows affected (0.01 sec)

mysql> drop table user;
Query OK, 0 rows affected (0.00 sec)

mysql> drop database gangshen;
Query OK, 0 rows affected (0.01 sec)

mysql> flush logs;
Query OK, 0 rows affected (0.01 sec)
```

所有场景中均按照以上步骤操作，只有系统变量 binlog_format 以及 transaction_isolation 不同。

注意：在 READ-UNCOMMITTED 以及 READ-COMMITTED 隔离级别下，将系统变量 binlog_format 设置为 statement 时，执行 DML 操作会有如下报错信息：

```
ERROR 1665 (HY000): Cannot execute statement: impossible to write to binary log since
BINLOG_FORMAT = STATEMENT and at least one table uses a storage engine limited to row-based
logging. InnoDB is limited to row-logging when transaction isolation level is READ COMMITTED
or READ UNCOMMITTED.
    Error (Code 1665): Cannot execute statement: impossible to write to binary log since
BINLOG_FORMAT = STATEMENT and at least one table uses a storage engine limited to row-based
logging. InnoDB is limited to row-logging when transaction isolation level is READ COMMITTED
or READ UNCOMMITTED.
    Error (Code 1015): Can't lock file (errno: 170 - It is not possible to log this statement)
```

错误信息的大意为：对于 InnoDB 存储引擎，在 READ-COMMITTED 以及 READ- UNCOMMITTED 隔离级别下，二进制日志格式不能设置为 statement，MySQL 为了保证数据复制时的正确性，设定了该限制。

那么，为什么在 READ-COMMITTED 以及 READ-UNCOMMITTED 隔离级别下，二进制日志格式不能设置为 statement？

因为在 READ-UNCOMMITTED 隔离级别下可以看到其他事务还未提交的数据（即脏读），在 READ-COMMITTED 隔离级别下，一个先开启的事务在自身还未提交时可以看到其他后开启的事务已提交的数据（即不可重复读），存在脏读、不可重复读以及幻读的问题，会导致 MySQL 在主从复制时发生主从库数据不一致的情况。

假设一个场景：在主库 master 上存在一张表 t，t 表中有(a,b)两个字段，有(1,2)、(2,1)两行记录。这时有两个会话连上 master，会话 1 显式开启一个事务（使用 BEGIN 或者 START TRANSACTION 语句），然后执行更新操作 "UPDATE t SET a=5 WHERE b = 2;" 执行完之后不提交事务。此时，会话 2 显式开启一个事务（使用 BEGIN 或者 START TRANSACTION 语句），执行更新操作 "UPDATE t SET b = 2 WHERE a = 2;"，接着会话 2 先提交事务，会话 1 后提交。此时主库 master 上 t 表的记录为(5,2)、(2,2)。事务写入二进制日志并在从库 slave 上回放，由于二进制日志在事务提交时才会记录，所以会话 2 的事务先写入二进制日志中，会话 1 的事务后写入。

在从库 slave 上回放二进制日志时，会先回放会话 2 的事务，再回放会话 1 的。如果 binlog_format = statement，在从库上等于先执行 "UPDATE t SET b = 2 WHERE a = 2"，后执行 "UDPATE t SET a = 5 WHERE b = 2"，则在从库中最终 t 表上的记录为(5,2)和(5,2)，这样主从库的数据就不一致了。

MySQL 为了避免这种情况导致主从库数据不一致，设定在 READ-COMMITTED 以及 READ-UNCOMMITTED 隔离级别下，当系统变量 binlog_format 设置为 statement 时，执行 DML 操作会报错。

在 REPEATABLE-READ 以及 SERIALIZABLE 隔离级别下，MySQL 在锁定更新的记录行之外还加了 GAP Lock，解决了不可重复读的问题，关于 GAP Lock 是如何解决不可重复读的问题的，这里不展开介绍，感兴趣的读者可以自行研究。

27.3 二进制日志内容解析

参照 26.3.1.2 节中的方式，对各个场景下产生的二进制日志进行解析：

```
[root@localhost ~]# mysqlbinlog -vv mysql-bin.000032
```

经过分析，发现演示过程中使用的常规 DDL 操作，在不同场景下产生的二进制日志形式没有区别，都是以 Query_log_event 事件形式记录在二进制日志中。所以，本章只对 READ-UNCOMMITTED 隔离级别下，binlog_format = row 的场景产生的二进制日志进行解释，对于其余场景的二进制日志不重复解释，感兴趣的读者可以自己动手操作。

关于 READ-UNCOMMITTED 隔离级别下，binlog_format = row 的场景产生的二进制日志中，Format_description_event、Gtid_log_event、Previous_Gtid_log_event 等事件内容的解析，请参考第 26 章内容，本章只对常规 DDL 操作进行解释。

```
......
# at 190
#190615 23:54:25 server id 330657 end_log_pos 251 GTID last_committed=0    sequence_number=1   rbr_only=no
SET @@SESSION.GTID_NEXT= '16b154ef-7a12-11e9-b850-0800276d49d5:151'/*!*/;
# at 251
#190615 23:54:25 server id 330657 end_log_pos 358 Query thread_id=8 exec_time=0 error_code=0
SET TIMESTAMP=1560657265/*!*/;
SET @@session.pseudo_thread_id=8/*!*/;
SET @@session.foreign_key_checks=1, @@session.sql_auto_is_null=0, @@session.unique_checks=1, @@session.autocommit=1/*!*/;
SET @@session.sql_mode=1073741824/*!*/;
SET @@session.auto_increment_increment=2, @@session.auto_increment_offset=1/*!*/;
/*!\C utf8 *//*!*/;
SET @@session.character_set_client=33,@@session.collation_connection=33,@@session.collation_server=83/*!*/;
SET @@session.lc_time_names=0/*!*/;
SET @@session.collation_database=DEFAULT/*!*/;
create database gangshen
/*!*/;
```

"create database gangshen" 语句产生的二进制日志，包含一个 Gtid_log_event 和一个 Query_log_event 事件。

```
# at 358
#190615 23:54:25 server id 330657 end_log_pos 419 GTID last_committed=1    sequence_number=2   rbr_only=no
SET @@SESSION.GTID_NEXT= '16b154ef-7a12-11e9-b850-0800276d49d5:152'/*!*/;
# at 419
#190615 23:54:25 server id 330657 end_log_pos 571 Query thread_id=8 exec_time=0 error_code=0
```

```
use `gangshen`/*!*/;
SET TIMESTAMP=1560657265/*!*/;
create table user(id int primary key auto_increment,name varchar(20))
/*!*/;
```

"create table user(id int primary key auto_increment, name varchar(20))"语句产生的二进制日志，包含一个 Gtid_log_event 和一个 Query_log_event 事件。

```
# at 967
#190615 23:54:25 server id 330657  end_log_pos 1028 GTID last_committed=3   sequence_number=4   rbr_only=no
    SET @@SESSION.GTID_NEXT= '16b154ef-7a12-11e9-b850-0800276d49d5:154'/*!*/;
# at 1028
#190615 23:54:25 server id 330657  end_log_pos 1158 Query    thread_id=8 exec_time=0 error_code=0
SET TIMESTAMP=1560657265/*!*/;
alter table user add column address varchar(50)
/*!*/;
```

"alter table user add column address varchar(50)"语句产生的二进制日志，包含一个 Gtid_log_event 和一个 Query_log_event 事件。

```
# at 1158
#190615 23:54:25 server id 330657  end_log_pos 1219 GTID last_committed=4   sequence_number=5   rbr_only=no
    SET @@SESSION.GTID_NEXT= '16b154ef-7a12-11e9-b850-0800276d49d5:155'/*!*/;
# at 1219
#190615 23:54:25 server id 330657 end_log_pos 1353 Query    thread_id=8 exec_time=0 error_code=0
SET TIMESTAMP=1560657265/*!*/;
alter table user modify column address varchar(100)
/*!*/;
```

"alter table user modify column address varchar(100)"语句产生的二进制日志，包含一个 Gtid_log_event 和一个 Query_log_event 事件。

```
# at 1353
#190615  23:54:25  server  id  330657  end_log_pos  1414  GTID  last_committed=5 sequence_number=6   rbr_only=no
    SET @@SESSION.GTID_NEXT= '16b154ef-7a12-11e9-b850-0800276d49d5:156'/*!*/;
# at 1414
#190615 23:54:25 server id 330657 end_log_pos 1533 Query    thread_id=8 exec_time=0 error_code=0
SET TIMESTAMP=1560657265/*!*/;
alter table user drop column address
/*!*/;
```

"alter table user drop column address"语句产生的二进制日志，包含一个 Gtid_log_event 和一个 Query_log_event 事件。

```
# at 1533
```

```
    #190615 23:54:25 server id 330657 end_log_pos 1594 GTID last_committed=6    sequence_
number=7    rbr_only=no
    SET @@SESSION.GTID_NEXT= '16b154ef-7a12-11e9-b850-0800276d49d5:157'/*!*/;
    # at 1594
    #190615 23:54:25 server id 330657 end_log_pos 1685 Query    thread_id=8 exec_time=0
error_code=0
    SET TIMESTAMP=1560657265/*!*/;
    truncate table user
    /*!*/;
```

"truncate table user"语句产生的二进制日志,包含一个 Gtid_log_event 和一个 Query_log_event 事件。

```
    # at 1685
    #190615 23:54:25 server id 330657 end_log_pos 1746 GTID last_committed=7    sequence_
number=8    rbr_only=no
    SET @@SESSION.GTID_NEXT= '16b154ef-7a12-11e9-b850-0800276d49d5:158'/*!*/;
    # at 1746
    #190615 23:54:25 server id 330657 end_log_pos 1872 Query    thread_id=8 exec_time=0
error_code=0
    SET TIMESTAMP=1560657265/*!*/;
    drop table `user` /* generated by server */
    /*!*/;
```

"drop table `user`"语句产生的二进制日志,包含一个 Gtid_log_event 和一个 Query_log_event 事件。

```
    # at 1872
    #190615 23:54:25 server id 330657 end_log_pos 1933 GTID last_committed=8    sequence_
number=9    rbr_only=no
    SET @@SESSION.GTID_NEXT= '16b154ef-7a12-11e9-b850-0800276d49d5:159'/*!*/;
    # at 1933
    #190615 23:54:25 server id 330657 end_log_pos 2027 Query    thread_id=8 exec_time=0
error_code=0
    SET TIMESTAMP=1560657265/*!*/;
    drop database gangshen
    /*!*/;
```

"drop database gangshen"语句产生的二进制日志,包含一个 Gtid_log_event 和一个 Query_log_event 事件。

从解析结果来看,常规 DDL 操作在二进制日志中都是以 Query_log_event 事件的形式记录的,Query_log_event 事件中记录了执行的原始语句形式,即在数据库中执行的语句,在二进制日志中也是以相同形式记录的。如果开启了 GTID,每个 DDL 操作都会分配一个 GTID,表示一个事务。

因为 DDL 操作都是对整表或者整库进行操作的,且执行完会自动提交,不受 BEGIN、COMMIT 等事务控制语句的控制,所以 DDL 操作以 Query_log_event 事件的形式存储,不会出现 27.2 节中提到的在 READ-UNCOMMITTED 和 READ-COMMITTED 隔离级别下

binlog_format 设置为 statement 时，允许执行 DML 操作导致的主从库数据不一致情况。

27.4 小结

下面对本章内容做一个小结：

- 常规 DDL 操作在二进制日志中是以 Query_log_event 事件的形式存储的，不受事务隔离级别以及系统变量 binlog_format 的影响。

- 在开启 GTID 的情况下，MySQL 会为每个 DDL 操作分配一个 GTID，表示一个事务。

- DDL 操作在执行完之后会自动提交，不受 BEGIN、COMMIT 等事务控制语句的控制。

- 在显式开启事务（使用 BEGIN 或者 START TRANSACTION 语句）的情况下，在事务中执行 DDL 操作后，会自动提交该事务。

- 在 READ-UNCOMMITTED 和 READ-COMMITTED 隔离级别下，当系统变量 binlog_format 设置为 statement 时，MySQL 不会执行 DML 操作，如果执行的话会有错误提示，这是为了确保主从库数据的一致性。

第 28 章　为何二进制日志中同一个事务的事件时间点会乱序

对于这个问题，相信很多人都只是大概知道其中的原理，并没有做过具体的案例分析。本章将以一个常见的 UPDATE 语句更新数据的事务作为案例，详细分析其中的过程。

28.1　操作环境信息

1. 操作系统版本

    CentOS release 6.5（Final）

2. MySQL 版本

    MySQL 5.7.20

3. MySQL 的关键参数

    innodb_flush_log_at_trx_commit=1

    sync_binlog=1

    binlog_format=row

    gtid_mode=on

    enforce_gtid_consistency=on

4. 演示过程所需数据

    利用 sysbench 制造一份用于测试的基础数据（可自行造数，只要一张表中有一条数据即可）。

28.2　验证前的准备

登录数据库，手动滚动二进制日志文件：

```
mysql> flush binary logs;select now();
Query OK, 0 rows affected (0.01 sec)
```

```
+---------------------+
| now()               |
+---------------------+
| 2018-10-04 15:31:55 |    # 滚动二进制日志文件的时间点为 2018-10-04 15:31:55
+---------------------+
1 row in set (0.00 sec)
```

解析最新的二进制日志文件内容及其对应的时间点信息：

```
[root@localhost binlog]# mysqlbinlog -vv mysql-bin.000460
mysqlbinlog: [Warning] unknown variable 'loose_default-character-set=utf8'
/*!50530 SET @@SESSION.PSEUDO_SLAVE_MODE=1*/;
/*!50003 SET @OLD_COMPLETION_TYPE=@@COMPLETION_TYPE,COMPLETION_TYPE=0*/;
DELIMITER /*!*/;
# at 4
#181004 15:31:55 server id 3306111  end_log_pos 123 CRC32 0x9cc61928  Start: binlog v 4, server v 5.7.20-log created 181004 15:31:55
# Warning: this binlog is either in use or was not closed properly.
BINLOG '
68G1Ww9/cjIAdwAAAHsAAAABAAQANS43LjIwLWxvZwAAAAAAAAAAAAAAAAAAAAAAAAAA
AAAAAAAAAAAAAAAAAAAAAEzgNAAgAEgAEBAQEEgAAXwAEGggAAAAICAgCAAAACgoKKioAEjQA
ASgZxpw=
'/*!*/;
# at 123
#181004 15:31:55 server id 3306111  end_log_pos 194 CRC32 0xc34422d7  Previous-GTIDs
# ec123678-5e26-11e7-9d38-000c295e08a0:182318-199336
SET @@SESSION.GTID_NEXT= 'AUTOMATIC' /* added by mysqlbinlog */ /*!*/;
DELIMITER ;
# End of log file
/*!50003 SET COMPLETION_TYPE=@OLD_COMPLETION_TYPE*/;
/*!50530 SET @@SESSION.PSEUDO_SLAVE_MODE=0*/;
```

从上述二进制日志的解析内容来看，当前最新的二进制日志文件中，除了文件头的事件之外，并没有其他用户数据产生的事件，而且文件头的所有事件的时间点都为 2018-10-04 15:31:55，这个时间点也是在数据库中手动执行二进制日志文件滚动语句的时间点。也就是说，二进制日志文件头的事件在二进制日志文件滚动时就已产生，且写入最新的二进制日志文件，这些事件与用户数据无关。

28.3 验证过程

接下来，我们模拟用户对数据执行 UPDATE 操作。为了验证的需要，我们使用了显式事务，并且在事务执行过程中人为加入了停顿，以便更清晰地看到二进制日志文件中的事件是如何被记录的。

首先，登录数据库，显式开启一个事务：

```
# 查询测试数据，以便后续编写 UPDATE 语句
```

```
mysql> select * from sbtest1 limit 1;
+----+---------+---------------------------------------------+-----+
| id | k | c | pad |
+----+---------+---------------------------------------------+-----+
| 21 | 2483477 | 09279210219-37745839908-56185699327-79477158641-86711242956-61449540392-
42622804506-61031512845-36718422840-11028803849 | xxx |
+----+---------+---------------------------------------------+-----+
1 row in set (0.00 sec)

# 显式开启一个事务
mysql> begin;select now();
Query OK, 0 rows affected (0.00 sec)

+---------------------+
| now()               |
+---------------------+
| 2018-10-04 15:33:02 |    # BEGIN 语句执行的时间点为 2018-10-04 15:33:02
+---------------------+
1 row in set (0.00 sec)
```

这个时候，二进制日志文件中还未记录新事务的二进制日志。等待几十秒之后，再在数据库中执行 UPDATE 语句修改 k 列值（这里是单行修改，且 sbtest1 表的 ID 列为主键，所以 UPDATE 语句的执行时间是极短的，可以忽略不计）：

```
mysql> update sbtest1 set k=k+1 where id=21;select now();
Query OK, 1 row affected (0.00 sec)
Rows matched: 1  Changed: 1  Warnings: 0

+---------------------+
| now()               |
+---------------------+
| 2018-10-04 15:35:21 |    # UPDATE 语句执行的时间点为 2018-10-04 15:35:21
+---------------------+
1 row in set (0.00 sec)
```

此时，二进制日志文件中仍未记录新事务的二进制日志（这是因为事务并没有提交。而在事务未提交的状态下，是不会有二进制日志被写入二进制日志文件的，只有事务执行提交时，二进制日志才会被写入二进制日志文件）。对该显式事务执行 COMMIT 语句，提交事务：

```
mysql> commit;select now();
Query OK, 0 rows affected (0.01 sec)

+---------------------+
| now()               |
+---------------------+
| 2018-10-04 15:36:42 |    # COMMIT 语句执行的时间点为 2018-10-04 15:36:42
+---------------------+
1 row in set (0.00 sec)
```

现在，我们解析最新的二进制日志文件，查看其是如何记录该事务产生的事件数据的（注意，二进制日志文件中每个事件的时间点信息是在生成事件的时候产生的），为了方便大家阅读，后续将对二进制日志分段说明。

（1）解析二进制日志文件。

```
[root@localhost binlog]# mysqlbinlog -vv mysql-bin.000460
……
```

（2）以下内容为二进制日志文件头，事件时间点为手工滚动二进制日志文件的时间点 2018-10-04 15:31:55，这里并未有任何变化。

```
mysqlbinlog: [Warning] unknown variable 'loose_default-character-set=utf8'
/*!50530 SET @@SESSION.PSEUDO_SLAVE_MODE=1*/;
/*!50003 SET @OLD_COMPLETION_TYPE=@@COMPLETION_TYPE,COMPLETION_TYPE=0*/;
DELIMITER /*!*/;
# at 4
#181004 15:31:55 server id 3306111  end_log_pos 123 CRC32 0x9cc61928  Start: binlog v 4, server v 5.7.20-log created 181004 15:31:55
# Warning: this binlog is either in use or was not closed properly.
BINLOG '
68G1Ww9/cjIAdwAAAHsAAAABAAQANS43LjIwLWxvZwAAAAAAAAAAAAAAAAAAAAAAAAAA
AAAAAAAAAAAAAAAAAAAAEzgNAAgAEgAEBAQEEgAAXwAEGggAAAAICAgCAAAACgoKKioAEjQA
ASgZxpw=
'/*!*/;
# at 123
#181004 15:31:55 server id 3306111  end_log_pos 194 CRC32 0xc34422d7  Previous-GTIDs
# ec123678-5e26-11e7-9d38-000c295e08a0:182318-199336
```

（3）"at 194"处为事务的第 1 个事件，用于记录事务的 GTID，时间点为 15:36:42。对照我们在数据库中执行事务相关的操作语句时的时间记录，可以得知，该时间点与执行 COMMIT 语句的时间点相同（均为 15:36:42）。

```
# at 194
#181004 15:36:42 server id 3306111  end_log_pos 259 CRC32 0x7971213c  GTID    last_committed=0    sequence_number=1    rbr_only=yes
/*!50718 SET TRANSACTION ISOLATION LEVEL READ COMMITTED*//*!*/;
SET @@SESSION.GTID_NEXT= 'ec123678-5e26-11e7-9d38-000c295e08a0:199337'/*!*/;
```

（4）"at 259"处为事务的第 2 个事件，用于记录事务的 BEGIN 语句，时间点为 15:35:21。根据我们进行事务操作时记录的时间信息，该时间为执行 UPDATE 语句的时间点（15:35:21），并不是执行 BEGIN 语句的时间点（15:33:02），说明该事件是在执行 UPDATE 语句时产生的，并不是在执行 BEGIN 语句时产生的。

```
# at 259
#181004 15:35:21 server id 3306111  end_log_pos 338 CRC32 0x4d3af767  Query    thread_id=3 exec_time=0 error_code=0
SET TIMESTAMP=1538638521/*!*/;
SET @@session.pseudo_thread_id=3/*!*/;
SET @@session.foreign_key_checks=1, @@session.sql_auto_is_null=0, @@session.unique_checks=1, @@session.autocommit=1/*!*/;
SET @@session.sql_mode=1436549152/*!*/;
```

```
    SET @@session.auto_increment_increment=2, @@session.auto_increment_offset=1/*!*/;
    /*!\C utf8 *//*!*/;
    SET
@@session.character_set_client=33,@@session.collation_connection=33,@@session.collation
_server=83/*!*/;
    SET @@session.lc_time_names=0/*!*/;
    SET @@session.collation_database=DEFAULT/*!*/;
    BEGIN
    /*!*/;
```

（5）"at 338" 处为事务的第 3 个事件，用于记录事务执行的原始 SQL 语句（需要启用参数 binlog_rows_query_log_events = ON），时间点仍然是执行 UPDATE 语句的时间点（15:35:21），说明该事件也是在执行 UPDATE 语句时产生的。

```
    # at 338
    #181004 15:35:21 server id 3306111  end_log_pos 398 CRC32 0x066200c5 Rows_query
    # update sbtest1 set k=k+1 where id=21
```

（6）"at 398" 处为事务的第 4 个事件，用于记录事务对应的操作表在内存中的映射 ID，时间点仍然是执行 UPDATE 语句的时间点（15:35:21），说明该事件也是在执行 UPDATE 语句时产生的。

```
    # at 398
    #181004 15:35:21 server id 3306111  end_log_pos 457 CRC32 0x020d3fcf Table_map:
`sbtest`.`sbtest1` mapped to number 143
```

（7）"at 457" 处为事务的第 5 个事件，用于记录事务的完整操作语句和数据记录，也就是真正在执行 UPDATE 语句修改数据（时间点仍然是 15:35:21），说明该事件也是在执行 UPDATE 语句时产生的。

```
    # at 457
    #181004 15:35:21 server id 3306111  end_log_pos 761 CRC32 0xdcfd94ba Update_rows: table
id 143 flags: STMT_END_F

    BINLOG '
    ucK1Wx1/cjIAPAAAAI4BAACAACR1cGRhdGUgc2J0ZXN0MSBzZXQgaz1rKzEgd2hlcmUgaWQ9MjHF
    AGIG
    ucK1WxN/cjIAOwAAAMkBAAAAAI8AAAAAAAEABnNidGVzdAAHc2J0ZXN0MQAEAwP+/gTuaP60AM8/
    DQI=
    ucK1Wx9/cjIAMAEAAPkCAAAAAI8AAAAAAAEAAgAE///wFQAAABX1JQB3ADA5Mjc5MjEwMjE5LTM3
    NzQ1ODM5OTA4LTU2MTg1Njk5MzI3LTc5Dc3MTU4NjQxLTg2NzExMjQyOTU1LTYxNDQ0Mzky
    LTQyNjIyODA0NTA2LTYxMDMxNTEyODQ1LTM2NzE4NDIyODQwLTExMDA4ODA5A3h4ePAVAAAA
    FuUlAHcAMDkyNzkyMTAyMTktMzc3NDU4Mzk5MDgtNTYxODU2OTkzMjctNzkwNzcxNTg2NDEtODY3
    MTEyNDI5NTYtNjE0NDAzOTItNDI2MjI4MDQ1MDYtNjEwMzE1MTI4NDUtMzY3MTg0MjI4NDAt
    MTEwMMjg4MDM4NDkDeHh4upT93A==
    '/*!*/;
    ### UPDATE `sbtest`.`sbtest1`
    ### WHERE
    ###   @1=21 /* INT meta=0 nullable=0 is_null=0 */
```

```
    ### @2=2483477 /* INT meta=0 nullable=0 is_null=0 */
    ###
@3='09279210219-37745839908-56185699327-79477158641-86711242956-61449540392-42622804506
-61031512845-36718422840-11028803849' /* STRING(360) meta=61032 nullable=0 is_null=0 */
    ### @4='xxx' /* STRING(180) meta=65204 nullable=0 is_null=0 */
    ### SET
    ### @1=21 /* INT meta=0 nullable=0 is_null=0 */
    ### @2=2483478 /* INT meta=0 nullable=0 is_null=0 */
    ###
@3='09279210219-37745839908-56185699327-79477158641-86711242956-61449540392-42622804506
-61031512845-36718422840-11028803849' /* STRING(360) meta=61032 nullable=0 is_null=0 */
    ### @4='xxx' /* STRING(180) meta=65204 nullable=0 is_null=0 */
```

（8）"at 761"处为事务的第6个事件，用于记录事务的 COMMIT 语句，时间点为15:36:42。根据我们进行事务操作时记录的时间信息，该时间点为执行 COMMIT 语句的时间点（15:36:42），说明该事件是在执行 COMMIT 语句时产生的。

```
# at 761
#181004 15:36:42 server id 3306111  end_log_pos 792 CRC32 0x3ad66423 Xid = 30
COMMIT/*!*/;
SET @@SESSION.GTID_NEXT= 'AUTOMATIC' /* added by mysqlbinlog */ /*!*/;
DELIMITER ;
# End of log file
/*!50003 SET COMPLETION_TYPE=@OLD_COMPLETION_TYPE*/;
/*!50530 SET @@SESSION.PSEUDO_SLAVE_MODE=0*/;
```

在上述二进制日志文件的解析内容中，用户事务的事件有 Gtid_log_event（记录事务的GTID）、Query_log_event（记录事务的 BEGIN 语句）、Rows_query_log_event（在基于 row 的复制的二进制日志中，记录事务的原始 SQL 语句）、Table_map_event（记录事务操作表在内存中的映射 ID）、Update_rows_log_event（记录事务真正操作的语句和修改的具体数据）、Xid_event（记录事务的 COMMIT 语句）几种类型。其中：

- Gtid_log_event 和 Xid_event 类型的事件是在事务执行 COMMIT 语句时产生的。

- Query_log_event、Rows_query_log_event、Table_map_event、Update_rows_log_event 类型的事件是在事务执行 UPDATE 语句时产生的。

- 执行 BEGIN 语句时未产生任何事件。

提示：本章以一个显式启动的事务（显式启动的事务需要显式提交）为例，剖析一个事务产生的二进制日志事件组（Binlog Event Group）。由于显式提交事务的 BEGIN 语句、DML 操作以及 COMMIT 语句之间存在时差，因此会造成事件乱序，但如果使用自动提交的事务，则不存在事件时间点乱序的问题。

如果只需要查看一个事务有哪些二进制日志事件类型，除了直接使用 mysqlbinlog 命令解析二进制日志文件之外，还可以使用 mysql 命令行客户端解析二进制日志文件来查看，如图 28-1 所示（这是一个 INSERT 语句事务）。

```
admin@localhost : (none):40: > show binlog events in 'mysql-bin.000002';
+------------------+-----+----------------+-----------+-------------+-------------------------------------------------------------------+
| Log_name         | Pos | Event_type     | Server_id | End_log_pos | Info                                                              |
+------------------+-----+----------------+-----------+-------------+-------------------------------------------------------------------+
| mysql-bin.000002 |   4 | Format_desc    |   3306102 |         124 | Server ver: 8.0.19, Binlog ver: 4                                 |
| mysql-bin.000002 | 124 | Previous_gtids |   3306102 |         155 |                                                                   |
| mysql-bin.000002 | 155 | Gtid           |   3306102 |         232 | SET @@SESSION.GTID_NEXT= 'bc86fafe-3844-11ea-8adc-0025905b06da:1' |
| mysql-bin.000002 | 232 | Query          |   3306102 |         351 | create database sbtest /* xid=40 */                               |
| mysql-bin.000002 | 351 | Gtid           |   3306102 |         428 | SET @@SESSION.GTID_NEXT= 'bc86fafe-3844-11ea-8adc-0025905b06da:2' |
| mysql-bin.000002 | 428 | Query          |   3306102 |         552 | create table sbtest.test (id int) /* xid=41 */                    |
| mysql-bin.000002 | 552 | Gtid           |   3306102 |         631 | SET @@SESSION.GTID_NEXT= 'bc86fafe-3844-11ea-8adc-0025905b06da:3' |
| mysql-bin.000002 | 631 | Query          |   3306102 |         707 | BEGIN                                                             |
| mysql-bin.000002 | 707 | Rows_query     |   3306102 |         764 | # insert into sbtest.test values(1)                               |
| mysql-bin.000002 | 764 | Table_map      |   3306102 |         816 | table_id: 110 (sbtest.test)                                       |
| mysql-bin.000002 | 816 | Write_rows     |   3306102 |         856 | table_id: 110 flags: STMT_END_F                                   |
| mysql-bin.000002 | 856 | Xid            |   3306102 |         887 | COMMIT /* xid=42 */                                               |
+------------------+-----+----------------+-----------+-------------+-------------------------------------------------------------------+
12 rows in set (0.01 sec)
```

图 28-1

第 29 章　复制 AUTO_INCREMENT 字段

设计 MySQL 的表结构时，一般建议使用 AUTO_INCREMENT 字段作为表的主键，而不是使用 UUID 作为主键，原因是使用前者作为主键能保证数据行是按顺序写入的。如果采用随机写入的方式，InnoDB 在写入数据时会产生大量的随机 I/O 操作，并且会频繁做数据页的分裂操作。AUTO_INCREMENT 字段主键在插入性能以及抑制碎片空间的产生方面都比较有优势。但是它的值是由 MySQL Server 产生的，那么在复制拓扑中，MySQL 是如何保证主从库之间 AUTO_INCREMENT 字段数据的一致性的呢？本章讨论在 MySQL 中是如何保证 AUTO_INCREMENT 字段被正确复制的。

29.1　操作环境信息

1. 操作系统版本

   Red Hat Enterprise Linux Server release 7.4（Maipo）

2. 数据库版本

   5.7.22-log MySQL Community Server（GPL）

3. 目的

   通过操作演示在不同事务隔离级别以及系统变量 binlog_format 的不同设置下，AUTO_INCREMENT 字段值是如何被记录到二进制日志并正确复制到从库的。

4. 数据库系统变量
   - gtid_mode=on
   - enforce_gtid_consistency=true

29.2　复制 AUTO_INCREMENT 字段的操作示例

本次测试分别在四种 MySQL 事务隔离级别以及三种二进制日志格式下进行，演示 AUTO_INCREMENT 字段在不同场景下如何被正确复制。由于操作的重复性，这里不给出所有场景下的操作记录。下面列出本次测试所用的语句，感兴趣的读者可以自己动手操作：

```
flush logs;
```

```
create database gangshen;
use gangshen;
create table user(id int primary key auto_increment,name varchar(20));
insert into user(id,name) values(1,'zhangsan'),(2,'lisi'),(3,'wangwu');
insert into user(name) values('zhaoliu');
drop database gangshen;
flush logs;
```

在不同的事务隔离级别以及二进制日志格式下，分别执行上述语句，并对对应产生的二进制日志进行解析，进而分析 MySQL 是如何保证 AUTO_INCREMENT 字段被正确复制的。

在测试过程中，当事务隔离级别为 READ-UNCOMMITTED 或 READ-COMMITTED 时，如果将系统变量 binlog_format 设置为 statement，执行 DML 语句时会出现 ERROR 1665 报错，报错原因在 28.2 节中已经做了解释，不再赘述。

29.3　对二进制日志的解析及解释

参照 26.3.1.2 节介绍的方式对 29.2 节中各个场景下产生的二进制日志进行解析，分析复制数据的正确性，主要关注 INSERT 语句对应的二进制日志是如何记录及复制的。通过对所有场景产生的二进制日志的分析，下面主要从三种二进制日志格式的维度对结果进行解释，至于具体二进制日志中事件内容的解释，参见第 26 章 "二进制日志文件的基本组成"，本节不再赘述。

29.3.1　基于 row 的复制中 AUTO_INCREMENT 字段的复制

基于 row 的复制是 MySQL 中最安全（保证主从库数据一致性）的复制格式，因为主库二进制日志是以 row 格式记录的。当系统变量 binlog_format 设置为 row 时，二进制日志记录的是实际每一行记录的变更情况。关于复制格式的细节，参见第 3 章 "复制格式详解"。在基于 row 的复制中，INSERT 语句在插入带 AUTO_INCREMENT 字段的表时，由 MySQL 产生的自增值被明确记录到二进制日志，因此 AUTO_INCREMENT 字段都能被正确地复制。

下面给出事务隔离级别为 READ-UNCOMMITTED 时，AUTO_INCREMENT 字段在 row 格式二进制日志中记录的形式：

```
1. ……
2. # at 694
3. #190701 12:44:39 server id 330657 end_log_pos 784 Rows_query
4. # insert into user(id,name) values(1,'zhangsan'),(2,'lisi'),(3,'wangwu')
5. # at 784
6. #190701 12:44:39 server id 330657 end_log_pos 834 Table_map: `gangshen`.`user` mapped to number 138
7. # at 834
8. #190701 12:44:39 server id 330657 end_log_pos 901 Write_rows: table id 138 flags: STMT_END_F
```

```
 9. ……
10. ### INSERT INTO `gangshen`.`user`
11. ### SET
12. ###   @1=1 /* INT meta=0 nullable=0 is_null=0 */
13. ###   @2='zhangsan' /* VARSTRING(60) meta=60 nullable=1 is_null=0 */
14. ### INSERT INTO `gangshen`.`user`
15. ### SET
16. ###   @1=2 /* INT meta=0 nullable=0 is_null=0 */
17. ###   @2='lisi' /* VARSTRING(60) meta=60 nullable=1 is_null=0 */
18. ### INSERT INTO `gangshen`.`user`
19. ### SET
20. ###   @1=3 /* INT meta=0 nullable=0 is_null=0 */
21. ###   @2='wangwu' /* VARSTRING(60) meta=60 nullable=1 is_null=0 */
22. # at 901
23. ……
24. # at 1121
25. #190701 12:44:39 server id 330657  end_log_pos 1171 Table_map: `gangshen`.`user` mapped to number 138
26. # at 1171
27. #190701 12:44:39 server id 330657  end_log_pos 1215 Write_rows: table id 138 flags: STMT_END_F
28. ……
29. ### INSERT INTO `gangshen`.`user`
30. ### SET
31. ###   @1=4 /* INT meta=0 nullable=0 is_null=0 */
32. ###   @2='zhaoliu' /* VARSTRING(60) meta=60 nullable=1 is_null=0 */
33. # at 1215
34. #190701 12:44:39 server id 330657  end_log_pos 1242 Xid = 212
35. COMMIT/*!*/;
36. # at 1242
37.
38. ……
```

其中，第 2~22 行的内容为"insert into user(id,name) values(1,'zhangsan'),(2,'lisi'),(3,'wangwu')"语句产生的二进制日志，即 INSERT 语句中指定了 AUTO_INCREMENT 字段的值；第 24~35 行的内容为"insert into user(name) values('zhaoliu')"语句产生的二进制日志，即 INSERT 语句中未指定 AUTO_INCREMENT 字段的值，由 MySQL 自己生成。

查看第 12、16、20 以及 31 行的内容可知，AUTO_INCREMENT 字段无论是由 INSERT 语句指定值还是由 MySQL 生成值，id 字段对应的值都被记录到 Write_rows_log_event 事件中，所以在从库上可以正确地重放二进制日志且能保证主从库数据的一致性。

29.3.2　基于 statement 的复制中 AUTO_INCREMENT 字段的复制

基于 statement 的复制在记录二进制日志时，将 DML 操作的原始变更语句记录到二进制日志中。如果 INSERT 语句在插入有 AUTO_INCREMENT 字段的表时没有指定自增值，且主库和从库对 auto_increment_offset 和 auto_increment_increment 变量的设置不同，MySQL 会如何保证复制的正确性呢？

在本章的演示过程中，当事务隔离级别为 READ-UNCOMIITTED 以及 READ-COMMITTED 时，将系统变量 binlog_format 设置为 statement，在执行 DML 操作时会出现 ERROR 1665 错误，DML 语句无法执行。所以，我们只需要关注在 REPEATABLE- READ 以及 SERIALIZABLE 隔离级别下 statement 格式的二进制日志如何保证 AUTO_INCREMENT 字段的正确复制即可。下面给出在 REPEATABLE-READ 隔离级别下，对应操作记录产生的二进制日志的解析结果：

```
1. ......
2. # at 705
3. #190706 22:18:30 server id 330657 end_log_pos 853 Query thread_id=3 exec_time=0 error_code=0
4. SET TIMESTAMP=1562465910/*!*/;
5. insert into user(id,name) values(1,'zhangsan'),(2,'lisi'),(3,'wangwu')
6. /*!*/;
7. # at 853
8. #190706 22:18:30 server id 330657 end_log_pos 880 Xid = 66
9. COMMIT/*!*/;
10. # at 880
11. #190706 22:18:30 server id 330657 end_log_pos 941 GTID last_committed=3 sequence_number=4 rbr_only=no
12. SET @@SESSION.GTID_NEXT= '16b154ef-7a12-11e9-b850-0800276d49d5:361'/*!*/;
13. # at 941
14. #190706 22:18:30 server id 330657 end_log_pos 1024 Query thread_id=3 exec_time=0 error_code=0
15. SET TIMESTAMP=1562465910/*!*/;
16. BEGIN
17. /*!*/;
18. # at 1024
19. # at 1052
20. #190706 22:18:30 server id 330657 end_log_pos 1052 Intvar
21. SET INSERT_ID=4/*!*/;
22. #190706 22:18:30 server id 330657 end_log_pos 1170 Query thread_id=3 exec_time=0 error_code=0
23. SET TIMESTAMP=1562465910/*!*/;
24. insert into user(name) values('zhaoliu')
25. /*!*/;
26. # at 1170
27. #190706 22:18:30 server id 330657 end_log_pos 1197 Xid = 67
28. COMMIT/*!*/;
```

其中，第 2~9 行内容为 "insert into user(id,name) values(1,'zhangsan'),(2,'lisi'),(3,'wangwu')" 语句产生的二进制日志，即 INSERT 语句中指定了 AUTO_INCREMENT 字段的值；第 10~28 行的内容为 "insert into user(name) values('zhaoliu')" 语句产生的二进制日志，即 INSERT 语句中未指定 AUTO_INCREMENT 字段的值，由 MySQL 自己生成该值。

通过分析可以得知，当 INSERT 语句指定了 AUTO_INCREMENT 字段的值时，二进制日志中以 Query_log_event 事件记录对应的 INSERT 语句，因为 INSERT 语句中指定了 AUTO_INCREMENT 字段的值，当其被复制到从库时，从库也会插入相同的值。当 INSERT 语句中未指定 AUTO_INCREMENT 字段的值，需要靠 MySQL 自己生成时，在主库上 MySQL 将生成的值以 Intvar_event 事件的形式记录下来，当其被复制到从库时，会先执行 SET INSERT_ID 语句，设置下一个会生成的 AUTO_INCREMENT 字段值为多少，然后执行 INSERT 语句，能保证从库插入与主库一样的 AUTO_INCREMENT 字段值。所以，在基于 statement 的复制中，AUTO_INCREMENT 字段也能被正确复制。

29.3.3 混合复制中 AUTO_INCREMENT 字段的复制

使用混合复制时（系统变量 binlog_format 设置为 mixed），默认情况下会使用 statement 格式记录二进制日志，如果 MySQL 判断无法使用 statement 格式，或者使用 statement 格式不安全时，会采用 row 格式。

对于本章的操作示例，在 READ-UNCOMMITTED 以及 READ-COMMITTED 事务隔离级别下，由于无法使用 statement 格式记录二进制日志，所以在这两种事务隔离级别下，混合复制的二进制日志将 INSERT 语句以 row 格式记录，其具体内容及形式与基于 row 的复制的一致，可以参考 29.3.1 节。在 REPEATABLE-READ 和 SERIALIZABLE 事务隔离级别下，可以使用 statement 格式记录二进制日志，所以在这两种事务隔离级别下，混合复制的二进制日志将以 statement 格式记录 INSERT 语句，其具体内容与基于 statement 的复制的一致，可以参考 29.3.2 节。

29.4 使用 AUTO_INCREMENT 字段时的注意事项

前面分析了 AUTO_INCREMENT 字段在 MySQL 中是如何被正确复制的，但是在演示操作过程中，还需要注意以下几点：

- 系统变量 auto_increment_offset 设置的是 AUTO_INCREMENT 字段产生的起始值，系统变量 auto_increment_increment 设置的是 AUTO_INCREMENT 字段在生成自增值时的步长，即两个自增值之间的差值。
- 查看 MySQL 表中的下一个自增值时有以下几种方式：
  - 通过 SHOW CREATE TABLE 语句查看表结构的定义。
  - 通过 SHOW TABLE STATUS 语句输出中的 AUTO_INCREMENT 字段查看。
  - 通过语句 SELECT auto_increment FROM information_schema.tables WHERE table_schema = 'XXX' AND table_name = 'XXX'查看。
- 在 MySQL 中，单表最多支持一个字段设置 AUTO_INCREMENT 属性，如果想要两个及以上字段实现自增，可以通过触发器或者用程序代码实现。

- AUTO_INCREMENT 字段必须有索引，而且一般在 AUTO_INCREMENT 字段上设置唯一索引。
- 通过存储过程或者触发器对 AUTO_INCREMENT 字段进行更新或者插入操作时，会出现如下报错：

```
Note (Code 1592): Unsafe statement written to the binary log using statement format since BINLOG_FORMAT = STATEMENT. Statement is unsafe because it invokes a trigger or a stored function that inserts into an AUTO_INCREMENT column. Inserted values cannot be logged correctly.
```

- 使用 ALTER TABLE 语句向表中添加 AUTO_INCREMENT 字段，可能导致主从库之间的数据行排序不同，因为行编号的顺序取决于表使用的特定存储引擎和插入行的顺序。如果主从库之间的数据行排序必须相同，则在分配 AUTO_INCREMENT 字段编号之前必须对行排序。例如，向只包含 col1 和 col2 列的表 t1 添加一个 AUTO_INCREMENT 字段，根据 MySQL 官方文档可使用如下方式来操作：

```
# 先创建 t2 表，表结构参考 t1
create table t2 like t1;
# 然后对 t2 表添加一个 AUTO_INCREMENT 字段
alter table t2 add id int auto_increment primary key;
# 对 t1 表中的数据全表扫描并对 t1 表的所有字段排序之后，再插入 t2 表（注意，为了保证主从库之间的数据行排序相同，必须使用 ORDER BY 语句对 t1 表中的所有数据排序）
insert into t2 select * from t1 order by col1,col2;

# 创建 t2 表之后，删除 t1 表，然后将 t2 表更名为 t1，如下:
drop table t1;
alter table t2 rename t1;
```

注意：前面给出的 CREATE TABLE ... LIKE 建表语句中，如果碰到外键定义则外键的定义会被忽略，DATA DIRECTORY 和 INDEX DIRECTORY 表选项也会被忽略。如果表定义包含任何这些特性，请直接参考 t1 的建表语句，添加 AUTO_INCREMENT 字段来创建 t2 表。

29.5 小结

通过本章的示例，我们可以明确地知道 AUTO_INCREMENT 字段在不同事务隔离级别、不同复制格式下都能被正确地从主库复制到从库，能保证主从库数据的一致性。

第 30 章　复制 CREATE ... IF NOT EXISTS 语句

MySQL 中有 CREATE ... IF NOT EXISTS 语句，对于程序创建库或者表很方便：如果库或者表不存在，则创建；反之，则不创建。无论主库中是否存在某个库，使用 CREATE DATABASE IF NOT EXISTS 语句创建库的时候，该语句都能被正确地复制到从库。同样，无论主库中是否存在某张表，CREATE TABLE IF NOT EXISTS 语句（除 CREATE TABLE IF NOT EXISTS ... SELECT 外）都能被正确复制到从库。本章我们就来看看 CREATE DATABASE IF NOT EXISTS 和 CREATE TABLE IF NOT EXISTS 语句是如何被正确复制到从库的。

30.1　操作环境信息

1. 操作系统版本

   Red Hat Enterprise Linux Server release 7.4（Maipo）

2. 数据库版本

   5.7.22-log MySQL Community Server（GPL）

3. 目的

   演示在不同事务隔离级别以及系统变量 binlog_format 的不同设置下，CREATE ... IF NOT EXISTS 语句是如何被记录到二进制日志，并被正确复制到从库的。

4. 数据库系统变量

   - gtid-mode = on
   - enforce-gtid-consistency = true
   - sql_mode = 'ONLY_FULL_GROUP_BY, STRICT_TRANS_TABLES, NO_ZERO_IN_DATE, NO_ZERO_DATE, ERROR_FOR_DIVISION_BY_ZERO, NO_AUTO_CREATE_USER, NO_ENGINE_SUBSTITUTION'

30.2 复制 CREATE ... IF NOT EXISTS 语句的操作演示

本次操作分别在四种事务隔离级别以及三种二进制日志记录格式下进行，演示 CREATE ... IF NOT EXISTS 语句在不同场景下如何被正确复制。由于操作的重复性，在此就不列出各个场景下的操作记录，只列出操作语句，感兴趣的读者可以自己动手跟随操作。这些语句如下：

```
flush logs;
drop database gangshen;
create database if not exists gangshen;
create database if not exists gangshen;
use gangshen;
create table if not exists test(id int primary key auto_increment,name varchar(20));
create table if not exists test(id int primary key auto_increment,name varchar(20));
create table if not exists test_new like test;
create table if not exists test_new like test;
```

由于使用 CREATE ... IF NOT EXISTS 语句，当库或表存在时不创建，不存在则创建，所以在我们的操作语句中 CREATE ... IF NOT EXISTS 都会执行两遍，以便观察当库或表不存在以及存在的情况下记录的二进制日志是否有所不同。

下面给出事务隔离级别为 READ-UNCOMMITTED，以及系统变量 binlog_format 设置为 statement 的场景下数据库的操作记录，供参考：

```
mysql> set session transaction_isolation='read-uncommitted';
Query OK, 0 rows affected (0.00 sec)

mysql> set session binlog_format='statement';
Query OK, 0 rows affected (0.00 sec)

mysql> show variables where variable_name in ('transaction_isolation','binlog_format')\G
*************************** 1. row ***************************
Variable_name: binlog_format
        Value: STATEMENT
*************************** 2. row ***************************
Variable_name: transaction_isolation
        Value: READ-UNCOMMITTED
2 rows in set (0.01 sec)

mysql> flush logs;
Query OK, 0 rows affected (0.02 sec)

mysql> drop database gangshen;
Query OK, 2 rows affected (0.03 sec)

mysql> create database if not exists gangshen;
```

```
Query OK, 1 row affected (0.00 sec)

mysql> create database if not exists gangshen;
Query OK, 1 row affected, 1 warning (0.00 sec)

mysql> use gangshen;
Database changed

mysql> create table if not exists test(id int primary key auto_increment,name varchar(20));
Query OK, 0 rows affected (0.02 sec)

mysql> create table if not exists test(id int primary key auto_increment,name varchar(20));
Query OK, 0 rows affected, 1 warning (0.00 sec)

mysql> create table if not exists test_new like test;
Query OK, 0 rows affected (0.02 sec)

mysql> create table if not exists test_new like test;
Query OK, 0 rows affected, 1 warning (0.00 sec)
```

30.3 二进制日志解析结果的解释

参照 26.3.1.2 节中二进制日志的解析方式，对 30.2 节各场景下产生的二进制日志进行解析，分析解析结果，确定 MySQL 是如何正确复制 CREATE ... IF NOT EXISTS 语句的。通过分析，我们发现不同场景下产生的二进制日志相同。所以，这里只给出事务隔离级别为 READ-UNCOMMITTED，系统变量 binlog_format 设置为 statement 的场景下产生的二进制日志解析结果，如下所示：

```
......
# at 412
#190810 15:15:29 server id 3306244  end_log_pos 528    Query    thread_id=2 exec_time=0 error_code=0
SET TIMESTAMP=1565421329/*!*/;
create database if not exists gangshen
/*!*/;
......
# at 589
#190810 15:15:29 server id 3306244  end_log_pos 705    Query    thread_id=2 exec_time=0 error_code=0
SET TIMESTAMP=1565421329/*!*/;
create database if not exists gangshen
/*!*/;
......
# at 766
#190810 15:15:29 server id 3306244  end_log_pos 927    Query    thread_id=2 exec_time=0
```

```
  error_code=0
     USE `gangshen`/*!*/;
     SET TIMESTAMP=1565421329/*!*/;
     create table if not exists test(id int primary key auto_increment,name varchar(20))
     /*!*/;
     ......
     # at 988
     #190810 15:15:29 server id 3306244  end_log_pos 1149   Query    thread_id=2 exec_time=0
  error_code=0
     SET TIMESTAMP=1565421329/*!*/;
     create table if not exists test(id int primary key auto_increment,name varchar(20))
     /*!*/;
     ......
     # at 1210
     #190810 15:15:29 server id 3306244  end_log_pos 1322   Query    thread_id=2 exec_time=0
  error_code=0
     SET TIMESTAMP=1565421329/*!*/;
     create table if not exists test_new like test
     /*!*/;
     ......
     # at 1383
     #190810 15:15:32 server id 3306244  end_log_pos 1495   Query    thread_id=2 exec_time=0
  error_code=0
     SET TIMESTAMP=1565421332/*!*/;
     create table if not exists test_new like test
     ......
```

通过二进制日志解析结果我们可以看到：

- CREATE ... IF NOT EXISTS 语句（除 CREATE TABLE IF NOT EXISTS ... SELECT 外）无论在何种事务隔离级别、系统变量 binlog_format 为何值的情况下，都是以 Query_log_event 事件的形式记录在二进制日志中的。
- CREATE ... IF NOT EXISTS 语句（除 CREATE TABLE IF NOT EXISTS ... SELECT 外）无论是否真实触发过创建库或表的操作，都会被记录到二进制日志中，所以在二进制日志的解析结果中看到每个语句都出现了两次。

30.4 小结

从操作的演示结果看，对于 CREATE ... IF NOT EXISTS 语句（除 CREATE TABLE IF NOT EXISTS ... SELECT 外），MySQL 能保证将其复制到从库。理论上 CREATE ... IF NOT EXISTS 语句有幂等性，多次执行也不会报错，可以放心使用它建库和建表。

另外，DROP ... IF EXISTS 语句在复制中的同步情况与 CREATE ... IF NOT EXISTS 语句的类似，本书不再对 DROP ... IF EXISTS 语句进行说明，感兴趣的读者可以自行验证。对于 DROP ... IF EXISTS 语句，MySQL 能保证将其复制到从库（即使主库上不存在要删除的数据库、表、视图），这是为了保证主从库中都不遗留要被删除的对象。

第 31 章　复制 CREATE TABLE ... SELECT 语句

第 30 章解释了 CREATE ... IF NOT EXISTS 语句(除 CREATE TABLE IF NOT EXISTS ... SELECT 之外)是如何被正确复制的,本章我们来看第 30 章未涉及的 CREATE TABLE IF NOT EXISTS ... SELECT 以及 CREATE TABLE ... SELECT 语句是如何被正确复制的。

31.1　操作环境信息

1. 操作系统版本

    Red Hat Enterprise Linux Server release 7.4(Maipo)

2. 数据库版本

    5.7.22-log MySQL Community Server(GPL)

3. 目的

    演示在不同事务隔离级别以及系统变量 binlog_format 的不同设置下,CREATE TABLE ... SELECT 语句是如何被记录到二进制日志,并被正确复制到从库的。

4. 数据库系统变量的设置

    - gtid-mode = off
    - enforce-gtid-consistency = off
    - sql_mode = 'ONLY_FULL_GROUP_BY, STRICT_TRANS_TABLES, NO_ZERO_IN_DATE, NO_ZERO_DATE, ERROR_FOR_DIVISION_BY_ZERO, NO_AUTO_CREATE_USER, NO_ENGINE_SUBSTITUTION'

31.2　复制 CREATE TABLE ... SELECT 语句的操作示例

本次操作分别在四种事务隔离级别以及三种二进制日志格式下进行,演示 CREATE TABLE ... SELECT 语句在不同场景下如何被正确复制。由于操作的重复性,在此就不列出各场景下的操作记录,仅列出操作语句,感兴趣的读者可以自己跟随这些语句进行操作。这些语句如下:

```
    flush logs;
    drop database gangshen;
    create database gangshen;
    use gangshen;
    create table my_customer (id smallint primary key auto_increment,first_name
varchar(45),last_name varchar(45)) select customer_id,first_name,last_name from sakila.
customer where customer_id <6;
    create table if not exists my_customer_new (id smallint primary key auto_increment,
first_name varchar(45),last_name varchar(45)) select customer_id,first_name,last_name from
sakila.customer where customer_id < 6;
    create table if not exists my_customer_new (id smallint primary key auto_increment,
first_name varchar(45),last_name varchar(45)) select customer_id,first_name,last_name from
sakila.customer where customer_id < 6;
    select * from my_customer;
    select * from my_customer_new;
```

本次演示过程中需要注意的关键点：

- 本次演示我们利用了 MySQL 官方提供的 sakila 测试库，它包含一些基本的表结构以及数据。可以在 MySQL 官网下载 sakila 测试库。

- GTID 不支持 CREATE TABLE ... SELECT 语句，在开启 GTID 的情况下，执行 CREATE TABLE ... SELECT 语句会得到报错提示：

```
ERROR 1786 (HY000): Statement violates GTID consistency: CREATE TABLE ... SELECT.
```

- 在 READ-UNCOMMITTED 和 READ-COMMITTED 事务隔离级别下，当系统变量 binlog_format 设置为 statement 时，CREATE TABLE ... SELECT 语句也无法执行，会得到报错提示：

```
ERROR 1665 (HY000): Cannot execute statement: impossible to write to binary log since
BINLOG_FORMAT = STATEMENT and at least one table uses a storage engine limited to row-based
logging. InnoDB is limited to row-logging when transaction isolation level is READ COMMITTED
or READ UNCOMMITTED.
```

下面给出在事务隔离级别为 READ-UNCOMMITTED，系统变量 binlog_format 设置为 row 的场景下数据库的操作记录，供参考：

```
mysql> set session transaction_isolation='read-uncommitted';
Query OK, 0 rows affected (0.00 sec)

mysql> set session binlog_format='row';
Query OK, 0 rows affected (0.00 sec)

mysql> show variables where variable_name in ('transaction_isolation','binlog_
format')\G
*************************** 1. row ***************************
Variable_name: binlog_format
        Value: ROW
*************************** 2. row ***************************
```

```
    Variable_name: transaction_isolation
            Value: READ-UNCOMMITTED
2 rows in set (0.01 sec)

mysql> flush logs;
Query OK, 0 rows affected (0.01 sec)

mysql> drop database gangshen;
Query OK, 0 rows affected (0.00 sec)

mysql> create database gangshen;
Query OK, 1 row affected (0.00 sec)

mysql> use gangshen;
Database changed

mysql> create table my_customer (id smallint primary key auto_increment,first_name
varchar(45),last_name varchar(45)) select customer_id,first_name,last_name from sakila.
customer where customer_id <6;
Query OK, 5 rows affected (0.04 sec)
Records: 5  Duplicates: 0  Warnings: 0

mysql> create table if not exists my_customer_new (id smallint primary key auto_
increment,first_name varchar(45),last_name varchar(45)) select customer_id,first_name,
last_name from sakila.customer where customer_id < 6;
Query OK, 5 rows affected (0.02 sec)
Records: 5  Duplicates: 0  Warnings: 0

mysql> create table if not exists my_customer_new (id smallint primary key auto_
increment,first_name varchar(45),last_name varchar(45)) select customer_id,first_name,
last_name from sakila.customer where customer_id < 6;
Query OK, 0 rows affected, 1 warning (0.00 sec)

mysql> select * from my_customer;
+----+-------------+------------+-----------+
| id | customer_id | first_name | last_name |
+----+-------------+------------+-----------+
|  1 |           1 | MARY       | SMITH     |
|  2 |           2 | PATRICIA   | JOHNSON   |
|  3 |           3 | LINDA      | WILLIAMS  |
|  4 |           4 | BARBARA    | JONES     |
|  5 |           5 | ELIZABETH  | BROWN     |
+----+-------------+------------+-----------+
5 rows in set (0.00 sec)

mysql> select * from my_customer_new;
+----+-------------+------------+-----------+
| id | customer_id | first_name | last_name |
```

```
+----+-------------+-------------+-----------+
| 1  | 1           | MARY        | SMITH     |
| 2  | 2           | PATRICIA    | JOHNSON   |
| 3  | 3           | LINDA       | WILLIAMS  |
| 4  | 4           | BARBARA     | JONES     |
| 5  | 5           | ELIZABETH   | BROWN     |
+----+-------------+-------------+-----------+
5 rows in set (0.00 sec)
```

31.3 二进制日志的解析及解释

参照 26.3.1.2 节的二进制日志解析方式，对 31.2 节中各场景下产生的二进制日志进行解析，分析解析结果，确定 MySQL 是如何正确复制 CREATE TABLE ... SELECT 语句的。对解析结果可以分为两类进行解释：一类是系统变量 binlog_format 为 statement，另一类是系统变量 binlog_format 为 row 或 mixed。

31.3.1 statement 格式二进制日志的解析及解释

因为只有事务隔离级别为 REPEATABLE-READ 和 SERIALIZABLE 时，才支持以 statement 格式的二进制日志记录 CREATE TABLE ... SELECT 语句，而且由于 statement 格式的二进制日志记录的内容是类似的，所以这里只给出 REPEATABLE-READ 隔离级别下二进制日志的解析结果。

```
......
# at 412
#190811 15:47:22 server id 3306244  end_log_pos 514   Query      thread_id=9 exec_time=0 error_code=0
SET TIMESTAMP=1565509642/*!*/;
create database gangshen
/*!*/;
......
# at 575
# at 603
#190811 15:47:22 server id 3306244  end_log_pos 603 Intvar
SET INSERT_ID=1/*!*/;
#190811 15:47:22 server id 3306244  end_log_pos 864   Query      thread_id=9 exec_time=0 error_code=0
USE `gangshen`/*!*/;
SET TIMESTAMP=1565509642/*!*/;
create table my_customer (id smallint primary key auto_increment,first_name varchar(45),last_name varchar(45)) select customer_id,first_name,last_name from sakila.customer where customer_id <6
/*!*/;
......
# at 925
# at 953
```

```
    #190811 15:47:22 server id 3306244 end_log_pos 953 Intvar
    SET INSERT_ID=1/*!*/;
    #190811 15:47:22 server id 3306244 end_log_pos 1233 Query    thread_id=9 exec_time=0
error_code=0
    SET TIMESTAMP=1565509642/*!*/;
    create table if not exists my_customer_new (id smallint primary key auto_increment,
first_name varchar(45),last_name varchar(45)) select customer_id,first_name,last_name from
sakila.customer where customer_id < 6
    /*!*/;
    ......
```

通过二进制日志的解析结果可以看到：

- CREATE TABLE ... SELECT 语句在 statement 格式的二进制日志中会以 Query_log_event 事件的形式记录，而且由于创建的目标表中含有 AUTO_INCREMENT 字段，所以会通过 Intvar_event 事件记录数据的自增值。

- 对于 CREATE TABLE IF NOT EXISTS...SELECT 语句，若目标表不存在，MySQL 会将其完整记录下来，若存在，则不会记录，这与 CREATE ... IF NOT EXISTS 语句（不含 CREATE TABLE ... IF NOT EXISTS ... SELECT）不同，后者不管目标表是否存在都会被 MySQL 记录下来。

31.3.2　row 格式和 mixed 格式二进制日志的解析及解释

当系统变量 binlog_format 设置为 row 或者 mixed 时，二进制日志记录的内容类似，所以此处只给出 REPEATABLE-READ 隔离级别下的二进制日志解析结果。

```
    ......
    # at 575
    #190811 15:49:04 server id 3306244 end_log_pos 913 Query    thread_id=9 exec_time=0
error_code=0
    use `gangshen`/*!*/;
    SET TIMESTAMP=1565509744/*!*/;
    create table `my_customer` (
      `id` smallint(6) not null auto_increment,
      `customer_id` smallint(5) unsigned not null default '0',
      `first_name` varchar(45) collate utf8_bin default null,
      `last_name` varchar(45) collate utf8_bin default null,
      primary key (`id`)
    )
    /*!*/;
    ......
    # at 1048
    #190811 15:49:04 server id 3306244 end_log_pos 1260 Rows_query
    # create table my_customer (id smallint primary key auto_increment,first_name varchar(45),
last_name varchar(45)) select customer_id,first_name,last_name from sakila.customer where
customer_id <6
    # at 1260
```

```
    #190811 15:49:04 server id 3306244 end_log_pos 1321 Table_map: `gangshen`.`my_customer`
mapped to number 206
    # at 1321
    #190811 15:49:04 server id 3306244 end_log_pos 1450 Write_rows: table id 206 flags:
STMT_END_F
    ......
    '/*!*/;
    ### INSERT INTO `gangshen`.`my_customer`
    ### SET
    ###   @1=1 /* SHORTINT meta=0 nullable=0 is_null=0 */
    ###   @2=1 /* SHORTINT meta=0 nullable=0 is_null=0 */
    ###   @3='MARY' /* VARSTRING(135) meta=135 nullable=1 is_null=0 */
    ###   @4='SMITH' /* VARSTRING(135) meta=135 nullable=1 is_null=0 */
    ### INSERT INTO `gangshen`.`my_customer`
    ### SET
    ###   @1=2 /* SHORTINT meta=0 nullable=0 is_null=0 */
    ###   @2=2 /* SHORTINT meta=0 nullable=0 is_null=0 */
    ###   @3='PATRICIA' /* VARSTRING(135) meta=135 nullable=1 is_null=0 */
    ###   @4='JOHNSON' /* VARSTRING(135) meta=135 nullable=1 is_null=0 */
    ### INSERT INTO `gangshen`.`my_customer`
    ### SET
    ###   @1=3 /* SHORTINT meta=0 nullable=0 is_null=0 */
    ###   @2=3 /* SHORTINT meta=0 nullable=0 is_null=0 */
    ###   @3='LINDA' /* VARSTRING(135) meta=135 nullable=1 is_null=0 */
    ###   @4='WILLIAMS' /* VARSTRING(135) meta=135 nullable=1 is_null=0 */
    ### INSERT INTO `gangshen`.`my_customer`
    ### SET
    ###   @1=4 /* SHORTINT meta=0 nullable=0 is_null=0 */
    ###   @2=4 /* SHORTINT meta=0 nullable=0 is_null=0 */
    ###   @3='BARBARA' /* VARSTRING(135) meta=135 nullable=1 is_null=0 */
    ###   @4='JONES' /* VARSTRING(135) meta=135 nullable=1 is_null=0 */
    ### INSERT INTO `gangshen`.`my_customer`
    ### SET
    ###   @1=5 /* SHORTINT meta=0 nullable=0 is_null=0 */
    ###   @2=5 /* SHORTINT meta=0 nullable=0 is_null=0 */
    ###   @3='ELIZABETH' /* VARSTRING(135) meta=135 nullable=1 is_null=0 */
    ###   @4='BROWN' /* VARSTRING(135) meta=135 nullable=1 is_null=0 */
    # at 1450
    #190811 15:49:04 server id 3306244 end_log_pos 1477 Xid = 2538
    COMMIT/*!*/;
    ......
    # at 1538
    #190811 15:49:04 server id 3306244 end_log_pos 1894 Query   thread_id=9 exec_time=0
error_code=0
    SET TIMESTAMP=1565509744/*!*/;
    create table if not exists `my_customer_new` (
      `id` smallint(6) not null auto_increment,
      `customer_id` smallint(5) unsigned not null default '0',
```

```
    `first_name` varchar(45) collate utf8_bin default null,
    `last_name` varchar(45) collate utf8_bin default null,
    primary key (`id`)
  )
  /*!*/;
  ......
  # at 2260
  #190811 15:49:04 server id 3306244 end_log_pos 2325 Table_map: `gangshen`.`my_customer_
new` mapped to number 207
  # at 2325
  #190811 15:49:04 server id 3306244 end_log_pos 2454 Write_rows: table id 207 flags:
STMT_END_F
  ......
  '/*!*/;
  ### INSERT INTO `gangshen`.`my_customer_new`
  ### SET
  ###   @1=1 /* SHORTINT meta=0 nullable=0 is_null=0 */
  ###   @2=1 /* SHORTINT meta=0 nullable=0 is_null=0 */
  ###   @3='MARY' /* VARSTRING(135) meta=135 nullable=1 is_null=0 */
  ###   @4='SMITH' /* VARSTRING(135) meta=135 nullable=1 is_null=0 */
  ### INSERT INTO `gangshen`.`my_customer_new`
  ### SET
  ###   @1=2 /* SHORTINT meta=0 nullable=0 is_null=0 */
  ###   @2=2 /* SHORTINT meta=0 nullable=0 is_null=0 */
  ###   @3='PATRICIA' /* VARSTRING(135) meta=135 nullable=1 is_null=0 */
  ###   @4='JOHNSON' /* VARSTRING(135) meta=135 nullable=1 is_null=0 */
  ### INSERT INTO `gangshen`.`my_customer_new`
  ### SET
  ###   @1=3 /* SHORTINT meta=0 nullable=0 is_null=0 */
  ###   @2=3 /* SHORTINT meta=0 nullable=0 is_null=0 */
  ###   @3='LINDA' /* VARSTRING(135) meta=135 nullable=1 is_null=0 */
  ###   @4='WILLIAMS' /* VARSTRING(135) meta=135 nullable=1 is_null=0 */
  ### INSERT INTO `gangshen`.`my_customer_new`
  ### SET
  ###   @1=4 /* SHORTINT meta=0 nullable=0 is_null=0 */
  ###   @2=4 /* SHORTINT meta=0 nullable=0 is_null=0 */
  ###   @3='BARBARA' /* VARSTRING(135) meta=135 nullable=1 is_null=0 */
  ###   @4='JONES' /* VARSTRING(135) meta=135 nullable=1 is_null=0 */
  ### INSERT INTO `gangshen`.`my_customer_new`
  ### SET
  ###   @1=5 /* SHORTINT meta=0 nullable=0 is_null=0 */
  ###   @2=5 /* SHORTINT meta=0 nullable=0 is_null=0 */
  ###   @3='ELIZABETH' /* VARSTRING(135) meta=135 nullable=1 is_null=0 */
  ###   @4='BROWN' /* VARSTRING(135) meta=135 nullable=1 is_null=0 */
  # at 2454
  #190811 15:49:04 server id 3306244 end_log_pos 2481 Xid = 2539
  COMMIT/*!*/;
  ......
```

通过二进制日志解析结果可以看到，在 row 或者 mixed 二进制日志格式下：

- CREATE TABLE ... SELECT 语句会被拆分成 CREATE TABLE 语句以 Query_log_event 事件的形式记录，插入数据的部分以 Insert_rows_log_even 事件的形式记录。
- 对于 CREATE TABLE IF NOT EXISTS ... SELECT 语句，当目标表不存在时，MySQL 会将语句完整记录下来，当目标表存在时，则不会记录，与二进制日志为 statement 格式的情况相同。

31.4 使用 CREATE TABLE ... SELECT 语句时的注意事项

本章的操作演示虽然证明了 CREATE TABLE ... SELECT 语句能被正确复制到从库，但我们仍然建议不要使用该种方式创建新表，因为使用该语句时有许多地方都需要注意：

- 需要关闭 GTID 复制模式才能使用该语句。
- CREATE TABLE ... SELECT 语句执行时会触发事务隐式提交。
- 如果目标表不存在：
  - 在 statement 格式的二进制日志中，语句会以 Query_log_event 事件形式记录下来。
  - 在 row 或者 mixed 格式的二进制日志中，该语句会被拆分为两部分来记录：一部分是 CREATE TABLE 语句部分，以 Query_log_event 事件的形式记录；另一部分是插入数据的部分，以 Insert_rows_log_event 事件的形式记录。
- 如果目标表已经存在，则该语句不会被记录到二进制日志。

31.5 小结

从操作演示的结果看，对于 CREATE TABLE ... SELECT 语句，MySQL 能正确将其复制到从库，保证主从库数据一致性，前提是需要按照 31.4 节中的注意事项操作。

第 32 章　在主从复制中使用不同的表定义

正常情况下，我们使用 MySQL 主从复制时，都是为了保证主从库数据的一致性，但有时可能会遇到一些需求。比如，出于安全的考虑，上游数据库同步到下游数据库时需要过滤某些字段，或者中间库需要对字段类型进行调整，那么在 MySQL 中如果不同步表（结构）定义，人为地分别在主库和从库上创建不同的表定义，MySQL 复制时只负责数据同步，这样可行吗？本章就来介绍一下在主从复制中使用不同的表定义进行复制。

在主从复制架构中，源表（主库中的表）和目标表（从库中的表）不必相同。主库上表的字段数可以比从库的更多或更少。此外，主从库之间相同名称的表的字段也可以根据条件使用不同数据类型，但是在源表和目标表具有不同的表定义的情况下，主从库上的数据库名称和表名称必须相同。

32.1　操作环境信息

1. 操作系统版本

    Red Hat Enterprise Linux Server release 7.4（Maipo）

2. 数据库版本

    5.7.27-log MySQL Community Server（GPL）

3. 目的

    操作演示在不同事务隔离级别以及系统变量 binlog_format 的不同设置下，主从库使用不同的表定义，主库上的变更操作是如何被记录到二进制日志并被复制到从库的。

4. 数据库系统变量的设置

    - gtid-mode = on
    - enforce-gtid-consistency = true
    - binlog_row_image = FULL
    - sql_mode = 'ONLY_FULL_GROUP_BY,STRICT_TRANS_TABLES, NO_ZERO_IN_DATE, NO_ZERO_DATE, ERROR_FOR_DIVISION_BY_ZERO, NO_AUTO_CREATE_USER, NO_ENGINE_SUBSTITUTION'

32.2　主从库的表字段数不同时如何复制

首先介绍当主从库使用含不同数量的字段的表结构时如何进行复制，主要对系统变量 binlog_format 为 row 或者 statement 的情况进行分析。不过，需要注意的是，在 READ-UNCOMMIT 和 READ-COMMITTED 事务隔离级别下，如果系统变量 binlog_format 设置为 statement，当执行 DML 操作时，MySQL 会报错，其原因在前面的章节中已说明，不再赘述。所以本章对于 binlog_format 为 statement，只讨论 REPEATABLE-READ 和 SERIALIZABLE 隔离级别下的情况。

MySQL 支持主从库之间具有不同字段数的表的复制，但是有一定的限制条件：

- 在源表和目标表中都存在的字段，其定义的顺序必须相同（除了两个表的字段数相同以外，其字段的定义顺序也必须相同）。
- 源表与目标表之间有差异的字段，差异字段的定义顺序必须在相同字段的后面。
- 源表与目标表之间有差异的字段需要设置默认值。

32.2.1　源表字段数多于目标表字段数

下面，我们通过操作演示主从库之间使用不同数量的字段进行复制是可行的。先演示源表字段数多于目标表字段数的情况。由于操作步骤都是类似的，所以此处演示事务隔离级别为 REPEATABLE-READ，系统变量 binlog_format 设置为 row 和 statement 时，MySQL 主从复制是否可行。

32.2.1.1　操作步骤

（1）安装两个 MySQL 实例并搭建主从复制（具体步骤略）。

（2）在主库上创建库并创建表：

```
mysql> create database gangshen;
mysql> use gangshen;
## 因为不希望将主库的表结构同步到从库，所以这里在会话级别将 sql_log_bin 设置为 0，表示不记录二进制日志，建表语句也就无法同步到从库
mysql> set sql_log_bin=0;
mysql> create table t1(c1 int,c2 int,c3 int);
## 执行完建表语句后，恢复系统变量 sql_log_bin 的值，继续记录二进制日志
mysql> set sql_log_bin=1;
```

（3）在从库上根据需要创建表：

```
mysql> use gangshen;
mysql> set sql_log_binlog=0;
## 创建的表名需要与主库上源表表名一致
mysql> create table t1(c1 int,c2 int);
mysql> set sql_log_binlog=1;
```

（4）将主库系统变量 binlog_format 设置为 row，并执行插入语句，看数据能否正常同步：

```
mysql> set binlog_format=row;
mysql> insert into t1(c1,c2,c3) values(1,1,1);
mysql> insert into t1 values(2,2,2);
```

（5）检查从库的数据以及复制的状态：

```
mysql> select * from gangshen.t1;
mysql> show slave status\G
```

（6）将主库系统变量 binlog_format 设置为 statement，并执行 INSERT 语句，看数据能否正常同步：

```
mysql> set binlog_format=statement;
mysql> insert into t1 (c1,c2) values(3,3);
mysql> insert into t1 (c1,c2,c3) values(4,4,4);
mysql> insert into t1 values(5,5,5);
```

（7）检查从库的数据以及复制的状态：

```
mysql> select * from gangshen.t1;
mysql> show slave status\G
```

32.2.1.2　演示结果及解释

当 binlog_format 为 row 时，语句都能正常复制，源表比目标表多出来的字段都被忽略了。

当 binlog_format 为 statement 时，如果主库上执行的语句含有的字段只涉及源表与目标表之间相同的字段，就能正常复制；如果主库上执行的语句含有源表与目标表之间存在差异的字段，则复制时会报错。对于上述操作，"Insert into t1(c1,c2) values(3,3)"语句能正常复制，而"insert into t1(c1,c2,c3) values(4,4,4)"和"insert into t1 values(5,5,5)"语句会导致主从复制时报错。

当系统变量 binlog_format 为 row 时，语句在二进制日志中都是以字段值的形式记录的：

```
......
# insert into t1(c1,c2,c3) values(1,1,1)
......
### INSERT INTO `gangshen`.`t1`
### SET
###   @1=1 /* INT meta=0 nullable=1 is_null=0 */
###   @2=1 /* INT meta=0 nullable=1 is_null=0 */
###   @3=1 /* INT meta=0 nullable=1 is_null=0 */
......
# INSERT INTO t1 VALUES(2,2,2)
......
### INSERT INTO `gangshen`.`t1`
### SET
###   @1=2 /* INT meta=0 nullable=1 is_null=0 */
```

```
### @2=2 /* INT meta=0 nullable=1 is_null=0 */
### @3=2 /* INT meta=0 nullable=1 is_null=0 */
......
```

在以 row 格式记录的二进制日志中，"@1 @2 @3"代表字段的顺序，与 INFORMATION_SCHEMA 库 COLUMNS 表中记录的字段顺序一致，依次建立对应关系。当源表字段数多于目标表字段数时，多余的字段被 MySQL 忽略了，目标表中含有多少个字段，则依次取对应数量的值进行应用。但是需要注意，以上的情况有一个前提，即系统变量 binlog_row_image 为 FULL，若不为 FULL，由于 MySQL 不会记录变更记录行的所有字段，可能会有问题。这种情况比较复杂，在此就不展开介绍了。

当系统变量 binlog_format 为 statement 时，语句在二进制日志中是以原始语句的形式记录的：

```
......
SET TIMESTAMP=1568356269/*!*/;
insert into t1 (c1,c2) values(3,3)
......
#190913  2:31:09 server id 33062  end_log_pos 1533  Query thread_id=2 exec_time=0 error_code=0
SET TIMESTAMP=1568356269/*!*/;
insert into t1 (c1,c2,c3) values(4,4,4)
......
#190913  2:31:09 server id 33062  end_log_pos 1820  Query thread_id=2 exec_time=0 error_code=0
SET TIMESTAMP=1568356269/*!*/;
insert into t1 values(5,5,5)
```

对于 statement 格式的二进制日志，从库也是按照执行语句的方式应用的，所以当语句无法在从库上执行时，主从复制就会报错。

32.2.2 目标表字段数多于源表字段数

本节接着演示目标表字段数多于源表字段数的情况。同样，这里只演示事务隔离级别为 REPEATABLE-READ，系统变量 binlog_format 设置为 row 和 statement 时，MySQL 主从复制是否可行。

32.2.2.1 操作步骤

（1）安装两个 MySQL 实例并搭建主从复制（具体操作步骤略）。

（2）在主库上创建库并创建表：

```
mysql> create database gangshen;
mysql> use gangshen;
## 因为不希望将主库的表结构同步到从库，所以在会话级别将 sql_log_bin 设置为 0，表示不记录二进制日志，建表语句也就无法同步到从库
mysql> set sql_log_bin=0;
mysql> create table t1(c1 int,c2 int);
```

```
## 执行完建表语句后,恢复系统变量 sql_log_bin 的值,继续记录二进制日志
mysql> set sql_log_bin=1;
```

(3)在从库上根据需要创建表:

```
mysql> use gangshen;
mysql> set sql_log_binlog=0;
## 创建的表名需要与主库上源表的表名一致
mysql> create table t1(c1 int,c2 int,c3 int);
mysql> set sql_log_binlog=1;
```

(4)将主库系统变量 binlog_format 设置为 row,并执行插入语句,看数据能否正常同步:

```
mysql> set binlog_format=row;
mysql> insert into t1(c1,c2) values(1,1);
mysql> insert into t1 values(2,2);
```

(5)检查从库的数据以及复制的状态:

```
mysql> select * from gangshen.t1;
mysql> show slave status\G
```

(6)将主库系统变量 binlog_format 设置为 statement,并执行 INSERT 语句,看数据能否正常同步:

```
mysql> set binlog_format=statement;
mysql> insert into t1 (c1,c2) values(3,3);
mysql> insert into t1 (c1,c2,c3) values(4,4,4);
mysql> insert into t1 values(5,5,5);
```

(7)检查从库的数据以及复制的状态:

```
mysql> select * from gangshen.t1;
mysql> show slave status\G
```

32.2.2.2 演示结果及解释

系统变量 binlog_format 为 row 时,语句都能正常复制,目标表与源表有差异的字段都被复制为默认值。

系统变量 binlog_format 为 statement 时,如果主库上执行的语句含有的字段只涉及源表与目标表相同的字段,就能正常复制;如果含有源表与目标表有差异的字段,则会报错。对于上述操作,"insert into t1(c1,c2) values(3,3)"语句能正常复制,"insert into t1(c1,c2,c3) values(4,4,4)"和"insert into t1 values(5,5,5)"语句会导致主从复制报错。

其原因与 32.2.1.2 节中解释的相同,只是在系统变量 binlog_format 为 row 时,目标表为比源表多出的字段的值设置了默认值。

32.3 不同类型字段的复制

在理想情况下,源表与目标表中相同的字段具有相同的数据类型定义,在前面演示的例子中,只是源表与目标表的字段数量不一致,相同字段的数据类型是一样的。但其实如

果满足某些条件，源表和目标表中相同字段定义的数据类型并不需要是相同的。

32.3.1 属性提升

通常，主库目标表中给定数据类型的字段可以复制到从库中相同数据类型、相同定义长度的字段，或更长定义长度的字段中。例如，从源表的 CHAR(10)字段复制到目标表的 CHAR(10)字段，或从源表的 CHAR(10)字段复制到目标表的 CHAR(25)字段，都不会出现任何问题。在某些情况下，也可以从一种数据类型的字段复制到另一种不同数据类型的字段。当源表中的数据类型复制后变为更长的定义长度的相同数据类型时，称为属性提升。当系统变量 binlog_format 为 statement 和 row 时，MySQL 复制中的属性提升也不同，下文会讨论，但属性提升与主库或从库使用的存储引擎无关。

使用基于 statement 的复制格式时，如果在主库上运行的语句能够在从库上成功运行，数据就能被复制到从库，即如果主库执行的语句涉及的字段与从库中给定字段的类型兼容，语句就可以被正确复制。例如，TINYINT 字段的任何值可以成功插入到 BIGINT 字段中。因此，如果主库使用 TINYINT 字段类型，从库使用 BIGINT 字段类型，数据是可以被正确复制的，因为 TINYINT 字段的值范围不可能超过 BIGINT 字段的。或者主库表结构为"CREATE TABLE t1(id INT, gmt_create DATETIME)"，从库表结构为"CREATE TABLE t1(id INT,gmt_create VARCHAR(50))"，当在主库上执行"INSERT INTO t1(id ,gmt_create) VALUES(1, '2019-09-13 03:19:00')语句"时，在 statement 格式的二进制日志中记录的是语句，因为该语句在从库上也能执行，所以能正确复制。主库的 DATETIME 字段类型也就被转换为从库的 VARCHAR 字段类型。

基于 row 的复制格式支持较小数据类型和较大数据类型之间的属性提升（从小数据类型到大数据类型）和属性降级（从大数据类型到小数据类型），此外还可以设置是否允许有损转换或无损转换，下文会阐述。

32.3.2 有损转换与无损转换

如果目标字段的数据类型无法完整表示要插入的源数据值，则必须选择转换方式。如果允许转换，而且会截断（或以其他方式修改）源值，以适应目标数据字段的类型，这种数据类型转换称为有损转换；如果允许转换但不需要截断源值或做类似的修改，以适应目标字段中数据类型，这种数据类型转换就称为无损转换。

类型转换的模式由系统变量 slave_type_conversions 控制，其默认值为空，不允许动态修改，必须重启 MySQL 进程才能使改后的值生效。slave_type_conversions 的有效值和对应的转换行为如表 32-1 所示。

表 32-1

有效值（模式）	行为
ALL_LOSSY	此模式允许有损转换。注意，此模式不允许无损转换，仅允许有损转换或者不需要转换的情况。例如，允许将 INT 字段转换为 TINYINT 字段（有损转换），但不允许 TINYINT 字段转换为 INT 字段（无损转换）。在该模式下，如果遇到无损转换，会导致复制发生错误而中止
ALL_NON_LOSSY	此模式仅允许不需要截断，或不需要对源值进行其他特殊处理的转换的情况，即允许目标数据类型具有比源数据类型范围更宽的转换（此模式和是否允许有损转换无关，有损转换由 ALL_LOSSY 模式控制）。如果仅设置 ALL_NON_LOSSY，但不设置 ALL_LOSSY，则在尝试复制一些可能导致数据丢失的转换时会导致复制报错而中止。例如：INT 到 TINYINT、CHAR(25)到 VARCHAR(20)的转换会导致从库的复制中止并报错
ALL_LOSSY ALL_NON_LOSSY	此模式允许所有支持的类型转换，包括有损和无损的转换
ALL_SIGNED	在此模式下，将提升后的整数类型视为有符号值（默认行为）
ALL_UNSIGNED	在此模式下，将提升后的整数类型视为无符号值
ALL_SIGNED ALL_UNSIGNED	在此模式下，如果可能，将提升后的整数类型视为有符号值，否则视为无符号值
[empty]	slave_type_conversions 的默认值，空串。未设置该变量时，即为默认值。在此模式下，不允许属性的提升或降级，这意味着源表和目标表中的所有字段必须具有相同的类型，否则会导致从库复制报错而中止

由于篇幅限制，在此不具体列出哪些数据类型之间可以相互转换，哪些数据类型之间不可以，有兴趣的读者可以自行参考 MySQL 官网中的相关内容。

32.4 小结

最后，对本章内容进行总结：

- 在一定限制条件下，MySQL 主从复制中源表和目标表可以采用不同的表（结构）定义。

- 如果源表与目标表的字段数量不同，则：
  - row 格式的二进制日志能保证数据正常复制。
  - statement 格式的二进制日志需要确认语句能否在从库执行，如果能在从库执行，则能保证数据正常复制，否则不行。

- 如果源表与目标表字段类型不同，在复制时需要考虑是否存在字段类型的属性提升，以及类型的转换是否属于有损/无损转换。

第 33 章　复制中的调用功能

日常工作中，DBA 不可避免会遇到使用存储过程、触发器、自定义函数等的场景，因此，本章我们来讲一讲在 MySQL 的复制中是如何保证存储过程、触发器、自定义函数等的正确性的。

33.1　操作环境信息

1. 操作系统版本

   Red Hat Enterprise Linux Server release 7.4（Maipo）

2. 数据库版本

   5.7.27-log MySQL Community Server（GPL）

3. 目的

   演示在不同事务隔离级别以及系统变量 binlog_format 的不同设置下，MySQL 中的调用功能（存储过程、触发器、自定义函数等）是如何被记录到二进制日志，并被正确复制到从库的。

4. 数据库系统变量的设置

   - gtid-mode = on
   - enforce-gtid-consistency = true
   - event-scheduler = on
   - sql_mode = 'ONLY_FULL_GROUP_BY, STRICT_TRANS_TABLES, NO_ZERO_IN_DATE, NO_ZERO_DATE, ERROR_FOR_DIVISION_BY_ZERO, NO_AUTO_CREATE_USER, NO_ENGINE_SUBSTITUTION'

33.2　复制中的调用功能操作示例

本次操作分别在四种事务隔离级别以及三种二进制日志格式下进行，演示存储过程、触发器、自定义函数等在不同场景下如何被正确地复制。

注意：在操作之前需要确认系统变量 event-scheduler 是否被设置为 on，否则即使创建了事件（event）[1]，MySQL 也不会调用。

33.2.1　在 READ-COMMITTED 隔离级别、基于 row 的复制场景下数据库的操作记录

下面给出在事务隔离级别为 READ-COMMITTED，系统变量 binlog_format 设置为 row 的场景下数据库的操作记录，供参考：

```
mysql> set session transaction_isolation='read-committed';
Query OK, 0 rows affected (0.00 sec)

mysql> set session binlog_format='row';
Query OK, 0 rows affected (0.00 sec)

mysql> show variables where variable_name in ('transaction_isolation', 'binlog_format')\G
*************************** 1. row ***************************
Variable_name: binlog_format
        Value: ROW
*************************** 2. row ***************************
Variable_name: transaction_isolation
        Value: READ-COMMITTED
2 rows in set (0.00 sec)

mysql> drop database if exists gangshen;
Query OK, 2 rows affected (0.01 sec)

mysql> flush logs;
Query OK, 0 rows affected (0.02 sec)

mysql> create database gangshen;
Query OK, 1 row affected (0.00 sec)

mysql> use gangshen
Database changed

mysql> create table tb1(id int primary key auto_increment,name varchar(20),gmt_create datetime);
Query OK, 0 rows affected (0.01 sec)

mysql> create table tb2(id int primary key auto_increment,gmt_create datetime);
Query OK, 0 rows affected (0.03 sec)
```

---

[1] 本章中所说的 event 是指计划任务中的 event，请注意和二进制日志中的 event 概念区分。

```
mysql> delimiter //

mysql> create procedure simpleproc ()
    -> begin
    -> insert into tb1(name,gmt_create) values('procedure_test',now());
    -> end//
Query OK, 0 rows affected (0.00 sec)

mysql> delimiter ;

mysql> call simpleproc();
Query OK, 1 row affected (0.00 sec)

mysql> select * from tb1;
+----+----------------+---------------------+
| id | name           | gmt_create          |
+----+----------------+---------------------+
|  1 | procedure_test | 2019-09-15 22:02:21 |
+----+----------------+---------------------+
1 row in set (0.00 sec)

mysql> drop procedure simpleproc;
Query OK, 0 rows affected (0.01 sec)

mysql> create function hello (s char(20))
    -> returns char(50) deterministic
    -> return concat('hello, ',s,'!');
Query OK, 0 rows affected (0.00 sec)

mysql> select hello('world');
+----------------+
| hello('world') |
+----------------+
| Hello, world!  |
+----------------+
1 row in set (0.00 sec)

mysql> drop function hello;
Query OK, 0 rows affected (0.00 sec)

mysql> create event event_test on schedule every 5 second do insert into tb1(name,
gmt_create) values('event_test',now());
Query OK, 0 rows affected (0.00 sec)

mysql> select sleep(15);
+-----------+
| sleep(15) |
+-----------+
```

```
| 0 |
+-----------+
1 row in set (15.00 sec)

mysql> select * from tb1;
+----+---------------+---------------------+
| id | name          | gmt_create          |
+----+---------------+---------------------+
| 1  | procedure_test | 2019-09-15 22:02:21 |
| 3  | event_test    | 2019-09-15 22:02:21 |
| 5  | event_test    | 2019-09-15 22:02:26 |
| 7  | event_test    | 2019-09-15 22:02:31 |
+----+---------------+---------------------+
4 rows in set (0.00 sec)

mysql> drop event event_test;
Query OK, 0 rows affected (0.01 sec)

mysql> create trigger trigger_test after insert on tb1 for each row insert into
tb2(gmt_create) values(now());
Query OK, 0 rows affected (0.00 sec)

mysql> insert into tb1(name,gmt_create) values('trigger',now());
Query OK, 1 row affected (0.00 sec)

mysql> select * from tb1;
+----+---------------+---------------------+
| id | name          | gmt_create          |
+----+---------------+---------------------+
| 1  | procedure_test | 2019-09-15 22:02:21 |
| 3  | event_test    | 2019-09-15 22:02:21 |
| 5  | event_test    | 2019-09-15 22:02:26 |
| 7  | event_test    | 2019-09-15 22:02:31 |
| 9  | trigger       | 2019-09-15 22:02:36 |
+----+---------------+---------------------+
5 rows in set (0.01 sec)

mysql> select * from tb2;
+----+---------------------+
| id | gmt_create          |
+----+---------------------+
| 1  | 2019-09-15 22:02:36 |
+----+---------------------+
1 row in set (0.00 sec)

mysql> drop trigger trigger_test;
Query OK, 0 rows affected (0.00 sec)
```

33.2.2 在 READ-COMMITTED 隔离级别、基于 statement 的复制场景下数据库的操作记录

在事务隔离级别为 READ-COMMITTED，系统变量 binlog_format 设置为 statement 的场景下，具体操作过程以及执行命令后输出的结果与 binlog_format 为 row 的基本一致，除了在调用存储过程以及触发触发器的时候。

调用存储过程时报 ERROR 1665 错误：

```
mysql> call simpleproc();
ERROR 1665 (HY000): Cannot execute statement: impossible to write to binary log since
BINLOG_FORMAT = STATEMENT and at least one table uses a storage engine limited to row-based
logging. InnoDB is limited to row-logging when transaction isolation level is READ COMMITTED
or READ UNCOMMITTED.
    Error (Code 1665): Cannot execute statement: impossible to write to binary log since
BINLOG_FORMAT = STATEMENT and at least one table uses a storage engine limited to row-based
logging. InnoDB is limited to row-logging when transaction isolation level is READ COMMITTED
or READ UNCOMMITTED.
    Error (Code 1015): Can't lock file (errno: 170 - It is not possible to log this statement)
```

触发触发器时也会报 ERROR 1665 错误：

```
mysql> insert into tb1(name,gmt_create) values('trigger',now());
ERROR 1665 (HY000): Cannot execute statement: impossible to write to binary log since
BINLOG_FORMAT = STATEMENT and at least one table uses a storage engine limited to row-based
logging. InnoDB is limited to row-logging when transaction isolation level is READ COMMITTED
or READ UNCOMMITTED.
    Error (Code 1665): Cannot execute statement: impossible to write to binary log since
BINLOG_FORMAT = STATEMENT and at least one table uses a storage engine limited to row-based
logging. InnoDB is limited to row-logging when transaction isolation level is READ COMMITTED
or READ UNCOMMITTED.
    Error (Code 1015): Can't lock file (errno: 170 - It is not possible to log this statement)
```

33.3 二进制日志的解析及解释

参照 26.3.1.2 节的二进制日志解析方式，对 33.2.1 和 33.2.2 节中各个场景下产生的二进制日志进行解析，分析解析结果，确定 MySQL 是如何正确复制存储过程、触发器、自定义函数的。对结果可以分为两类进行解释：一类是系统变量 binlog_format 为 statement，另一类是系统变量 binlog_format 为 row 或 mixed。

33.3.1 row 和 mixed 格式二进制日志的解析及解释

当系统变量 binlog_format 设置为 row 或者 mixed 时，二进制日志记录的内容类似，所以，此处只给出隔离级别为 READ-COMMITTED、基于 row 的复制中操作记录生成的二进

制日志解析结果。

```
......
# at 805
#190915 10:02:21 server id 33062  end_log_pos 866   GTID    last_committed=3    sequence_number=4    rbr_only=no
SET @@SESSION.GTID_NEXT= '6385db95-cbc2-11e9-9f05-080027136405:289'/*!*/;
# at 866
#190915 10:02:21 server id 33062  end_log_pos 1098  Query   thread_id=34    exec_time=0    error_code=0
SET TIMESTAMP=1568556141/*!*/;
create definer=`root`@`localhost` procedure `simpleproc`()
begin
insert into tb1(name,gmt_create) values('procedure_test',now());
end
/*!*/;
```

以上为二进制日志中记录的创建存储过程的 CREATE PROCEDURE 语句内容，是以 Quyery_log_event 事件的形式记录的。但是细心的读者会发现，在二进制日志中记录的内容多了 "DEFINER = root@localhost"，这是因为 MySQL 在二进制日志中记录 CREATE PROCEDURE、CREATE FUNCTION、CREATE EVENT、CREATE TRIGGER 语句时，会将执行这些语句的用户也记录下来，在有需要时我们可以通过解析二进制日志查到是哪个用户创建了存储过程、触发器等。

```
# at 1098
#190915 10:02:21 server id 33062  end_log_pos 1159  GTID    last_committed=4    sequence_number=5    rbr_only=yes
/*!50718 SET TRANSACTION ISOLATION LEVEL READ COMMITTED*//*!*/;
SET @@SESSION.GTID_NEXT= '6385db95-cbc2-11e9-9f05-080027136405:290'/*!*/;
# at 1159
#190915 10:02:21 server id 33062  end_log_pos 1244  Query   thread_id=34    exec_time=0    error_code=0
SET TIMESTAMP=1568556141/*!*/;
SET @@session.time_zone='+08:00'/*!*/;
BEGIN
/*!*/;
# at 1244
#190915 10:02:21 server id 33062  end_log_pos 1327  Rows_query
# insert into tb1(name,gmt_create) values('procedure_test',now())
# at 1327
#190915 10:02:21 server id 33062  end_log_pos 1378  Table_map: `gangshen`.`tb1` mapped to number 141
# at 1378
#190915 10:02:21 server id 33062  end_log_pos 1434  Write_rows: table id 141 flags: STMT_END_F
### INSERT INTO `gangshen`.`tb1`
### SET
###   @1=1 /* INT meta=0 nullable=0 is_null=0 */
```

```
    ###   @2='procedure_test' /* VARSTRING(60) meta=60 nullable=1 is_null=0 */
    ###   @3='2019-09-15 22:02:21' /* DATETIME(0) meta=0 nullable=1 is_null=0 */
    # at 1434
    #190915 10:02:21 server id 33062  end_log_pos 1461   Xid = 393
    COMMIT/*!*/;
```

以上为调用存储过程时产生的操作产生的二进制日志,在基于 row 的复制或者混合(mixed)复制中,存储过程产生的变更操作会以 row 格式记录在二进制日志中。

```
    # at 1461
    #190915 10:02:21 server id 33062  end_log_pos 1522   GTID    last_committed=5    sequence_number=6    rbr_only=no
    SET @@SESSION.GTID_NEXT= '6385db95-cbc2-11e9-9f05-080027136405:291'/*!*/;
    # at 1522
    #190915 10:02:21 server id 33062  end_log_pos 1630   Query   thread_id=34    exec_time=0  error_code=0
    SET TIMESTAMP=1568556141/*!*/;
    drop procedure simpleproc
    /*!*/;
```

以上为二进制日志中记录的删除存储过程(DROP PROCEDURE 语句)的内容,是以 Query_log_event 事件的形式记录的,但是与 CREATE PROCEDURE 语句不同,前者在二进制日志中不会记录执行语句的用户。

```
    # at 1630
    #190915 10:02:21 server id 33062  end_log_pos 1691   GTID    last_committed=6    sequence_number=7    rbr_only=no
    SET @@SESSION.GTID_NEXT= '6385db95-cbc2-11e9-9f05-080027136405:292'/*!*/;
    # at 1691
    #190915 10:02:21 server id 33062  end_log_pos 1948   Query   thread_id=34    exec_time=0  error_code=0
    SET TIMESTAMP=1568556141/*!*/;
    CREATE DEFINER=`root`@`localhost` FUNCTION `hello`(s CHAR(20)) RETURNS char(50) CHARSET utf8 COLLATE utf8_bin
        DETERMINISTIC
    RETURN CONCAT('Hello, ',s,'!')
    /*!*/;
    # at 1948
    #190915 10:02:21 server id 33062  end_log_pos 2009   GTID    last_committed=7    sequence_number=8    rbr_only=no
    SET @@SESSION.GTID_NEXT= '6385db95-cbc2-11e9-9f05-080027136405:293'/*!*/;
    # at 2009
    #190915 10:02:21 server id 33062  end_log_pos 2111   Query   thread_id=34    exec_time=0  error_code=0
    SET TIMESTAMP=1568556141/*!*/;
    DROP FUNCTION hello
    /*!*/;
```

以上为二进制日志中记录的 CREATE FUNCTION 以及 DROP FUNCTION 语句的内容,

它们都是以 Query_log_event 事件的形式记录的。

```
    # at 2111
    #190915 10:02:21 server id 33062  end_log_pos 2172  GTID    last_committed=8    sequence_
number=9   rbr_only=no
    SET @@SESSION.GTID_NEXT= '6385db95-cbc2-11e9-9f05-080027136405:294'/*!*/;
    # at 2172
    #190915 10:02:21 server id 33062  end_log_pos 2425  Query    thread_id=34    exec_time=0
error_code=0
    SET TIMESTAMP=1568556141/*!*/;
    CREATE DEFINER=`root`@`localhost` event event_test ON SCHEDULE EVERY 5 SECOND DO INSERT
INTO tb1(name,gmt_create) VALUES('event_test',now())
    /*!*/;
```

以上为二进制日志中记录的创建事件（CREATE EVENT 语句）的内容，是以 Query_log_event 事件的形式记录的，并且记录了执行该语句的用户。

```
    # at 2425
    #190915 10:02:21 server id 33062  end_log_pos 2486  GTID    last_committed=8    sequence_
number=10   rbr_only=yes
    /*!50718 SET TRANSACTION ISOLATION LEVEL READ COMMITTED*//*!*/;
    SET @@SESSION.GTID_NEXT= '6385db95-cbc2-11e9-9f05-080027136405:295'/*!*/;
    # at 2486
    #190915 10:02:21 server id 33062  end_log_pos 2571  Query    thread_id=62    exec_time=0
error_code=0
    SET TIMESTAMP=1568556141/*!*/;
    SET @@session.sql_auto_is_null=1/*!*/;
    BEGIN
    /*!*/;
    # at 2571
    #190915 10:02:21 server id 33062  end_log_pos 2650  Rows_query
    # INSERT INTO tb1(name,gmt_create) VALUES('event_test',now())
    # at 2650
    #190915 10:02:21 server id 33062  end_log_pos 2701  Table_map: `gangshen`.`tb1` mapped
to number 141
    # at 2701
    #190915 10:02:21 server id 33062  end_log_pos 2753  Write_rows: table id 141 flags:
STMT_END_F
    '/*!*/;
    ### INSERT INTO `gangshen`.`tb1`
    ### SET
    ###   @1=3 /* INT meta=0 nullable=0 is_null=0 */
    ###   @2='event_test' /* VARSTRING(60) meta=60 nullable=1 is_null=0 */
    ###   @3='2019-09-15 22:02:21' /* DATETIME(0) meta=0 nullable=1 is_null=0 */
    # at 2753
    #190915 10:02:21 server id 33062  end_log_pos 2780  Xid = 401
    COMMIT/*!*/;
    ……
    # at 3135
```

```
    #190915 10:02:31 server id 33062  end_log_pos 3196   GTID    last_committed=11
sequence_number=12 rbr_only=yes
    /*!50718 SET TRANSACTION ISOLATION LEVEL READ COMMITTED*//*!*/;
    SET @@SESSION.GTID_NEXT= '6385db95-cbc2-11e9-9f05-080027136405:297'/*!*/;
    # at 3196
    #190915 10:02:31 server id 33062  end_log_pos 3281  Query   thread_id=64   exec_time=0
error_code=0
    SET TIMESTAMP=1568556151/*!*/;
    BEGIN
    /*!*/;
    # at 3281
    #190915 10:02:31 server id 33062  end_log_pos 3360  Rows_query
    # INSERT INTO tb1(name,gmt_create) VALUES('event_test',now())
    # at 3360
    #190915 10:02:31 server id 33062  end_log_pos 3411  Table_map: `gangshen`.`tb1` mapped
to number 141
    # at 3411
    #190915 10:02:31 server id 33062  end_log_pos 3463  Write_rows: table id 141 flags:
STMT_END_F
    ### INSERT INTO `gangshen`.`tb1`
    ### SET
    ###   @1=7 /* INT meta=0 nullable=0 is_null=0 */
    ###   @2='event_test' /* VARSTRING(60) meta=60 nullable=1 is_null=0 */
    ###   @3='2019-09-15 22:02:31' /* DATETIME(0) meta=0 nullable=1 is_null=0 */
    # at 3463
    #190915 10:02:31 server id 33062  end_log_pos 3490  Xid = 404
    COMMIT/*!*/;
```

以上为调用事件（event）时产生的操作所对应的二进制日志，在基于 row 的复制或者混合复制中，产生的操作都以 row 格式记录在二进制日志中。

```
    # at 3490
    #190915 10:02:36 server id 33062  end_log_pos 3551   GTID    last_committed=12
sequence_number=13 rbr_only=no
    SET @@SESSION.GTID_NEXT= '6385db95-cbc2-11e9-9f05-080027136405:298'/*!*/;
    # at 3551
    #190915 10:02:36 server id 33062 end_log_pos 3661  Query   thread_id=34   exec_time=0
error_code=0
    SET TIMESTAMP=1568556156/*!*/;
    SET @@session.sql_auto_is_null=0/*!*/;
    DROP EVENT event_test
    /*!*/;
```

以上为在二进制日志中记录的 DROP EVENT 语句的内容，是以 Query_log_event 事件的形式记录的。

```
    # at 3661
    #190915 10:02:36 server id 33062  end_log_pos 3722   GTID    last_committed=13
sequence_number=14 rbr_only=no
    SET @@SESSION.GTID_NEXT= '6385db95-cbc2-11e9-9f05-080027136405:299'/*!*/;
```

```
    # at 3722
    #190915 10:02:36 server id 33062  end_log_pos 3954  Query   thread_id=34   exec_time=0
error_code=0
    SET TIMESTAMP=1568556156.529062/*!*/;
    CREATE DEFINER=`root`@`localhost` trigger trigger_test AFTER INSERT ON tb1 FOR EACH ROW
INSERT INTO tb2(gmt_create) VALUES(now())
    /*!*/;
```

以上为二进制日志中记录的创建触发器（CREATE TRIGGER 语句）的内容，以 Query_log_event 事件的形式记录，并且记录了执行该语句的用户的信息。

```
    # at 3954
    #190915 10:02:36 server id 33062  end_log_pos 4015  GTID    last_committed=14
sequence_number=15 rbr_only=yes
    /*!50718 SET TRANSACTION ISOLATION LEVEL READ COMMITTED*//*!*/;
    SET @@SESSION.GTID_NEXT= '6385db95-cbc2-11e9-9f05-080027136405:300'/*!*/;
    # at 4015
    #190915 10:02:36 server id 33062  end_log_pos 4100  Query   thread_id=34   exec_time=0
error_code=0
    SET TIMESTAMP=1568556156/*!*/;
    BEGIN
    /*!*/;
    # at 4100
    #190915 10:02:36 server id 33062  end_log_pos 4176  Rows_query
    # INSERT INTO tb1(name,gmt_create) values('trigger',now())
    # at 4176
    #190915 10:02:36 server id 33062  end_log_pos 4227  Table_map: `gangshen`.`tb1` mapped
to number 142
    # at 4227
    #190915 10:02:36 server id 33062  end_log_pos 4275  Table_map: `gangshen`.`tb2` mapped
to number 143
    # at 4275
    #190915 10:02:36 server id 33062  end_log_pos 4324  Write_rows: table id 142
    # at 4324
    #190915 10:02:36 server id 33062  end_log_pos 4365  Write_rows: table id 143 flags:
STMT_END_F
    ### INSERT INTO `gangshen`.`tb1`
    ### SET
    ###   @1=9 /* INT meta=0 nullable=0 is_null=0 */
    ###   @2='trigger' /* VARSTRING(60) meta=60 nullable=1 is_null=0 */
    ###   @3='2019-09-15 22:02:36' /* DATETIME(0) meta=0 nullable=1 is_null=0 */
    ### INSERT INTO `gangshen`.`tb2`
    ### SET
    ###   @1=1 /* INT meta=0 nullable=0 is_null=0 */
    ###   @2='2019-09-15 22:02:36' /* DATETIME(0) meta=0 nullable=1 is_null=0 */
    # at 4365
    #190915 10:02:36 server id 33062  end_log_pos 4392  Xid = 408
    COMMIT/*!*/;
```

以上为触发器产生的操作对应的二进制日志，在基于 row 的复制或者混合复制中，触发器产生的操作都以 row 格式记录在二进制日志中。

```
    # at 4392
    #190915 10:02:36 server id 33062  end_log_pos 4453    GTID    last_committed=15
sequence_number=16    rbr_only=no
    SET @@SESSION.GTID_NEXT= '6385db95-cbc2-11e9-9f05-080027136405:301'/*!*/;
    # at 4453
    #190915 10:02:36 server id 33062  end_log_pos 4561   Query    thread_id=34    exec_time=0
error_code=0
    SET TIMESTAMP=1568556156/*!*/;
    DROP TRIGGER trigger_test
    /*!*/;
    ......
```

以上为二进制日志中记录的 DROP TRIGGER 语句的内容，以 Query_log_event 事件的形式记录。

33.3.2　statement 格式二进制日志的解析及解释

通过 33.2.2 节中的操作记录可以看到，在 READ-COMMITTED 隔离级别下，系统变量 binlog_format 为 statement 时，CREATE EVENT、ALTER EVENT、DROP EVENT、CREATE PROCEDURE、DROP PROCEDURE、CREATE FUNCTION、DROP FUNCTION、CREATE TRIGGER、DROP TRIGGER 等语句能正常执行，但是调用存储过程、函数、触发器时会报 ERROR 1665 错误。

因为在 READ-UNCOMMITTED 以及 READ-COMMITTED 隔离级别下，当 binlog_format 为 statement 时，MySQL 不支持 DML 操作，存储过程、触发器等包含 DML 操作，所以调用它们的过程中会出错。

这里只给出 binlog_format 为 statement，且隔离级别为 REPEATABLE-READ 的场景下二进制日志的解析结果。

```
    # at 805
    #190915 10:12:15 server id 33062  end_log_pos 866    GTID    last_committed=3    sequence_
number=4    rbr_only=no
    SET @@SESSION.GTID_NEXT= '6385db95-cbc2-11e9-9f05-080027136405:355'/*!*/;
    # at 866
    #190915 10:12:15 server id 33062  end_log_pos 1098   Query    thread_id=34    exec_time=0
error_code=0
    SET TIMESTAMP=1568556735/*!*/;
    CREATE DEFINER=`root`@`localhost` PROCEDURE `simpleproc`()
    BEGIN
    INSERT INTO tb1(name,gmt_create) values('procedure_test',now());
    END
    /*!*/;
```

以上为二进制日志中记录的创建存储过程（CREATE PROCEDURE 语句）的内容，以 Query_log_event 事件的形式记录。

```
# at 1098
#190915 10:12:15 server id 33062  end_log_pos 1159  GTID  last_committed=4  sequence_number=5  rbr_only=no
   SET @@SESSION.GTID_NEXT= '6385db95-cbc2-11e9-9f05-080027136405:356'/*!*/;
# at 1159
#190915 10:12:15 server id 33062  end_log_pos 1255  Query  thread_id=34  exec_time=0  error_code=0
   SET TIMESTAMP=1568556735/*!*/;
   SET @@session.time_zone='+08:00'/*!*/;
   BEGIN
/*!*/;
# at 1255
# at 1283
#190915 10:12:15 server id 33062  end_log_pos 1283  Intvar
   SET INSERT_ID=1/*!*/;
#190915 10:12:15 server id 33062  end_log_pos 1437  Query  thread_id=34  exec_time=0  error_code=0
   SET TIMESTAMP=1568556735/*!*/;
   INSERT INTO tb1(name,gmt_create) VALUES('procedure_test',now())
/*!*/;
# at 1437
#190915 10:12:15 server id 33062  end_log_pos 1464  Xid = 526
COMMIT/*!*/;
```

以上为调用存储过程产生的操作对应的二进制日志，在基于 statement 的复制中，存储过程产生的影响以 statement 格式记录在二进制日志中。

```
# at 1464
#190915 10:12:15 server id 33062  end_log_pos 1525  GTID  last_committed=5  sequence_number=6  rbr_only=no
   SET @@SESSION.GTID_NEXT= '6385db95-cbc2-11e9-9f05-080027136405:357'/*!*/;
# at 1525
#190915 10:12:15 server id 33062  end_log_pos 1633  Query  thread_id=34  exec_time=0  error_code=0
   SET TIMESTAMP=1568556735/*!*/;
   DROP PROCEDURE simpleproc
/*!*/;
```

以上为二进制日志中记录的 DROP PROCEDURE 语句的内容，是以 Query_log_event 事件的形式记录的。

```
# at 1633
#190915 10:12:15 server id 33062  end_log_pos 1694  GTID  last_committed=6  sequence_number=7  rbr_only=no
   SET @@SESSION.GTID_NEXT= '6385db95-cbc2-11e9-9f05-080027136405:358'/*!*/;
# at 1694
#190915 10:12:15 server id 33062  end_log_pos 1951  Query  thread_id=34  exec_time=0
```

```
error_code=0
    SET TIMESTAMP=1568556735/*!*/;
    CREATE DEFINER=`root`@`localhost` FUNCTION `hello`(s CHAR(20)) RETURNS char(50) CHARSET
utf8 COLLATE utf8_bin
        DETERMINISTIC
    RETURN CONCAT('Hello, ',s,'!')
    /*!*/;
    # at 1951
    #190915 10:12:15 server id 33062  end_log_pos 2012   GTID    last_committed=7    sequence_
number=8    rbr_only=no
    SET @@SESSION.GTID_NEXT= '6385db95-cbc2-11e9-9f05-080027136405:359'/*!*/;
    # at 2012
    #190915 10:12:15 server id 33062  end_log_pos 2114   Query    thread_id=34    exec_time=0
error_code=0
    SET TIMESTAMP=1568556735/*!*/;
    DROP FUNCTION hello
```

以上为二进制日志中记录的 CREATE FUNCTION 和 DROP FUNCTION 语句的内容，都是以 Query_log_event 事件的形式记录的。

```
    /*!*/;
    # at 2114
    #190915 10:12:15 server id 33062  end_log_pos 2175   GTID    last_committed=8    sequence_
number=9    rbr_only=no
    SET @@SESSION.GTID_NEXT= '6385db95-cbc2-11e9-9f05-080027136405:360'/*!*/;
    # at 2175
    #190915 10:12:15 server id 33062  end_log_pos 2428   Query    thread_id=34    exec_time=0
error_code=0
    SET TIMESTAMP=1568556735/*!*/;
    CREATE DEFINER=`root`@`localhost` EVENT event_test ON SCHEDULE EVERY 5 SECOND DO INSERT
INTO tb1(name,gmt_create) VALUES('event_test',now())
    /*!*/;
```

以上为二进制日志中记录的 CREATE EVENT 语句的内容，是以 Query_log_event 事件的形式记录的。

```
    # at 2428
    #190915 10:12:15 server id 33062  end_log_pos 2489   GTID    last_committed=8    sequence_
number=10    rbr_only=yes
    /*!50718 SET TRANSACTION ISOLATION LEVEL READ COMMITTED*//*!*/;
    SET @@SESSION.GTID_NEXT= '6385db95-cbc2-11e9-9f05-080027136405:361'/*!*/;
    # at 2489
    #190915 10:12:15 server id 33062  end_log_pos 2574   Query    thread_id=76    exec_time=0
error_code=0
    SET TIMESTAMP=1568556735/*!*/;
    SET @@session.sql_auto_is_null=1/*!*/;
    BEGIN
    /*!*/;
    # at 2574
    #190915 10:12:15 server id 33062  end_log_pos 2653   Rows_query
```

```
# INSERT INTO tb1(name,gmt_create) VALUES('event_test',now())
# at 2653
#190915 10:12:15 server id 33062  end_log_pos 2704  Table_map: `gangshen`.`tb1` mapped to number 153
# at 2704
#190915 10:12:15 server id 33062  end_log_pos 2756  Write_rows: table id 153 flags: STMT_END_F
### INSERT INTO `gangshen`.`tb1`
### SET
###   @1=3 /* INT meta=0 nullable=0 is_null=0 */
###   @2='event_test' /* VARSTRING(60) meta=60 nullable=1 is_null=0 */
###   @3='2019-09-15 22:12:15' /* DATETIME(0) meta=0 nullable=1 is_null=0 */
# at 2756
#190915 10:12:15 server id 33062  end_log_pos 2783  Xid = 534
COMMIT/*!*/;
……
# at 3138
#190915 10:12:25 server id 33062  end_log_pos 3199  GTID    last_committed=11 sequence_number=12 rbr_only=yes
/*!50718 SET TRANSACTION ISOLATION LEVEL READ COMMITTED*//*!*/;
SET @@SESSION.GTID_NEXT= '6385db95-cbc2-11e9-9f05-080027136405:363'/*!*/;
# at 3199
#190915 10:12:25 server id 33062  end_log_pos 3284  Query   thread_id=78   exec_time=0   error_code=0
SET TIMESTAMP=1568556745/*!*/;
BEGIN
/*!*/;
# at 3284
#190915 10:12:25 server id 33062  end_log_pos 3363  Rows_query
# INSERT INTO tb1(name,gmt_create) VALUES('event_test',now())
# at 3363
#190915 10:12:25 server id 33062  end_log_pos 3414  Table_map: `gangshen`.`tb1` mapped to number 153
# at 3414
#190915 10:12:25 server id 33062  end_log_pos 3466  Write_rows: table id 153 flags: STMT_END_F
### INSERT INTO `gangshen`.`tb1`
### SET
###   @1=7 /* INT meta=0 nullable=0 is_null=0 */
###   @2='event_test' /* VARSTRING(60) meta=60 nullable=1 is_null=0 */
###   @3='2019-09-15 22:12:25' /* DATETIME(0) meta=0 nullable=1 is_null=0 */
# at 3466
#190915 10:12:25 server id 33062  end_log_pos 3493  Xid = 537
COMMIT/*!*/;
```

以上为调用事件产生的操作对应的二进制日志。你可能会奇怪，系统变量 binlog_format 不是设置为 statement 吗？为什么对应的操作是以 row 格式记录的？对此笔者在操作过程中也感觉比较迷惑，但实际操作的结果确实是以 row 格式记录的。MySQL 官方文档上也没说明原因，只能猜测为了保证主从复制时数据的一致性，所以以 row 格式记录二进制日志。

```
# at 3493
#190915 10:12:30 server id 33062  end_log_pos 3554  GTID  last_committed=12  sequence_number=13  rbr_only=no
SET @@SESSION.GTID_NEXT= '6385db95-cbc2-11e9-9f05-080027136405:364'/*!*/;
# at 3554
#190915 10:12:30 server id 33062  end_log_pos 3664  Query  thread_id=34  exec_time=0  error_code=0
SET TIMESTAMP=1568556750/*!*/;
SET @@session.sql_auto_is_null=0/*!*/;
DROP EVENT event_test
/*!*/;
```

以上为二进制日志中记录的 DROP EVENT 语句的内容，是以 Query_log_event 事件的形式记录的。

```
# at 3664
#190915 10:12:30 server id 33062  end_log_pos 3725  GTID  last_committed=13  sequence_number=14  rbr_only=no
SET @@SESSION.GTID_NEXT= '6385db95-cbc2-11e9-9f05-080027136405:365'/*!*/;
# at 3725
#190915 10:12:30 server id 33062  end_log_pos 3957  Query  thread_id=34  exec_time=0  error_code=0
SET TIMESTAMP=1568556750.039449/*!*/;
CREATE DEFINER=`root`@`localhost` TRIGGER trigger_test AFTER INSERT ON tb1 FOR EACH ROW INSERT INTO tb2(gmt_create) VALUES(now())
/*!*/;
```

以上为二进制日志中记录的 CREATE TRIGGER 语句的内容，是以 Query_log_event 事件的形式记录的，并且记录了执行该语句的用户的信息。

```
# at 3957
#190915 10:12:30 server id 33062  end_log_pos 4018  GTID  last_committed=14  sequence_number=15  rbr_only=no
SET @@SESSION.GTID_NEXT= '6385db95-cbc2-11e9-9f05-080027136405:366'/*!*/;
# at 4018
#190915 10:12:30 server id 33062  end_log_pos 4114  Query  thread_id=34  exec_time=0  error_code=0
SET TIMESTAMP=1568556750/*!*/;
BEGIN
/*!*/;
# at 4114
# at 4142
#190915 10:12:30 server id 33062  end_log_pos 4142  Intvar
SET INSERT_ID=9/*!*/;
#190915 10:12:30 server id 33062  end_log_pos 4289  Query  thread_id=34  exec_time=0  error_code=0
SET TIMESTAMP=1568556750/*!*/;
insert into tb1(name,gmt_create) values('trigger',now())
```

```
/*!*/;
# at 4289
#190915 10:12:30 server id 33062  end_log_pos 4316  Xid = 541
COMMIT/*!*/;
```

以上为触发器产生的操作对应的二进制日志，在基于 statement 的复制中，触发器产生的操作都以 statement 格式记录到二进制日志。

```
# at 4316
#190915 10:12:30 server id 33062  end_log_pos 4377   GTID      last_committed=15
sequence_number=16 rbr_only=no
SET @@SESSION.GTID_NEXT= '6385db95-cbc2-11e9-9f05-080027136405:367'/*!*/;
# at 4377
#190915 10:12:30 server id 33062 end_log_pos 4485  Query    thread_id=34    exec_time=0
error_code=0
SET TIMESTAMP=1568556750/*!*/;
drop trigger trigger_test
......
```

以上为二进制日志中记录的 DROP TRIGGER 语句的内容，以 Query_log_event 事件的形式记录。

33.4 小结

根据以上演示操作过程以及结果，总结存储过程、函数、触发器、事件在 MySQL 复制中的注意事项：

- CREATE EVENT、ALTER EVENT、DROP EVENT、CREATE PROCEDURE、DROP PROCEDURE、CREATE FUNCTION、DROP FUNCTION、CREATE TRIGGER、DROP TRIGGER 语句都以 Query_log_event 事件记录在二进制日志中并复制。
- 使用以上语句执行创建、删除、修改操作时产生的影响（对数据的 DML 操作之类）在基于 row 的复制或者混合复制中是以 row 格式记录的，在基于 statement 的复制中是以 statement 格式记录的。但需要注意的是，对于事件，其产生的影响即使在系统变量 binlog_format 为 statement 的情况下，也是以 row 格式记录的。
- 在执行 CREATE EVENT 和 ALTER EVENT 语句时，事件在从库上的状态都被设置为 SLAVESIDE_DISABLED，表示在从库上即使配置了事件，但是也不会调用，该状态可以在 information_schema.events 表的 STATUS 字段中查到。

第 34 章　复制 LIMIT 子句

DBA 一般会建议开发人员在使用 MySQL 时不要使用大事务，比如在对某一张表清理上百万行数据时，不建议直接用一条 DELETE 语句清理完所有数据，而是建议使用 LIMIT 子句，小批量、多次清理。LIMIT 子句对符合条件的结果集限定返回的行数，但是无法明确是哪几行。在 MySQL 的复制中，对于 LIMIT 子句这种具有不确定性的子句如何保证复制的一致性呢？本章我们就来看看 LIMIT 子句是如何被正确复制的。

34.1　操作环境信息

1. 操作系统版本

    Red Hat Enterprise Linux Server release 7.4（Maipo）

2. 数据库版本

    5.7.27-log MySQL Community Server（GPL）

3. 目的

    操作演示在不同事务隔离级别以及系统变量 binlog_format 的不同设置下，带 LIMIT 子句的 DML 语句是如何被记录到二进制日志中，并被正确复制到从库的。

4. 数据库系统变量的设置

    - gtid-mode = on
    - enforce-gtid-consistency = true
    - binlog_row_image = FULL
    - sql_mode = 'ONLY_FULL_GROUP_BY, STRICT_TRANS_TABLES, NO_ZERO_IN_DATE, NO_ZERO_DATE, ERROR_FOR_DIVISION_BY_ZERO, NO_AUTO_CREATE_USER, NO_ENGINE_SUBSTITUTION'

34.2　复制 LIMIT 子句的操作示例

本次演示分别在四种事务隔离级别以及三种二进制日志格式下进行，演示 LIMIT 子句在不同场景下如何被正确地复制。由于这些操作都是类似的，在此就不列出各场景下的操作记录，只列出操作语句，感兴趣的读者可以自己动手跟着操作。这些操作语句如下：

```
drop database if exists gangshen;
create database gangshen;
use gangshen;
create table tb1(id int primary key auto_increment,name varchar(20),gmt_create datetime default now());
create table tb2(id int primary key auto_increment,name varchar(20),gmt_create datetime default now());
insert into tb1(name) values("liuyi");
insert into tb2(name) values("chener");
insert into tb1(name) values("zhangsan");
insert into tb1(name) values("lisi");
insert into tb1(name) values("wangwu");
insert into tb1(name) values("zhaoliu");
insert into tb1(name) values("sunqi");
insert into tb1(name) values("zhouba");
insert into tb1(name) values("wujiu");
insert into tb1(name) values("zhengshi");
flush logs;
update tb1 set name = "update_name" limit 3;
delete from tb1 limit 3;
insert into tb2(name,gmt_create) select name,gmt_create from tb1 limit 3;
flush logs;
```

在操作过程中有几点需要注意：

- 在 READ-UNCOMMITTED 和 READ-COMMITTED 事务隔离级别下，当系统变量 binlog_format 设置为 statement 时，DML 语句无法执行，会报 ERROR 1665 错误。

- 在 REPEATABLE-READ 和 SERIALIZABLE 事务隔离级别下，系统变量 binlog_format 设置为 statement，当执行带 LIMIT 子句的 DML 语句时，MySQL 会有告警提示（因为无法保证查询结果的确定性），但是该语句能正常执行：

```
Note (Code 1592): Unsafe statement written to the binary log using statement format since BINLOG_FORMAT = STATEMENT. The statement is unsafe because it uses a LIMIT clause. This is unsafe because the set of rows included cannot be predicted.
```

34.3 二进制日志的解析及解释

参照 26.3.1.2 节的二进制日志解析方式，对 34.2 节中各场景下产生的二进制日志进行解析，分析解析的结果，确定 MySQL 是如何正确复制带 LIMIT 子句的 DML 语句的。通过分析，对结果可以分为两类进行解释：一类是系统变量 binlog_format 为 statement，另一类是系统变量 binlog_format 为 row 或 mixed。

34.3.1 statement 格式二进制日志的解析及解释

只有事务隔离级别为 REPEATABLE-READ 和 SERIALIZABLE 时，才支持以 statement 格式的二进制日志记录带 LIMIT 子句的 DML 语句。由于 statement 格式的二进制日志记录的内容是类似的，所以此处只给出 REPEATABLE-READ 隔离级别下的二进制日志解析结果。

```
......
# at 339
#190928  8:17:29 server id 33062  end_log_pos 465 Query thread_id=3 exec_time=0 error_code=0
use `gangshen`/*!*/;
SET TIMESTAMP=1569673049/*!*/;
update tb1 set name = "update_name" limit 3
/*!*/;
# at 465
#190928  8:17:29 server id 33062  end_log_pos 492 Xid = 399
COMMIT/*!*/;
```

statement 格式的二进制日志中记录的 UPDATE ... LIMIT 语句形式，与 MySQL 执行的原始语句一致。

```
# at 492
#190928  8:17:29 server id 33062  end_log_pos 553 GTID last_committed=1 sequence_number=2 rbr_only=no
SET @@SESSION.GTID_NEXT= '6385db95-cbc2-11e9-9f05-080027136405:737'/*!*/;
# at 553
#190928  8:17:29 server id 33062  end_log_pos 641 Query thread_id=3 exec_time=0 error_code=0
SET TIMESTAMP=1569673049/*!*/;
BEGIN
/*!*/;
# at 641
#190928  8:17:29 server id 33062  end_log_pos 747 Query thread_id=3 exec_time=0 error_code=0
SET TIMESTAMP=1569673049/*!*/;
delete from tb1 limit 3
/*!*/;
```

statement 格式的二进制日志中记录的 DELETE ... LIMIT 语句形式，与 MySQL 执行的原始语句一致。

```
# at 747
#190928  8:17:29 server id 33062  end_log_pos 774 Xid = 401
COMMIT/*!*/;
# at 774
#190928  8:17:29 server id 33062  end_log_pos 835 GTID last_committed=2 sequence_number=3 rbr_only=no
SET @@SESSION.GTID_NEXT= '6385db95-cbc2-11e9-9f05-080027136405:738'/*!*/;
# at 835
#190928  8:17:29 server id 33062  end_log_pos 923 Query thread_id=3 exec_time=0 error_code=0
SET TIMESTAMP=1569673049/*!*/;
BEGIN
```

```
/*!*/;
# at 923
# at 951
#190928 8:17:29 server id 33062 end_log_pos 951 Intvar
SET INSERT_ID=3/*!*/;
#190928 8:17:29 server id 33062 end_log_pos 1106 Query thread_id=3 exec_time=0 error_code=0
SET TIMESTAMP=1569673049/*!*/;
insert into tb2(name,gmt_create) select name,gmt_create from tb1 limit 3
/*!*/;
......
```

statement 格式的二进制日志中记录的 INSERT INTO ... SELECT ... LIMIT 语句形式，与 MySQL 执行的原始语句一致。

通过二进制日志的解析结果可以看到：

- 带有 LIMIT 子句的 DML 语句，在 statement 格式的二进制日志中会以 Query_log_event 事件的形式记录下来。

- 建议在基于 statement 的复制中，慎用带有 LIMIT 子句的 DML 语句，因为无法保证相同语句在主库和从库上查询出同样的数据，容易出现主从库数据不一致的情况。

34.3.2　row 格式和 mixed 格式二进制日志的解析及解释

当系统变量 binlog_format 设置为 row 或者 mixed 时，二进制日志记录的内容类似，所以这里只给出 REPEATABLE-READ 隔离级别下 row 格式二进制日志的解析结果。

```
# at 190
#190928 8:19:55 server id 33062 end_log_pos 251 GTID last_committed=0 sequence_number=1 rbr_only=yes
/*!50718 SET TRANSACTION ISOLATION LEVEL READ COMMITTED*//*!*/;
SET @@SESSION.GTID_NEXT= '6385db95-cbc2-11e9-9f05-080027136405:753'/*!*/;
......
# at 328
#190928 8:19:55 server id 33062 end_log_pos 391 Rows_query
# UPDATE tb1 SET name = "update_name" LIMIT 3
# at 391
#190928 8:19:55 server id 33062 end_log_pos 442 Table_map: `gangshen`.`tb1` mapped to number 141
# at 442
#190928 8:19:55 server id 33062 end_log_pos 590 Update_rows: table id 141 flags: STMT_END_F
......
### UPDATE `gangshen`.`tb1`
### WHERE
###   @1=1 /* INT meta=0 nullable=0 is_null=0 */
###   @2='liuyi' /* VARSTRING(60) meta=60 nullable=1 is_null=0 */
###   @3='2019-09-28 20:19:55' /* DATETIME(0) meta=0 nullable=1 is_null=0 */
### SET
```

```
    ### @1=1 /* INT meta=0 nullable=0 is_null=0 */
    ### @2='update_name' /* VARSTRING(60) meta=60 nullable=1 is_null=0 */
    ### @3='2019-09-28 20:19:55' /* DATETIME(0) meta=0 nullable=1 is_null=0 */
    ### UPDATE `gangshen`.`tb1`
    ### WHERE
    ###   @1=3 /* INT meta=0 nullable=0 is_null=0 */
    ###   @2='zhangsan' /* VARSTRING(60) meta=60 nullable=1 is_null=0 */
    ###   @3='2019-09-28 20:19:55' /* DATETIME(0) meta=0 nullable=1 is_null=0 */
    ### SET
    ###   @1=3 /* INT meta=0 nullable=0 is_null=0 */
    ###   @2='update_name' /* VARSTRING(60) meta=60 nullable=1 is_null=0 */
    ###   @3='2019-09-28 20:19:55' /* DATETIME(0) meta=0 nullable=1 is_null=0 */
    ### UPDATE `gangshen`.`tb1`
    ### WHERE
    ###   @1=5 /* INT meta=0 nullable=0 is_null=0 */
    ###   @2='lisi' /* VARSTRING(60) meta=60 nullable=1 is_null=0 */
    ###   @3='2019-09-28 20:19:55' /* DATETIME(0) meta=0 nullable=1 is_null=0 */
    ### SET
    ###   @1=5 /* INT meta=0 nullable=0 is_null=0 */
    ###   @2='update_name' /* VARSTRING(60) meta=60 nullable=1 is_null=0 */
    ###   @3='2019-09-28 20:19:55' /* DATETIME(0) meta=0 nullable=1 is_null=0 */
    # at 590
    #190928 8:19:55 server id 33062 end_log_pos 617 Xid = 427
    COMMIT/*!*/;
```

对于 UPDATE ... LIMIT 语句，在使用 row 格式的二进制日志时，MySQL 会将每一行变更都记录在二进制日志中。

```
    # at 617
    #190928 8:19:55 server id 33062 end_log_pos 678 GTID last_committed=1 sequence_number=2 rbr_only=yes
    /*!50718 SET TRANSACTION ISOLATION LEVEL READ COMMITTED*//*!*/;
    SET @@SESSION.GTID_NEXT= '6385db95-cbc2-11e9-9f05-080027136405:754'/*!*/;
    # at 678
    #190928 8:19:55 server id 33062 end_log_pos 755 Query thread_id=3 exec_time=0 error_code=0
    SET TIMESTAMP=1569673195/*!*/;
    BEGIN
    /*!*/;
    # at 755
    #190928 8:19:55 server id 33062 end_log_pos 798 Rows_query
    # delete from tb1 limit 3
    # at 798
    #190928 8:19:55 server id 33062 end_log_pos 849 Table_map: `gangshen`.`tb1` mapped to number 141
    # at 849
    #190928 8:19:55 server id 33062 end_log_pos 946 Delete_rows: table id 141 flags: STMT_END_F
    ......
    ### DELETE FROM `gangshen`.`tb1`
    ### WHERE
```

```
    ### @1=1 /* INT meta=0 nullable=0 is_null=0 */
    ### @2='update_name' /* VARSTRING(60) meta=60 nullable=1 is_null=0 */
    ### @3='2019-09-28 20:19:55' /* DATETIME(0) meta=0 nullable=1 is_null=0 */
    ### DELETE FROM `gangshen`.`tb1`
    ### WHERE
    ### @1=3 /* INT meta=0 nullable=0 is_null=0 */
    ### @2='update_name' /* VARSTRING(60) meta=60 nullable=1 is_null=0 */
    ### @3='2019-09-28 20:19:55' /* DATETIME(0) meta=0 nullable=1 is_null=0 */
    ### DELETE FROM `gangshen`.`tb1`
    ### WHERE
    ### @1=5 /* INT meta=0 nullable=0 is_null=0 */
    ### @2='update_name' /* VARSTRING(60) meta=60 nullable=1 is_null=0 */
    ### @3='2019-09-28 20:19:55' /* DATETIME(0) meta=0 nullable=1 is_null=0 */
    # at 946
    #190928 8:19:55 server id 33062  end_log_pos 973 Xid = 428
    COMMIT/*!*/;
```

对于 DELETE ... LIMIT 语句，在使用 row 格式的二进制日志时，MySQL 会将每一行变更的记录都记录在二进制日志中。

```
    # at 973
    #190928 8:19:55 server id 33062  end_log_pos 1034 GTID last_committed=2 sequence_number=3 rbr_only=yes
    /*!50718 SET TRANSACTION ISOLATION LEVEL READ COMMITTED*//*!*/;
    SET @@SESSION.GTID_NEXT= '6385db95-cbc2-11e9-9f05-080027136405:755'/*!*/;
    # at 1034
    #190928 8:19:55 server id 33062  end_log_pos 1111 Query  thread_id=3 exec_time=0 error_code=0
    SET TIMESTAMP=1569673195/*!*/;
    BEGIN
    /*!*/;
    # at 1111
    #190928 8:19:55 server id 33062  end_log_pos 1203 Rows_query
    # insert into tb2(name,gmt_create) select name,gmt_create from tb1 limit 3
    # at 1203
    #190928 8:19:55 server id 33062  end_log_pos 1254 Table_map: `gangshen`.`tb2` mapped to number 142
    # at 1254
    #190928 8:19:55 server id 33062  end_log_pos 1336 Write_rows: table id 142 flags: STMT_END_F
    ......
    ### INSERT INTO `gangshen`.`tb2`
    ### SET
    ### @1=3 /* INT meta=0 nullable=0 is_null=0 */
    ### @2='wangwu' /* VARSTRING(60) meta=60 nullable=1 is_null=0 */
    ### @3='2019-09-28 20:19:55' /* DATETIME(0) meta=0 nullable=1 is_null=0 */
    ### INSERT INTO `gangshen`.`tb2`
    ### SET
    ### @1=5 /* INT meta=0 nullable=0 is_null=0 */
    ### @2='zhaoliu' /* VARSTRING(60) meta=60 nullable=1 is_null=0 */
```

```
### @3='2019-09-28 20:19:55' /* DATETIME(0) meta=0 nullable=1 is_null=0 */
### INSERT INTO `gangshen`.`tb2`
### SET
### @1=7 /* INT meta=0 nullable=0 is_null=0 */
### @2='sunqi' /* VARSTRING(60) meta=60 nullable=1 is_null=0 */
### @3='2019-09-28 20:19:55' /* DATETIME(0) meta=0 nullable=1 is_null=0 */
# at 1336
#190928 8:19:55 server id 33062 end_log_pos 1363 Xid = 429
COMMIT/*!*/;
```

对于 INSERT INTO ... SELECT ... LIMIT 语句，在基于 row 的复制中，MySQL 将每一行的变更都记录在二进制日志中。

通过二进制日志的解析结果，可以看到：

- 在基于 row 的复制或者混合复制中，带有 LIMIT 子句的 DML 语句会以 row 格式记录在二进制日志中，记录了每一行的具体变更。
- 基于 row 的复制或者混合复制，能保证 LIMIT 子句复制的正确性。

34.4 小结

从操作演示的结果看，带有 LIMIT 子句的 DML 语句能被复制到从库，但是使用基于 statement 的复制无法保证主从库数据的一致性，而基于 row 的复制能保证，所以建议使用后者。

第 35 章　复制 LOAD DATA 语句

　　LOAD DATA 语句能很方便地将文件中的数据导入 MySQL，是开发人员平时习惯使用的一种导入数据的方式。本章我们就来看一看，在 MySQL 主从复制架构中是如何保证 LOAD DATA 语句正确地从主库复制到从库的。

35.1　操作环境信息

1. 操作系统版本

   Red Hat Enterprise Linux Server release 7.4（Maipo）

2. 数据库版本

   5.7.27-log MySQL Community Server（GPL）

3. 目的

   演示在不同事务隔离级别以及系统变量 binlog_format 的不同设置下，LOAD DATA 语句是如何被记录到二进制日志，并正确复制到从库的。

4. 数据库系统变量的设置

   - gtid-mode = on
   - enforce-gtid-consistency = true
   - binlog_row_image = FULL
   - secure_file_priv = "/tmp"
   - sql_mode = 'ONLY_FULL_GROUP_BY, STRICT_TRANS_TABLES, NO_ZERO_IN_DATE, NO_ZERO_DATE, ERROR_FOR_DIVISION_BY_ZERO, NO_AUTO_CREATE_USER, NO_ENGINE_SUBSTITUTION'

35.2　复制 LOAD DATA 语句的操作示例

35.2.1　准备演示数据

（1）建库：

```
mysql> create database gangshen;
Query OK, 1 row affected (0.01 sec)
```

（2）建表：

```
mysql> use gangshen
Database changed

mysql> create table if not exists test_load(id int unsigned not null primary key auto_increment,test varchar(100));
Query OK, 0 rows affected (0.02 sec)
```

（3）插入数据：

```
mysql> insert into test_load(test) values('1'),('2'),('null'),('4');
Query OK, 4 rows affected (0.01 sec)
Records: 4  Duplicates: 0  Warnings: 0

mysql> select * from test_load;
+----+------+
| id | test |
+----+------+
|  1 | 1    |
|  3 | 2    |
|  5 | null |
|  7 | 4    |
+----+------+
4 rows in set (0.00 sec)
```

（4）执行 SELECT ... INTO OUTFILE 语句，生成导入数据需要的数据文本文件：

```
mysql> select * from test_load into outfile "/tmp/test_load.txt";
Query OK, 4 rows affected (0.00 sec)
```

\# 使用 system 命令可以在 MySQL 命令行客户端中调用 OS 命令行命令。例如，这里为了方便，直接在 MySQL 命令行客户端中使用 cat 命令查看 /tmp/test_load.txt 文件（注意，文件过大时须慎用）

```
mysql> system cat /tmp/test_load.txt;
1	1
3	2
5	null
7	4
```

35.2.2 LOAD DATA 语句的操作

本次演示分别在四种事务隔离级别以及三种二进制日志格式下进行，演示 LOAD DATA 语句在不同场景下是如何被正确复制的。由于这些操作类似，这里就不列出各场景下的操作记录，只列出操作语句。感兴趣的读者可以自己动手操作。这些操作语句如下：

```
flush logs;
use gangshen;
truncate table test_load;
load data infile '/tmp/test_load.txt' into table test_load;
select * from test_load;
```

在操作过程中需要注意：在 READ-UNCOMMITTED 和 READ-COMMITTED 事务隔离级别下，当系统变量 binlog_format 设置为 statement 时，DML 语句无法执行，所以在执行 LOAD DATA 语句时，会报 ERROR 1665 错误。

35.3 二进制日志的解析及解释

参照 26.3.1.2 节中二进制日志的解析方式，对 35.2 节中各个场景下产生的二进制日志进行解析，分析解析结果，确定 MySQL 是如何正确复制 LOAD DATA 语句的。对结果可以分为两类进行解释：一类是系统变量 binlog_format 为 statement，另一类是系统变量 binlog_format 为 row 或者 mixed。

35.3.1 statement 格式二进制日志的解析及解释

因为只有事务隔离级别为 REPEATABLE-READ 和 SERIALIZABLE 时，才支持以 statement 格式的二进制日志记录 LOAD DATA 语句，而且两种情况下 statement 格式二进制日志记录的内容类似，所以在此只给出 REPEATABLE-READ 隔离级别下的解析结果。

```
......
# at 190
#191003 11:37:25 server id 33062 end_log_pos 251 GTID last_committed=0 sequence_number=1 rbr_only=no
SET @@SESSION.GTID_NEXT= '6385db95-cbc2-11e9-9f05-080027136405:855'/*!*/;
# at 251
#191003 11:37:25 server id 33062 end_log_pos 358 Query thread_id=2 exec_time=1 error_code=0
USE `gangshen`/*!*/;
......
# at 358
#191003 11:37:26 server id 33062 end_log_pos 419 GTID last_committed=1 sequence_number=2 rbr_only=no
SET @@SESSION.GTID_NEXT= '6385db95-cbc2-11e9-9f05-080027136405:856'/*!*/;
# at 419
#191003 11:37:26 server id 33062 end_log_pos 507 Query thread_id=2 exec_time=0
```

```
    error_code=0
       SET TIMESTAMP=1570117046/*!*/;
       BEGIN
       /*!*/;
       # at 507
       #191003 11:37:26 server id 33062  end_log_pos 549
       #Begin_load_query: file_id: 1 block_len: 19
       # at 549
       #191003  11:37:26 server  id 33062  end_log_pos  802  Execute_load_query  thread_id=2
    exec_time=0 error_code=0
       SET TIMESTAMP=1570117046/*!*/;
       LOAD DATA LOCAL INFILE '/tmp/SQL_LOAD_MB-1-0' INTO TABLE `test_load` FIELDS TERMINATED
    BY '\t' ENCLOSED BY '' ESCAPED BY '\\' LINES TERMINATED BY '\n' (`id`, `test`)
       /*!*/;
       # file_id: 1
       # at 802
       #191003 11:37:26 server id 33062  end_log_pos 829 Xid = 85
       COMMIT/*!*/;
       ......
```

从上面解析的二进制日志结果中可以看到：

- LOAD DATA 语句在内部被转换为 local 方式，走了 local 方式的流程，并且二进制日志中记录的是输入的原始 SQL 语句内容，以及默认的字段分隔符、转义符、行分隔符和导入表中的字段列表。

- LOAD DATA 语句在二进制日志中的关键字是 Begin_load_query 和 Execute_load_query。

- 原始的文件路径是 /tmp/test_load.txt，但是在二进制日志中是 /tmp/SQL_LOAD_MB-1-0。这是因为如果二进制日志为 statement 格式，从库在解析二进制日志文件的 Begin_load_query_log_event 类型事件时（如 Begin_load_query: file_id: 1 block_len: 19），把 file_id 读取出来作为文件名称索引，然后到 tmpdir 中（默认是在系统的 /tmp 目录下）查询以 SQL_LOAD_MB 为前缀名的文件。如果与 file_id 匹配，就看该文件名的最后一位数字是多少。例如文件名为 SQL_LOAD_MB-1-0，最后一位数字是 0，为了唯一标识这个文件，会把这个数字加 1（等于 1），生成新的文件名（如果没有发现与 file_id 匹配的文件，则最后一位数字以 0 开始）。当发现 Execute_load_query_log_event 类型的事件时，用这个新的文件名在 tmpdir 下创建一个新的临时文件，然后把从中继日志中提取的文本数据存放到这个临时文件中，并使用这个新文件名来生成 LOAD DATA 语句。

- LOAD DATA 语句导入的数据实际上存放在二进制日志中，但是 mysqlbinlog 解析 statement 格式二进制日志时并没有将数据展示出来。

- 当使用 mysqlbinlog 读取以 statement 格式记录的 LOAD DATA 语句时，会在 tmpdir 中创建一个本地临时文件。mysqlbinlog 解析完 LOAD DATA 语句的日志事件之后，

mysqlbinlog 或其他的 MySQL 程序不会自动删除这些临时文件，需要我们自行删除。例如，对于上述 REPETABLE-READ 事务隔离级别下 statement 格式的二进制日志，使用 mysqlbinlog 解析后，会在/tmp/目录下生成一个名为 SQL_LOAD_MB-1-0 的文件，该文件中保存的就是 LOAD DATA 导入的数据。

35.3.2　row 格式和 mixed 格式二进制日志的解析及解释

当系统变量 binlog_format 设置为 row 或者 mixed 时，二进制日志记录的内容类似，所以这里只给出 REPETABLE-READ 隔离级别下 row 格式的二进制日志的解析结果。

```
......
TRUNCATE table test_load
/*!*/;
# at 347
#191003 11:58:26 server id 33062 end_log_pos 408 GTID last_committed=1 sequence_number=2 rbr_only=yes
/*!50718 SET TRANSACTION ISOLATION LEVEL READ COMMITTED*//*!*/;
SET @@SESSION.GTID_NEXT= '6385db95-cbc2-11e9-9f05-080027136405:860'/*!*/;
# at 408
#191003 11:58:26 server id 33062 end_log_pos 485 Query thread_id=2 exec_time=0 error_code=0
SET TIMESTAMP=1570118306/*!*/;
BEGIN
/*!*/;
# at 485
#191003 11:58:26 server id 33062 end_log_pos 563 Rows_query
# LOAD DATA INFILE '/tmp/test_load.txt' INTO TABLE test_load
# at 563
#191003 11:58:26 server id 33062 end_log_pos 618 Table_map: `gangshen`.`test_load` mapped to number 118
# at 618
#191003 11:58:26 server id 33062 end_log_pos 684 Write_rows: table id 118 flags: STMT_END_F
......
### INSERT INTO `gangshen`.`test_load`
### SET
###   @1=1 /* INT meta=0 nullable=0 is_null=0 */
###   @2='1' /* VARSTRING(300) meta=300 nullable=1 is_null=0 */
### INSERT INTO `gangshen`.`test_load`
### SET
###   @1=3 /* INT meta=0 nullable=0 is_null=0 */
###   @2='2' /* VARSTRING(300) meta=300 nullable=1 is_null=0 */
### INSERT INTO `gangshen`.`test_load`
### SET
###   @1=5 /* INT meta=0 nullable=0 is_null=0 */
###   @2='null' /* VARSTRING(300) meta=300 nullable=1 is_null=0 */
### INSERT INTO `gangshen`.`test_load`
### SET
```

```
### @1=7 /* INT meta=0 nullable=0 is_null=0 */
### @2='4' /* VARSTRING(300) meta=300 nullable=1 is_null=0 */
# at 684
#191003 11:58:26 server id 33062  end_log_pos 711 Xid = 99
COMMIT/*!*/;
......
```

从上面的结果中可以看到，数据写入成功，LOAD DATA 语句以 row 格式记录在二进制日志中。

35.4　小结

在上面的操作演示过程中，在主库上成功执行的 LOAD DATA 语句，都能被正确地复制到从库，但是通常不建议在二进制日志为 statement 格式时使用 LOAD DATA 语句，这是不安全的。

使用 LOCAL 子句与不使用有如下差异：

- 如果 LOAD DATA 语句使用了 LOCAL 子句，则客户端使用 TCP 远程连接 MySQL 时，即使没有 file 权限也仍然能够导入文本文件。这是非常危险的，因为 LOCAL 子句是从客户端的主机读取文本文件，传送到服务端的/tmp 目录并保存为一个临时文件，再执行 LOAD DATA 语句的。另外，要使用 LOCAL 子句，还需要看服务端启动时是否关闭了 local_infile 选项（如果未指定该选项，则服务端默认开启），以及客户端连接时是否关闭了 local_infile 选项（如果未指定该选项，则客户端默认开启），只要服务端或客户端任意一端关闭了 local_infile，就不能使用 LOCAL 子句，否则会报错：ERROR 1148 (42000): The used command is not allowed with this MySQL version。

- 如果 LOAD DATA 语句不使用 LOCAL 子句，则用户必须有 file 权限才能导入文本文件（并且只能导入服务端的本地文本文件），如果没有 file 权限，可能会报没有 file 权限的错误，也可能报如下错误：ERROR 1045 (28000): Access denied for user 'test'@'%' (using password: YES)。

- 因为要限制客户端在没有 file 权限时使用 LOAD DATA 语句，所以需要在服务端使用 local_infile = OFF 来关闭 local_infile；不使用 LOCAL 子句时，如果用户没有 file 权限，显然就无法使用 LOAD DATA 语句，但是如果还想限制具有 file 权限的用户，可以设置系统变量 secure_file_priv 为 null，全面禁止使用 LOAD DATA 语句（不管是否使用了 LOCAL 子句都不允许执行 LOAD DATA 语句）。

第 36 章　系统变量 max_allowed_packet 对复制的影响

熟悉 MySQL 的人都知道，MySQL 有许多可设置的系统变量，用于控制其行为或者性能，比如系统变量 max_allowed_packet 可以控制客户端与服务端连接后，传输的数据包的最大尺寸。本章我们讲一讲系统变量 max_allowed_packet 对复制的影响，再介绍一些其他类似的系统变量。

36.1　系统变量简介

1. max_allowed_packet

它控制一个数据包或由 mysql_stmt_send_long_data() 函数发送的所有参数的最大尺寸，默认为 4 MB。要注意，此系统变量在客户端和服务端要设置为同样的值，比如在使用 mysqldump 进行备份时，如果要生成整表单条 INSERT 语句，max_allowed_packet 太小可能导致备份失败。一般建议将其设置为 32 MB 或 64 MB。

- 包消息缓冲区初始化为 net_buffer_length 定义的大小，但在需要时可以自动增加到 max_allowed_packet 定义的值。
- 如果使用大 BLOB 列或长字符串，则必须增加此值，把它设置为你想要使用的最大 BLOB 值。max_allowed_packet 的协议限制为 1 GB，其值应为 1024 的倍数，否则四舍五入取最接近的倍数。
- 通过更改 max_allowed_packet 的值来更改消息缓冲区大小时，如果客户端程序允许修改，则应该同时更改客户端的缓冲区大小。内置的客户端库中 max_allowed_packet 的默认值为 1 GB，但客户端程序可能会使用此系统变量定义另一个值，该自定义值将覆盖客户端库中的默认值。例如，mysql 和 mysqldump 分别定义 max_allowed_packet 的默认值为 16 MB 和 24 MB。它们还允许在命令行或选项文件中设置 max_allowed_packet 变量控制来客户端建立连接时的 max_allowed_packet 值。
- max_allowed_packet 是全局变量、会话变量、动态变量（注意，仅在全局级别是动态的，在会话级别是只读的，所以对于服务端而言，动态修改全局值会立即影响当前会话发送的包大小，此时会忽略会话变量的大小，而客户端变量的大小仍然是以会话变量为准的。因此，建议在动态修改这个值之后，断开连接，然后重连，避免出现不可预估的结果），单位为字节。MySQL 5.6.5 及其之前版本，max_allowed_packet 的默认值为 1 MB，从 MySQL 5.6.6 开始，默认值为 4 MB，最小值为 1 KB，最大值为 1 GB，整型值。

2. slave_max_allowed_packet

此变量设置从库 SQL 线程和 I/O 线程的最大数据包尺寸，以便在基于 row 的复制中，减少因为较大的事务更新产生的数据包大小超过 max_allowed_packet 设置的值，导致报错的问题。如果设置了此变量，会立即生效，包括正在运行的所有复制通道。

- 该变量的值必须是 1024 的正整数倍。如果设置的值不是 1024 的正整数倍，就向下舍入为大于该值的下一个 1024 的倍数值。例如，设置为 0，MySQL 会自动更正为 1024，同时发出设置值被截断的警告信息）。
- 也可以在启动时使用 --slave-max-allowed-packet 选项来设置此系统变量。
- 它是一个全局变量、动态变量，整型值，默认值为 1 GB，取值范围为 1 KB ~ 1GB。

3. net_buffer_length

此变量设置每个客户端会话线程相关的连接缓冲区和结果集缓冲区的初始大小。这两个缓冲区在连接创建伊始就被设置为 net_buffer_length 的值，后面根据需要，可动态增加到系统变量 max_allowed_packet 指定的大小，每个 SQL 语句执行完之后，结果集缓冲区自动缩至 net_buffer_length 指定的大小。

- 该变量通常不应该更改，但是如果内存很小，则可以将其设置得小一些。
- 它是一个全局变量、会话变量（在会话级别为只读变量）、动态变量，整型值，默认值为 16 KB，取值范围为 1024 字节~1048576 字节（1 KB~1 MB）。

36.2 操作环境信息

1. 主库 IP 地址

   172.20.10.3

2. 从库 IP 地址

   172.20.10.4

3. 操作系统版本

   Red Hat Enterprise Linux Server release 7.4（Maipo）

4. 数据库版本

   5.7.27-log MySQL Community Server（GPL）

5. 数据库系统变量的设置
- character_set = utf8
- transaction_isolation = REPEATABLE-READ
- binlog_format = ROW
- sql_mode ='ONLY_FULL_GROUP_BY, STRICT_TRANS_TABLES, NO_ZERO_IN_DATE, NO_ZERO_DATE, ERROR_FOR_DIVISION_BY_ZERO, NO_AUTO_CREATE_USER, NO_ENGINE_SUBSTITUTION'

36.3　max_allowed_packet 对复制的影响操作示例

本节演示系统变量 max_allowed_packet 的值设置过小时，对主库和从库的影响以及如何解决带来的问题。

36.3.1　max_allowed_packet 对主库的影响

（1）安装 MySQL，搭建主从复制（步骤省略）。

（2）建库：

```
mysql> create database gangshen;
Query OK, 1 row affected (0.01 sec)
```

（3）建表：

```
mysql> use gangshen;
Database changed
mysql> create table test(id int primary key auto_increment,name blob);
Query OK, 0 rows affected (0.02 sec)
```

（4）在全局级别将主库上的 max_allowed_packet、net_buffer_length 设置为最小值 1024，并重新连接：

```
mysql> set global net_buffer_length=1024;
Query OK, 0 rows affected (0.00 sec)

mysql> set global max_allowed_packet=1024;
Query OK, 0 rows affected (0.00 sec)

mysql> \q
Bye

# 重新连接 MySQL Server
mysql> show variables where variable_name in ('max_allowed_packet', 'net_buffer_length')\G
*************************** 1. row ***************************
Variable_name: max_allowed_packet
```

```
         Value: 1024
*************************** 2. row ***************************
Variable_name: net_buffer_length
         Value: 1024
2 rows in set (0.00 sec)
```

注意：在设置 max_allowed_packet 时，需要先将 net_buffer_length 设置为 1024，否则会提示设置失败。

（5）在主库上插入数据：

```
mysql> insert into test(name) values('这是一条十四个字符的演示语句这是一条十四个字符的演示语句这是一条十四个字符的演示语句这是一条十四个字符的演示语句这是一条十四个字符的演示语句这是一条十四个字符的演示语句这是一条十四个字符的演示语句这是一条十四个字符的演示语句这是一条十四个字符的演示语句这是一条十四个字符的演示语句这是一条十四个字符的演示语句这是一条十四个字符的演示语句这是一条十四个字符的演示语句这是一条十四个字符的演示语句这是一条十四个字符的演示语句这是一条十四个字符的演示语句这是一条十四个字符的演示语句这是一条十四个字');
ERROR 1153 (08S01): Got a packet bigger than 'max_allowed_packet' bytes
ERROR 2006 (HY000): MySQL server has gone away
No connection. Trying to reconnect...
Connection id:   54
Current database: gangshen
```

可以看到，当 max_allowed_packet 的值设置过小时，如果 MySQL 客户端传输的数据包过大，MySQL 服务端会报错（ERROR 1153），并重新建立连接。

（6）查看错误日志：

```
2019-10-03T17:01:21.819750-05:00 53 [Note] Aborted connection 53 to db: 'gangshen' user: 'root' host: 'localhost' (Got a packet bigger than 'max_allowed_packet' bytes)
```

在错误日志中也有提示信息，提示 id 为 53 的连接其数据包大小超过 max_allowed_packet 设置的值，导致连接断开。

36.3.2　max_allowed_packet 对从库的影响

（1）在从库上检查主从复制的状态是否正常：

```
mysql> show slave status\G
*************************** 1. row ***************************
……
            Slave_IO_Running: Yes
           Slave_SQL_Running: Yes
……
1 row in set (0.00 sec)
```

（2）在全局级别将从库上的 net_buffer_length、max_allowed_packet、slave_max_allowed_packet 设置为 1024：

```
mysql> stop slave;
```

```
Query OK, 0 rows affected (0.02 sec)

mysql> set global net_buffer_length=1024;
Query OK, 0 rows affected (0.00 sec)

mysql> set global max_allowed_packet=1024;
Query OK, 0 rows affected (0.00 sec)

mysql> set global slave_max_allowed_packet=1024;
Query OK, 0 rows affected (0.00 sec)

mysql> start slave;
Query OK, 0 rows affected (0.01 sec)
```

注意：要在停止从库复制线程之后设置系统变量，并重新启动从库复制线程，因为系统变量 slave_max_allowed_packet 只能在全局级别修改，修改之后需要重新建立复制关系才能生效。

（3）在主库上插入数据，并观察从库的复制状态：

```
# Master:172.20.10.3
mysql> insert into test(name) values('这是一条十四个字符的演示语句这是一条十四个字符的测 试语句这是一条十四个字符的演示语句这是一条十四个字符的演示语句这是一条十四个字符的演示语句这是一条十四个字符的演示语句这是一条十四个字符的演示语句这是一条十四个字符的演示语句这是一条十四个字符的演示语句这是一条十四个字符的演示语句这是一条十四个字符的演示语句这是一条十四个字符的演示语句这是一条十四个字符的演示语句这是一条十四个字符的演示语句这是一条十四个字符的演示语句这是一条十四个字符的演示语句这是一条十四个字符的演示语句这是一条十四个字符的演示语句这是一条十四个字符的演示语句这是一条十四个字符的演示语句这是一条十四个字符的演示语句这是一条十四个字符的演示语句这是一条十四个字符的演示语句这是一条十四个字符的演示语句这是一条十四个');
Query OK, 1 row affected (0.01 sec)

# Slave:172.20.10.4
mysql> show slave status\g
*************************** 1. row ***************************
......
           Slave_IO_Running: Yes
          Slave_SQL_Running: Yes
......
1 row in set (0.00 sec)

# Master:172.20.10.3
mysql> insert into test(name) select name from test;
query ok, 1 row affected (0.01 sec)
Records: 1  Duplicates: 0  Warnings: 0
# Slave:172.20.10.4
mysql> show slave status\g
*************************** 1. row ***************************
......
           Slave_IO_Running: Yes
          Slave_SQL_Running: Yes
```

```
......
1 row in set (0.00 sec)

# Master:172.20.10.3
mysql> insert into test(name) select name from test;
Query OK, 2 rows affected (0.01 sec)
Records: 2  Duplicates: 0  Warnings: 0

# Slave:172.20.10.4
mysql> show slave status\G
*************************** 1. row ***************************
......
          Slave_IO_Running: Yes
         Slave_SQL_Running: Yes
......
1 row in set (0.00 sec)

# Master:172.20.10.3
mysql> insert into test(name) select name from test;
Query OK, 4 rows affected (0.02 sec)
Records: 4  Duplicates: 0  Warnings: 0

# Slave:172.20.10.4
mysql> show slave status\G
*************************** 1. row ***************************
......
          Slave_IO_Running: Yes
         Slave_SQL_Running: Yes
......
1 row in set (0.00 sec)

# Master:172.20.10.3
mysql> insert into test(name) select name from test;
Query OK, 8 rows affected (0.01 sec)
Records: 8  Duplicates: 0  Warnings: 0

# Slave:172.20.10.4
mysql> show slave status\G
*************************** 1. row ***************************
......
          Slave_IO_Running: No
         Slave_SQL_Running: Yes
......
            Last_IO_Errno: 1153
            Last_IO_Error: Got a packet bigger than 'slave_max_allowed_packet' bytes
           Last_SQL_Errno: 0
           Last_SQL_Error:
......
1 row in set (0.00 sec)
```

可以看到，当主库上执行的 INSERT INTO 语句影响的行数达到一定数量时，主从库之间的复制会报错，提示超过了系统变量 slave_max_allowed_packet 设置的阈值。

（4）查看从库的错误日志：

```
2019-10-06T13:25:50.608359-05:00 37 [ERROR] Error reading packet from server for channel
'': Got packet bigger than 'max_allowed_packet' bytes (server_errno=2020)
2019-10-06T13:25:50.608385-05:00 37 [ERROR] Log entry on master is longer than
slave_max_allowed_packet (1024) on slave. If the entry is correct, restart the server with
a higher value of slave_max_allowed_packet
2019-10-06T13:25:50.608393-05:00 37 [ERROR] Slave I/O for channel '': Got a packet bigger
than 'slave_max_allowed_packet' bytes, Error_code: 1153
2019-10-06T13:25:50.608396-05:00 37 [Note] Slave I/O thread exiting for channel '', read
up to log 'mysql-bin.000106', position 808676
```

从错误日志中也可以看到，I/O 线程在传输数据包时，发现数据包大小超过了 slave_max_allowed_packet 设置的值，所以 I/O 线程中断。

（5）根据提示在从库上修改 slave_max_allowed_packet 值，并重新启动复制线程：

```
mysql> set global slave_max_allowed_packet=102400;
Query OK, 0 rows affected (0.00 sec)

mysql> start slave;
Query OK, 0 rows affected (0.00 sec)

mysql> show slave status\G
*************************** 1. row ***************************
......
        Slave_IO_Running: Yes
       Slave_SQL_Running: Yes
......
1 row in set (0.00 sec)
```

36.4 小结

从上面的操作演示过程可以看到，系统变量 max_allowed_packet 设置的是 MySQL 服务端和客户端之间传输的单个数据包大小的上限。如果要复制大型列的值（例如 BLOB、TEXT 类型字段），主库上设置的 max_allowed_packet 值太小，则主库会报告"Got a packet bigger than 'max_allowed_packet' bytes"的错误，并重新建立连接；如果从库上设置的 max_allowed_packet 或 slave_max_allowed_packet 值太小，则从库的 I/O 线程也会报错，导致从库中止 I/O 线程。

在基于 row 的复制中，需要将发生变更的数据的整行（整个表的所有列）发送到从库中，即在复制大型列时，需要将 max_allowed_packet 和 slave_max_allowed_packet 的值设置得足够大，以容纳要复制的表中的最大行。

在基于 statement 的复制中，也需要注意，执行的 SQL 语句不应过大。

系统变量 max_allowed_packet、slave_max_allowed_packet 影响的是 MySQL 协议中数据包的大小，由于数据包会层层封装以及压缩，所以数据包大小并不单纯指 SQL 语句或插入语句的大小。

第 37 章 复制临时表

与普通的表相比，相信许多读者对于临时表应该既陌生又熟悉，本书讲的是 MySQL 的复制，那么本章就来讲一讲临时表在 MySQL 中是如何复制的，以及在复制中遇到临时表该怎么办。

37.1 操作环境信息

1. IP 地址
   - 主库：172.20.10.3
   - 从库：172.20.10.4

2. 操作系统版本

   Red Hat Enterprise Linux Server release 7.4（Maipo）

3. 数据库版本：

   5.7.27-log MySQL Community Server（GPL）

4. 数据库系统变量的设置
   - character_set = utf8
   - sql_mode = 'ONLY_FULL_GROUP_BY, STRICT_TRANS_TABLES, NO_ZERO_IN_DATE, NO_ZERO_DATE, ERROR_FOR_DIVISION_BY_ZERO, NO_AUTO_CREATE_USER, NO_ENGINE_SUBSTITUTION'

37.2 复制临时表的操作示例

本节主要演示临时表在 MySQL 中是如何被复制的。为了方便，选择显式创建临时表的方式进行演示。

由于在 READ-UNCOMITTED 和 READ-COMITTED 隔离级别下，系统变量 binlog_format 设置为 statement 时无法执行 DML 语句，所以本次操作是在 REPEATABLE-READ 隔离级别下进行的。

37.2.1 基于 statement 的复制且隔离级别为 REPEATABLE-READ

（1）设置主库的隔离级别以及二进制日志格式：

```
mysql> set session transaction_isolation='repeatable-read';
Query OK, 0 rows affected (0.00 sec)

mysql> set session binlog_format='statement';
Query OK, 0 rows affected (0.00 sec)

mysql> show variables where variable_name in ('transaction_isolation','binlog_format')\G
*************************** 1. row ***************************
Variable_name: binlog_format
        Value: STATEMENT
*************************** 2. row ***************************
Variable_name: transaction_isolation
        Value: REPEATABLE-READ
2 rows in set (0.00 sec)
```

（2）在主库上创建临时表并插入数据：

```
mysql> use gangshen
Database changed

mysql> create temporary table test(id int primary key auto_increment,name varchar(20));
Query OK, 0 rows affected (0.01 sec)

mysql> insert into test(name) values('tmp_test');
Query OK, 1 row affected (0.00 sec)

mysql> select * from test;
+----+----------+
| id | name     |
+----+----------+
|  1 | tmp_test |
+----+----------+
1 row in set (0.00 sec)
```

（3）检查从库的数据以及复制的状态：

```
mysql> use gangshen
Database changed
mysql> show tables;
Empty set (0.00 sec)

mysql> show slave status\G
*************************** 1. row ***************************
......
            Slave_IO_Running: Yes
```

```
         Slave_SQL_Running: Yes
......
1 row in set (0.00 sec)
```

可以看到主从复制的状态正常,但是在从库上看不到临时表。

(4)在主库上删除临时表:

```
mysql> drop table test;
Query OK, 0 rows affected (0.01 sec)
```

(5)解析主库产生的二进制日志:

```
......
# at 251
#191013  4:21:28 server id 33062  end_log_pos 413       Query       thread_id=3
exec_time=0    error_code=0
use `gangshen`/*!*/;
......
create temporary table test(id int primary key auto_increment,name varchar(20))
/*!*/;
# at 413
#191013  4:21:28 server id 33062  end_log_pos 474       GTID    last_committed=1
sequence_number=2      rbr_only=no
SET @@SESSION.GTID_NEXT= '6385db95-cbc2-11e9-9f05-080027136405:1026'/*!*/;
# at 474
#191013  4:21:28 server id 33062  end_log_pos 562       Query       thread_id=3
exec_time=0    error_code=0
SET TIMESTAMP=1570954888/*!*/;
BEGIN
/*!*/;
# at 562
# at 590
#191013  4:21:28 server id 33062  end_log_pos 590       Intvar
SET INSERT_ID=1/*!*/;
#191013  4:21:28 server id 33062  end_log_pos 714       Query       thread_id=3
exec_time=0    error_code=0
SET TIMESTAMP=1570954888/*!*/;
insert into test(name) values('tmp_test')
/*!*/;
# at 714
#191013  4:21:28 server id 33062  end_log_pos 803       Query       thread_id=3
exec_time=0    error_code=0
SET TIMESTAMP=1570954888/*!*/;
COMMIT
/*!*/;
# at 803
#191013  4:25:31 server id 33062  end_log_pos 864       GTID    last_committed=2
sequence_number=3      rbr_only=no
SET @@SESSION.GTID_NEXT= '6385db95-cbc2-11e9-9f05-080027136405:1027'/*!*/;
# at 864
```

```
    #191013   4:25:31 server id 33062    end_log_pos 1000       Query      thread_id=3
exec_time=0    error_code=0
    SET TIMESTAMP=1570955131/*!*/;
    DROP TEMPORARY TABLE `test` /* generated by server */
    ......
```

从解析的内容可以看到,在主库上关于临时表的操作都被记录到二进制日志中,且符合 statement 格式。但需要注意的是,DROP TABLE 语句被 MySQL 改写为 DROP TEMPORARY TABLE 语句(由于临时表是会话独享的,会话各自创建的临时表互相不可见,因此,当一个会话创建临时表之后,连接断开时会自动删除临时表,并在二进制日志中写入 DROP TEMPORARY TABLE 语句)。不论采用何种二进制日志格式,该语句都会被改写,这是为了保证主库中不再需要的表在从库上也能被删除。

37.2.2 基于 row 的复制且隔离级别为 REPEATABLE-READ

(1)设置隔离级别以及二进制日志格式。与 37.2.1 节中的操作相同,在此不赘述。

(2)在主库上创建临时表并插入数据。与 37.2.1 节中的操作相同,在此不赘述。

(3)检查从库的数据以及复制的状态:

```
mysql> use gangshen
Database changed
mysql> show tables;
Empty set (0.00 sec)

mysql> show slave status\G
*************************** 1. row ***************************
           Slave_IO_Running: Yes
           Slave_SQL_Running: Yes
1 row in set (0.00 sec)
```

可以看到主从复制状态正常,但是在从库上看不到临时表。

(4)在主库上删除临时表:

```
mysql> drop table test;
Query OK, 0 rows affected (0.01 sec)
```

(5)解析主库产生的二进制日志:

```
    ......
    # at 190
    #191013  4:36:40 server id 33062  end_log_pos 251       GTID     last_committed=0
sequence_number=1      rbr_only=no
    SET @@SESSION.GTID_NEXT= '6385db95-cbc2-11e9-9f05-080027136405:1028'/*!*/;
    # at 251
    #191013  4:36:40 server id 33062  end_log_pos 397       Query     thread_id=3     exec_time=0
error_code=0
    USE `gangshen`/*!*/;
```

```
    SET TIMESTAMP=1570955800/*!*/;
    SET @@session.pseudo_thread_id=3/*!*/;
    SET @@session.foreign_key_checks=1, @@session.sql_auto_is_null=0, @@session.unique_
checks=1, @@session.autocommit=1/*!*/;
    SET @@session.sql_mode=1436549152/*!*/;
    SET @@session.auto_increment_increment=2, @@session.auto_increment_offset=1/*!*/;
    /*!\C utf8 *//*!*/;
    SET  @@session.character_set_client=33,@@session.collation_connection=33,@@session.
collation_server=83/*!*/;
    SET @@session.lc_time_names=0/*!*/;
    SET @@session.collation_database=DEFAULT/*!*/;
    DROP TEMPORARY TABLE IF EXISTS `test` /* generated by server */
    ......
```

通过解析结果看到，当二进制日志为 row 格式时，对于临时表，除了 DROP 操作，其余的相关操作都未被记录到二进制日志中。当基于 row 复制时，在其他事务隔离级别下 MySQL 对于临时表的操作记录的二进制日志内容相同，不再赘述。

37.2.3　混合复制且隔离级别为 REPEATABLE-READ

（1）设置隔离级别以及二进制日志格式。与 37.2.1 节操作相同，在此不赘述。

（2）在主库上创建临时表并插入数据。与 37.2.1 节操作相同，在此不赘述。

（3）检查从库的数据以及复制的状态：

```
mysql> show tables;
Empty set (0.00 sec)

mysql> show slave status\G
*************************** 1. row ***************************
            Slave_IO_Running: Yes
           Slave_SQL_Running: Yes
1 row in set (0.00 sec)
```

可以看到主从复制状态正常，但是在从库上查看不到临时表。

（4）在主库上删除临时表：

```
mysql> drop table test;
Query OK, 0 rows affected (0.01 sec)
```

（5）解析主库产生的二进制日志，可以看到混合复制时的二进制日志的解析结果与基于 statement 复制时的一致。

37.2.4　使用临时表时如何安全关闭从库

通过上面的操作演示可以看到，MySQL 复制临时表的过程中，在从库上是无法看到临时表的。下面的操作演示了如果临时表仍在使用中，关闭从库时会有什么问题。

（1）在主库上创建临时表并插入数据：

```
mysql> create temporary table test(id int primary key auto_increment,name varchar(20));
Query OK, 0 rows affected (0.00 sec)

mysql> insert into test(name) values('tmp_test');
Query OK, 1 row affected (0.01 sec)
```

（2）在从库上执行 STOP SLAVE SQL_THREAD 语句：

```
mysql> stop slave sql_thread;
Query OK, 0 rows affected, 1 warning (0.01 sec)

Warning (Code 3022): This operation may not be safe when the slave has temporary tables.
The tables will be kept open until the server restarts or until the tables are deleted by
any replicated DROP statement. Suggest to wait until slave_open_temp_tables = 0.

mysql> show status like '%Slave_open_temp_tables%';
+------------------------+-------+
| Variable_name          | Value |
+------------------------+-------+
| Slave_open_temp_tables | 1     |
+------------------------+-------+
1 row in set (0.00 sec)
```

可以看到当执行 STOP SLAVE SQL_THREAD 语句时，会有警告，提示从库上还有正在使用的临时表，建议等待 Slave_open_temp_tables 状态变量为 0。

（3）在主库上删除临时表：

```
mysql> drop table test;
Query OK, 0 rows affected (0.01 sec)
```

（4）在从库上重新启动并停止 SQL 线程：

```
mysql> start slave sql_thread;
Query OK, 0 rows affected (0.01 sec)

mysql> stop slave sql_thread;
Query OK, 0 rows affected (0.00 sec)

mysql> show status like '%Slave_open_temp_tables%';
+------------------------+-------+
| Variable_name          | Value |
+------------------------+-------+
| Slave_open_temp_tables | 0     |
+------------------------+-------+
1 row in set (0.00 sec)
```

可以看到，复制到 DROP TABLE 语句之后，再停止从库 SQL 线程，就没有出现警告信息了。

在使用临时表的过程中，如果从库的 mysqld 进程或者从库所在服务器被意外关闭或重启，则已复制的临时表及相关的更新数据将丢失。为了尽量避免发生这种情况，可以在可控的从库停机操作之前，按照如下步骤关闭从库：

- 在从库执行 STOP SLAVE SQL_THREAD 语句停止 SQL 线程。
- 用 SHOW STATUS 语句检查状态变量 Slave_open_temp_tables 的值。
- 如果 Slave_open_temp_tables 的值不为 0，则通过 START SLAVE SQL_THREAD 语句重新启动从库 SQL 线程，间隔数十秒，重复上述步骤（包括此步骤）。
- 当 Slave_open_temp_tables 的值为 0 时，执行命令 mysqladmin shutdown，停止从库的 mysqld 进程。

37.3　与临时表相关的其他注意事项

使用临时表时，还需要注意如下几点：

- 当 binlog_format = row 时，二进制日志中不会记录临时表的相关操作。也就是说，从库进程关闭时，不存在从库中的临时表数据丢失的说法。
- 在 MySQL 5.7 中，由于临时表被创建在 ibtmp 的表空间中，所以在磁盘上看不到具体的表文件，如果有需要，可以通过 INFORMATION_SCHEMA.INNODB_TEMP_TABLE_INFO 表查看。
- 当临时表未被删除或正确关闭之前，使用 mysqlbinlog 工具解析二进制日志会出现警告信息：Warning: this binlog is either in use or was not closed properly.

37.4　小结

MySQL 的临时表其实是一个比较复杂的话题，涉及的内容比较多，本章仅简单演示了临时表在 MySQL 中是如何复制的。关于临时表的更多细节，请感兴趣的读者自行深入了解。

第 38 章 复制中的事务不一致问题

在 MySQL 复制中，我们都会担心主从库数据一致性的问题。有些情况下，主从库数据不一致是由于主从库事务不一致而导致的。本章我们就来讲一讲 MySQL 复制中事务不一致的类型、原因以及结果。

38.1 事务不一致的场景类型

在 MySQL 复制中，从库的 I/O 线程负责从主库上拉取二进制日志，并将其存放到从库的中继日志中，然后由从库的 SQL 线程负责读取中继日志，对中继日志进行解析并重放，但是某些复制相关的配置项可能会导致从库重放事务的顺序与主库执行事务的顺序不一致。

在 MySQL 中事务存在以下几种不一致的情况：

- 事务半提交（half-applied transaction）：当一个事务的更新包含了不支持事务特性的存储引擎表，那么从库在重放主库事务时无法保证事务的原子性，可能导致事务部分提交成功，部分提交失败，这种情况称为事务半提交。
- 间隙（gap）：这种情况只有在多线程复制时会出现，指的是一组队列中的事务在从库上同时重放，队列中的某个事务尚未完全被应用时，其后面的事务已经应用完。为了避免出现间隙，可以在从库上设置 slave_preserve_commit_order = 1（设置该系统变量的前提是：slave_parallel_type = LOGICAL_CLOCK，开启二进制日志 log_bin = mysql-bin，开启 log_slave_updates）。需要注意的是，即使设置了 slave_preserve_commit_order = 1 也无法完全避免出现间隙，因为该系统变量无法保证非事务存储引擎表的 DML 操作顺序，因此在重放非事务存储引擎表的 DML 操作时，仍然可能出现间隙。
- 低水位情况（master log position lag）：这种情况也只会在多线程复制时出现。即使没有出现间隙，仍然可能会出现 Exec_master_log_pos 显示事务重放至位置点 N，但实际事务已经应用到位置点 N 的后面的情况。SHOW SLAVE STATUS 语句输出中的 Exec_master_log_pos 值，代表 SQL 线程已经应用完的事务的位置。在单线程复制场景中，这个位置之前的所有事务都已经被应用，而其后的事务还未被应用。但在多线程复制场景下，所有的 Worker 线程都在并行应用事务，每个 Worker 线程都有自己应用事务的位置，需要按照一定机制来计算所有 Worker 线程已经应用完的事务的位置（已经完成 checkpoint 的事务的位置），然后再用该位置更新 Exec_master_log_pos 的值。即在多线程复制场景下，可能 Exec_master_log_pos 值之后的一些事务实际上已经被应用完，如果还没有用这些被应用完的事务的位置更新 Exec_master_log_pos 值，就发生了意外，则产生记录的低水位位置不正确的情况。启用 slave_preserve_commit_order = 1 并不能防止出现低水位情况。

38.2 事务不一致的原因

前面讲了事务不一致的三种情况，接下来讲哪些场景可能会导致这些情况：

- 在从库的复制线程运行过程中，可能存在间隙和事务半提交的情况。
- mysqld 进程关闭，无论是干净地关闭还是不干净地关闭，都会中断应用中的事务，可能会留下间隙或者出现事务半提交的情况。
- KILL 复制线程（此处的复制线程指的是单线程复制时的 SQL 线程、多线程复制时的协调器线程）会中断应用中的事务，可能会留下间隙或者出现事务半提交的情况。
- applier 线程（Worker 线程）出错时，可能会出现间隙。如果错误发生在混合事务中（同一个事务更新了事务引擎和非事务引擎），则该事务处于半提交的状态。使用多线程从库时，未收到（或者说未产生）错误的 Worker 线程会完成队列中的事务，因此当其中一个线程发生错误时，要停止所有 Worker 线程可能需要一些时间。
- 在使用多线程复制的从库中，执行 STOP SLAVE 语句后，从库会等待填充完间隙，然后更新 Exec_master_log_pos 的值，确保不会留下间隙或低水位位置。但是，如果在执行完 STOP SLAVE 语句之前发生上述几种情况（即发生错误，或者另一个会话发出 KILL 命令杀死复制线程、或者 mysqld 被重新启动）则会产生问题。
- 如果中继日志中的最后一个事务只接收了一半，而且多线程从库的协调器线程已开始将事务调度给 Worker 线程，则在执行 STOP SLAVE 语句之后，等待 60 秒之后还未收到事务剩余的事件则会发生超时（因为 STOP SLAVE 语句在执行之后，只等待 60 秒，供事务完成提交或者回滚）。超时后，协调器线程放弃并中止正在执行的事务。如果是混合事务，可能就会变成半提交事务。
- 在使用单线程复制的从库中，执行 STOP SLAVE 语句时，如果正在执行的事务仅更新事务表，则该事务立即被回滚。如果正在执行的事务是混合的，则 STOP SLAVE 最多等待 60 秒以完成事务。超过 60 秒还仍未完成的事务会被终止，因此可能变成半提交的事务。

38.3 事务不一致的后果

如果复制通道存在间隙，那么最终可能导致以下几种后果：

（1）从库数据与主库数据不一致，从库中可能缺少了某些数据。

（2）SHOW SLAVE STATUS 中的字段 Exec_master_log_pos 的值只是一个"低水位"值，即在该位置之前的事务可以保证都已经提交了，但是在该位置之后的事务可能已经提交，也可能还没有提交。

（3）如果 applier 线程正在运行时，执行 CHANGE MASTER TO 语句将失败并显示错误，例如：ERROR 3021 (HY000): This operation cannot be performed with a running slave io thread; run STOP SLAVE IO_THREAD FOR CHANNEL '' first.

（4）如果使用--relay-log-recovery 选项启动 mysqld，则不会对该通道进行恢复（因为中继日志直接被清理，所以不会使用中继日志进行恢复），并打印警告信息。

（5）如果 mysqldump 与--dump-slave 选项一起使用，则不会记录间隙，因为备份文件生成的 CHANGE MASTR TO 语句中的 RELAY_LOG_POS 的位置是通过 SHOW SLAVE STATUS 语句获取的 Exec_master_log_pos 低水位位置值。在另一台服务器上应用备份文件并启动复制线程后，可能出现已经被应用过的事务通过复制被重复应用（因为拉取了一部分重复的二进制日志）。请注意，如果启用了 GTID，则可以自动跳过重复的事务；如果未启用，则不建议使用--dump-slave 选项。

如果复制通道存在低水位的情况，但不存在间隙，那么可能存在上述（2）～（5）场景描述的问题。

在 MySQL 复制中，关于主库位置点的信息会存放在 mysql.slave_worker_info 表中。START SLAVE [SQL THREAD]语句在执行时会从该表中查询信息，以应用正确的事务。即使在执行 START SLAVE 语句之前将 slave_parallel_workers 设置为 0，该表的信息也能保证事务正确性。同样，START SLAVE UNTILL SQL AFTER MTS GAPS 语句也能通过该表的信息填补复制间隙中缺失的事务。

38.4 小结

本章讲了 MySQL 复制中事务不一致的几种情况，分析其原因，并列举了后果。某些事务不一致的情况是可以通过合理地设置系统变量来避免的，但是由于涉及的因素比较多，没有做详细的介绍。请留意以下两点：

- RESET SLAVE 语句会清理中继日志并重置复制的位置。如果一个从库的复制存在间隙，而执行 RESET SLAVE 语句后关于间隙情况的信息丢失，就无法发现间隙问题了。
- 在使用 GTID 复制时，MySQL 会自动判断，如果某 GTID 的事务已经被应用，则跳过，利用 GTID 的特性可以简单有效地恢复主从复制。